# Computer-Aided Highway Engineering

T0134092

# Computer-Aided Highway Engineering

Sandipan Goswami

and

Pradip Sarkar

CRC Press
Taylor & Francis Group
Boca Raton London New York

CRC Press is an imprint of the
Taylor & Francis Group, an **informa** business

First edition published 2021
by CRC Press
6000 Broken Sound Parkway NW, Suite 300, Boca Raton, FL 33487-2742

and by CRC Press
2 Park Square, Milton Park, Abingdon, Oxon, OX14 4RN
First edition published by CRC Press 2021

CRC Press is an imprint of Taylor & Francis Group, LLC

© 2022 Taylor & Francis Group, LLC

ISBN: 978-0-367-49338-7 (hbk)
ISBN: 978-0-367-49397-4 (pbk)
ISBN: 978-1-003-04583-0 (ebk)

Typeset in Times
by SPi Technologies India Pvt Ltd (Straive)

# Dedication

*I, Sandipan Goswami, dedicate this work to my parents*

***Mr. Mrinal Kanti Goswami and Mrs. Snigdha Goswami***

*And also, to all my colleagues & friends in India and Malaysia who helped me learn and gather knowledge on the subject*

# Contents

While reading the following sections for computer-aided design in relevant chapters for:

- Processing of Various Survey Data
- New Highway Design with Uniform Cross Section by the design of Horizontal Alignment, Vertical Alignment, Cross Section Elements, Earthwork Estimation and finally the Design & Construction Drawings
- Design of At-Grade Intersections
- Design of Multi Level Grade Separated Interchanges
- Design of Flexible Pavement
- Design of Rigid Pavement
- Design of Highway Drainage

The readers are recommended that they should concurrently use the highway design software and the respective chapter wise data, by following step-by step instructions as given in "User's Guide for Hands On Practice".

The following tutorials are also provided in a separate folder of the software:

- The Downloading and Processing of Ground Data from Satellite
- Design of Highway Widening, Multiple Sections
- Design of Hill Roads and Car Racing Tracks
- Automatic Production of Multiple Drawings

All above tutorials are available for download from folder "Book Tutorials" from the website of the book at "www.techsoftglobal.com/download.php" and "www.roadbridgedesign.com".

# Preface

## INTRODUCTION

With the present advancement of internet and satellite technology, the road-building industry uses latest features of Hi-Tech facilities like Satellite/Google Earth based alignment design for 4/6/8 Lane National Highway or Expressway projects, GPS instruments, or Google Earth-based alignment designs. These remain very useful tools for the design of Highways and Hill road projects, processing Digital Terrain Models (DTM) using downloaded ground topography data from the internet by Shuttle Radar Topography Mission (SRTM, by using software Global Mapper) without carrying out any field survey and so on. This saves substantial design time.

The engineers and professionals will gain the most benefit from obtaining a technical environment complete with the latest advanced technology. Mastering this book along with the tutorials will make the reader as a complete "Highway Design Expert."

To assist the readers as much as possible, the trial version of "Highway Engineering And Design Software – Professional (HEADS Pro)," along with chapter tutorials containing design and field data, tutorial guides and various tutorial videos are provided in two separate folders, in the relevant section for the book, at the website www.techsoftglobal.com/download.php. Readers are strongly recommended to first download and install the software available in the folder "HEADS Pro Installation Setup." Next, watch the tutorial videos and finally process the tutorials available in the folder 'Book Tutorials' for the book.

Tutorials are provided for the special techniques of the design of Hill Roads requiring extra widening at curves and by using Bearing Line data, Slopes varying with depth of cut, Providing Gabion walls on cut side and Retaining walls on fill side.

Tutorials are also provided for the design of Vehicular Tunnels using Robot Total Station providing scan data for Bored Tunnels. The occurrences of Tunnels along the route of the highways are very common in many of the highway projects and the engineers are therefore provided with the best and straight forward professional design procedure in three-dimensional built-in CAD system.

## APPROACH

For the development of effective transport infrastructure in the developing countries most of the current projects are widening of old 2/4 Lane Highways to 4/6/8 Lane Expressways, where the existing road is strengthened with bituminous overlay and are widened with new construction on the side. Along the alignment of the route the side of widening changes from left to concentric to right, based on the site situation. In PPP/BOT mode projects there are requirements for route re-alignments, bypasses, underpasses, flyovers, service roads, bus bays, truck laybys, etc. These makes the design quite complex and is done by the highway and bridge engineers for the entire route with best possible alignment and profile, by ensuring most economic cost of construction. Finally, the best-quality sophisticated construction drawings are produced for construction.

Therefore, the demand is very high in the industry for the engineers having high level of professional skill and expertise to produce the best design, cost estimation and finally the construction drawings for the construction of highways, multi-lane and multi-section expressways, low-cost rural roads, hill roads, tunnels, pavements, traffic intersections, grade separated interchanges, and highway drainage.

The highway design project commonly starts with collecting and processing of topographic survey data from Total Station, Auto Level, GPS, Ground Elevation data downloaded from satellite by various ways by using programs for such purposes.

Improvement of rural roads and hill roads including bridges and culverts need specialized techniques as per the relevant standards of the country. The technology of alignment and profile optimization saves the cost of construction significantly, especially in the construction of low-cost rural roads.

The theory and process are explained for the full range of flexible and rigid pavement design facilities in AASHTO and IRC standard, strengthening of bituminous pavement wearing course by using geo synthetics by determining strength of various products for applied traffic load using layered system analysis, bituminous mix design for AC/BC, SDBC, DBM, BM for quality control at construction site, analysis of benkelman beam deflection test data for overlay design.

The theory and process are explained for the hydraulic design of highway drainage, with design report containing step-by-step calculations to check the adequacy of the cross section and profile.

In science and engineering, text books are commonly used as reference for various formulas, diagrams, tables, etc. There are very few books available on the market which helps the reader to become real-life professionals in science and technology, whereas in the field of computer science and programming' there are numerous books which help the readers to become professionals.

This book, *Computer-Aided Highway Engineering*, is aimed at creating professionals in the field of highway engineering. Highway projects are very expensive projects for a country and involve best possible contributions by professionals from various fields like, civil engineering with specialization in transportation engineering, structural engineering, geotechnical engineering, hydrologist, economist, financial analyst, estimator, surveyors, material experts, pavement specialist, etc. This book is mostly for the jobs done by the civil engineers, in planning, design, estimating and construction of a highway project.

In specific chapters, this book has relevant figures, tables, charts etc. to accompany theory and examples. The final aim is to enable the readers to use the field and design data as input in computerized processing and produce the deliverables as formatted output reports and design drawings for real-life projects.

To assist the readers to the maximum possible extent, a trial version of HEADS Pro software along with chapter tutorials containing design and field data, tutorial guides and various tutorial videos are provided in two separate DVDs. The highway engineering software HEADS Pro is based on the 'string modeling' concept and is recognized as extremely powerful software to design multi-section highways and expressways, hill roads, low-cost rural roads, intersections, multi-level grade separated interchanges, pavement design and drainage design with cost effective solutions. The features of this book by using the power of the software are highly demanded across the world and are highly effective in projects for new design or design checking. The procedures for design modifications and design audit are therefore well guided to the readers with the use of latest information technology and satellite technology for the present generation of engineers and professionals. Mastering computer-aided highway engineering will make a reader as a highly skilled professional of international standard to work with desired proficiency in any country across the world.

Tutorial videos are provided in the accompanying DVD to train the reader about facets of the design work. The learning of various processes of computer-aided highway engineering involves various design aspects which are listed below, these save the time and cost significantly for the highway design projects.

- Design of alignment for 4/6/8 Lane National Highway or Expressway projects by using Hi-Tech facilities like Google Earth based satellite imagery, Aerial photographs etc. Identify village roads in core-network and design the best alignment by using Google Earth and GPS instrument.
- Alignment design for hill road projects by using GPS instrument-based Bearing Line.
- Processing of survey data from Total Station, Auto Level, GPS and processing of ground elevation data downloaded from the internet by Shuttle Radar Topography Mission (SRTM) by using Global Mapper. The reader will gain the full benefit of a technical environment complete with the latest advanced technology.
- Design of uniform section highway and low-cost rural roads, multi-lane and multi-section highways and expressways, hill roads with extra widening at curves, cut slopes varying with depth of cut.
- Providing Gabion walls on cut side and retaining walls on fill side, at-grade traffic intersections, and multi-level grade separated interchanges with detail reports and highly sophisticated CAD design drawings of highest international standard.
- A complete set of construction drawings for PLAN, PROFILE, and CROSS SECTIONS are produced either for the complete length of the road or for every one kilometer (or as desired) length.
- Estimation report with detailed "Bill of Quantities."
- The occurrences of tunnels along the route of the highways are very common in most of the highway projects and the tutorial DVD is provided with the best design procedure in a three-dimensional built-in CAD system. Tunnel module provides complete process of design to the readers for the design of vehicular tunnels by using scan data taken inside the "Bored Tunnels" by "Robot Total Station."
- Design of flexible and rigid pavement design facilities in AASHTO and IRC Standard.
- Design of pavement overlay with analysis of Benkelman beam rebound deflection test data.

- Strengthening of bituminous pavement wearing course by using geo synthetics by determining strength of various products for applied traffic load using layered system analysis, the use of various geotextiles and fiberglass geogrids may be decided by obtaining the result on strength improvement given by detail step wise report.
- Bituminous mix design for AC/BC, SDBC, DBM, BM for quality control at construction site.
- Hydrological design of highway drainage.
- Land acquisition by maintaining land drawings and land records by HEADS Pro in database management system is highly useful for any highway project.
- Finished road levels on the cross section in bridges are generated at desired chainage intervals with normal camber/cross falls or with super elevations along the longitudinal profile of the bridge.
- Straight forward interactive design for main line alignment and interchange loops, ramps, and slip roads.

# Acknowledgment

*We thankfully acknowledge the kind co-operation extended by Mr. Cullen Bob of AASHTO, Public Works Department of Malaysia (JKR Malaysia).*

*We convey our heartfelt thanks to Dr. Gagandeep Singh and Mr. Lakshay Gaba, the wonderful people of CRC Press, Taylor and Francis Books India Pvt. Ltd. for their very effective co-operation and invaluable guidance, without whom this work wouldn't have been possible.*

# 1 Project Overview and Highway Engineering

## 1.1 GENERAL

Under normal practice, when a highway project is planned, it is awarded to a qualified engineering consultant to carry out the feasibility studies and to determine the viability of the project. The consultant carries out the required survey work and collects data with respect to existing classified traffic, engineering surveys on the existing road with bridges, the location of sources of suitable construction materials, land availability and environmental impact, law and order situation, exiting road network in the region, industrial prospects, and commercial transports, etc.

Based on the collected data, a preliminary design is done to estimate the cost of the project. The cost has two components, civil work cost (CWC) and the total project cost (TPC), which additionally includes the cost of interest of funding during construction, land acquisition, environmental restoration, and utilities relocation work, etc.

Next the economic analysis is done to obtain the internal rate of return (IRR). This enables the project authority to decide if the project should go ahead, if so, then the decision is taken on the mode of financing the project. The mode may be by either by government financing or by BOT concessionaires or by public private partnership (PPP). The financial analysis is carried out to get the profit returns to attract the private investors.

When a private equity firm places its first round bid for the toll-operate-transfer model, its team will have already spent months surveying traffic volumes on those stretches. This is normal, most road companies do not rely on the traffic estimates done by the project authority but research their own. Due to possible inaccuracy and misreporting, industry officials prefer to use their own research.

Road companies instead prefer to have their own team, they hire an agency to build their own traffic estimates. No one wants to take a financial decision based purely on the data provided by the project authority. Such big decisions need to be taken on one's own data. Road developer companies often find that the data shared by the project authority and their own data, does not differ by a huge margin.

Every government carries out feasibility studies for the road projects planned for construction. Next the governments try to attract private companies to the selected projects for their investment and earn revenue for a specified period either by commercial operation themselves or by the project authority. This is undertaken by the Public Private Partnership Appraisal Committee (PPPAC).

If the commercial operation is by the companies, then it is executed through a BOT/DBFOT in toll model, and if the commercial operation is by the project authority, then it is done through in annuity model. The builder company is called a

1

concessionaire and the annuity is an amount paid bi-annually by the project authority to the builder company for a fixed span of years, quoted by the builder company.

## 1.2 BUILD OPERATE AND TRANSFER (BOT) ANNUITY MODEL

The BOT annuity model is a PPP model for infrastructure projects, especially road projects. Under BOT annuity, a developer builds the highway, operates it for a specified duration, and then transfers it back to the government. The government starts making payments to the developer after the launch of the commercial operation of the project. Payment will be made on a six-month basis.

### 1.2.1 CONCESSIONS, BUILD OPERATE TRANSFER (BOT), AND DESIGN BUILD

#### 1.2.1.1 Operate (DBO) Projects

Concessions, BOT projects, and DBO projects are types of public private partnerships that are designed with output focus in mind. BOT and DBO projects typically involve significant design and construction as well as long term operations, for new build (greenfield) or projects involving significant refurbishment and extension (brownfield). The definitions below are of each type of agreement, as well as key features and examples of each project type. This page also includes links to checklists, toolkits, and sector-specific PPP information.

- Overview of Concessions, BOTs, and DBO Projects
- Key Features
  - o Concessions
  - o BOT Projects
- Contractual Structure

### 1.2.2 OVERVIEW OF CONCESSIONS, BOTs, AND DBO PROJECTS

A concession gives a concessionaire the long term right to use all utility assets conferred on the concessionaire, including responsibility for operations and some investment. Asset ownership remains with the authority, and the authority is typically responsible for the replacement of larger assets. Assets revert to the authority at the end of the concession period, including assets purchased by the concessionaire. In a concession the concessionaire typically obtains most of its revenues directly from the consumer and so it has a direct relationship with the consumer. A concession covers an entire infrastructure system (so may include the concessionaire taking over existing assets as well as building and operating new assets). The concessionaire will pay a concession fee to the authority which will usually be ring-fenced and put toward asset replacement and expansion. A concession is a specific term in civil law countries. To make it less confusing, in common law countries, projects that are more closely described as BOT projects are called concessions.

A BOT project is typically used to develop a discrete asset rather than a whole network and is generally entirely new or greenfield in nature (although refurbishment may be involved). In a BOT project the project company or operator generally obtains its revenues through a fee charged to the utility/government rather than tariffs charged to consumers. In common law countries a number of projects are called concessions, such as toll road projects, which are new build and have a number of similarities to BOTs.

In a DBO project, the public sector owns and finances the construction of new assets. The private sector designs, builds and operates the assets to meet certain agreed outputs. The documentation for a DBO is typically simpler than a BOT or concession as there are no financing documents and will typically consist of a turn-key construction contract plus an operating contract, or a section added to the turn-key contract covering operations. The operator is taking no or minimal financing risk on the capital and will typically be paid a sum for the design build of the plant, payable in installments on completion of construction milestones, and then an operating fee for the operating period. The operator is responsible for the design and the construction as well as operations and so if parts need to be replaced during the operation period prior to its assumed life span, the operator is likely to be responsible for replacement.

This section looks in greater detail at Concessions and BOT projects. It also looks at offtake/power purchase agreements, input supply/bulk supply agreements, and implementation agreements, which are used extensively in relation to BOT projects involving power plants.

This section does not address the complex array of finance documents typically found in a concession or BOT project.

### 1.2.3 KEY FEATURES

#### 1.2.3.1 Concessions

- A concession gives a private concessionaire responsibility not only for operation and maintenance of the assets but also for financing and managing all required investment.
- The concessionaire takes the risk for the condition of the assets and for investment.
- A concession may be granted an existing utility in relation to existing assets, or for extensive rehabilitation and extension of an existing asset (although often new build projects are called concessions).
- A concession is typically for a period of 25–30 years (i.e., long enough at least to fully amortize major initial investments).
- Asset ownership typically rests with the awarding authority and all rights in respect to those assets revert to the awarding authority at the end of the concession.
- The general public are usually the customer and main source of revenue for the concessionaire.
- Often the concessionaire will be operating the existing assets from the outset of the concession and so there will be immediate cash flow available to pay concessionaire, set aside for investment, service debt, etc.

- Unlike many management contracts, concessions are focused on outputs – i.e., the delivery of a service in accordance with performance standards. There is less focus on inputs – i.e., the concessionaire is left to determine how to achieve agreed performance standards, although there may be some requirements regarding frequency of asset renewal and consultation with the awarding authority or regulator on such key features as maintenance and renewal of assets, increase in capacity and asset replacement toward the end of the concession term.
- Some infrastructure services are deemed to be essential and some are monopolies. Limits will probably be placed on the concessionaire – by law, through the contract or through regulation – on tariff levels. The concessionaire will need assurances that it will be able to finance its obligations and still maintain a profitable rate of return and so appropriate safeguards will need to be included in the project or legislation. It will also need to know that the tariffs will be affordable and so will need to do due diligence on customers.
- In many countries there are sectors where the total collection of tariffs does not cover the cost of operation of the assets alone with further investment. In these cases, a clear basis of alternative cost recovery will need be set out in the concession, whether from general subsidies, taxation or loans from government, or other sources.
- The concept of a "concession" was first developed in France. As with "affermages," the framework for the concession is set out in the law and the contract contains provisions which are specific to the project. Emphasis is placed in the law on the public nature of the arrangement (because the concessionaire has a direct relationship with the consumer), and safeguards are enshrined in the law to protect the consumer. Similar legal frameworks have been incorporated into civil law systems elsewhere.
- Under French law the concessionaire has the obligation to provide continuity of services ("la continuité du service public"), to treat all consumers equally ("l'égalité des usagers") and to adapt the service according to changing needs ("l'adaptation du service"). In return, the concessionaire is protected against new concessions which would adversely affect the rights of the concessionaire. It is therefore important when considering concessions in civil law systems to understand what rights are already embodied in the law.
- Within the context of common law systems, the closest comparable legal structure is the BOT, which is typically considered for the purpose of constructing a facility or system.

### 1.2.3.2   BOT Projects

The following are considered from the planning to the execution stages of BOT projects:

- In a BOT project, the public sector grantor grants to a private company the right to develop and operate a facility or system for a certain period (the "project period"), in what would otherwise be a public sector project.
- It is usually a discrete, greenfield new build project.

- It is based on the operator's finances, who then own and construct the facility or system and operate it commercially for the project period, after which the facility is transferred to the authority.
- BOT is the typical structure for project finance. As it relates to new build, there is no revenue stream from the outset. Lenders are therefore anxious to ensure that project assets are ring-fenced within the operating project company and that all risks associated with the project are assumed and passed on to the appropriate actor. The operator is also prohibited from carrying out other activities. The operator is therefore usually a special purpose vehicle.
- The revenues are often obtained from a single "offtake purchaser" such as a utility or government, who purchases the project output from the project company (this is different from a pure concession where output is sold directly to consumers and end users). In the power sector, this will take the form of a power purchase agreement. For more, see **power purchase agreements**. There is likely to be a minimum payment that is required to be paid by the offtaker, provided that the operator can demonstrate that the facility can deliver the service based on the availability payment as well as a volumetric payment for quantities delivered above that level.
- The project company obtains financing for the project, and procures the design and construction of the works, and operates the facility during the concession period.
- The project company is a special purpose vehicle, its shareholders will often include companies with construction and/or operation experience, and with input supply and offtake purchase capabilities. It is also essential to include shareholders with experience in the management of the appropriate type of projects, such as working with diverse and multicultural partners, given the particular risks specific to these aspects of a BOT project. The offtake purchaser/utility will be anxious to ensure that the key shareholders remain in the project company for a period of time as the project is likely to have been awarded to it on the basis of their expertise and financial stability.
- The project company will coordinate the construction and operation in accordance with the requirements of the concession agreement. The offtaker will want to know the identity of the construction sub-contractor and the operator.
- The project company (and the lenders) in a project will be anxious to ensure it has a secure affordable source of raw materials and fuel. It will often enter into a bulk supply agreement for fuel.
- The revenues generated from the operation phase are intended to cover operating costs, maintenance, repayment of debt principal (which represents a significant portion of development and construction costs), financing costs (including interest and fees), and a return for the shareholders of the special purpose company.
- Lenders provide non-recourse or limited recourse financing and will, therefore, bear any residual risk along with the project company and its shareholders.
- The project company is assuming a lot of risk. It is anxious to ensure that those risks that stay with the grantor are protected. It is common for a project company to require some form of guarantee from the government and/or,

particularly in the case of power projects, commitments from the government that are incorporated into **implementation agreements**.

- In order to minimize such residual risk (as the lenders will only want, as far as possible, to bear a limited portion of the commercial risk of the project) the lenders will insist on passing the project company risk on to the other project participants through contracts, such as a construction contract, an operation and maintenance contract.

### 1.2.4   CONTRACTUAL STRUCTURE

The chart below shows the contractual structure of a typical BOT project or concession, including the lending agreements, the shareholder's agreement between the project company shareholders, and the subcontracts of the operating contract and the construction contract, which will typically be between the project company and a member of the project company consortium.

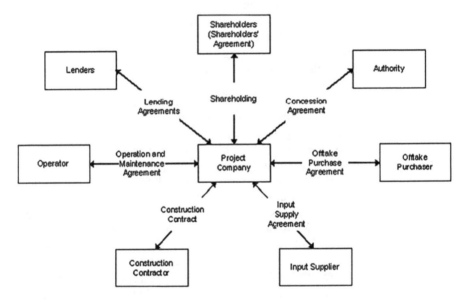

Each project will involve some variation of this contractual structure depending on its particular requirements – not all BOT projects will require a guaranteed supply of input, therefore a fuel/input supply agreement may not be necessary. The payment stream may be in part or completely through tariffs from the general public, rather than from an offtake purchaser.

## 1.3   THE HYBRID ANNUITY MODEL (HAM)

As the name suggests, HAM is a hybrid annuity model and is a mix of the engineering, procurement, and construction (EPC) and BOT models. Under the EPC model, the project authority pays a private builder to build roads. The private builder has no role in the road's ownership, toll collection or maintenance, rather it is taken care of by the government's project authority.

Under the BOT model though, private builders have an active role, they build, operate, and maintain the road for a specified number of years – say 10–15-years – before transferring the asset back to the government.

Under BOT, the private builders arrange all the finances for the project, while collecting toll revenue or annuity fee from the government, as agreed. The annuity fee arrangement is known as BOT annuity. Essentially, the toll revenue risk is taken by the government, while the private builder is paid a pre-fixed annuity for construction and maintenance of roads.

HAM combines EPC (40%) and BOT annuity (60%). On behalf of the government, the authority releases 40% of the TPC. It is given in five tranches linked to milestones. The balance has 60% arranged by the developer. Here, the developer builder usually invests no more than 20–25% of the project cost (rather than 40% or more, as they did before), while the remaining is raised as debt.

### 1.3.1 THE IMPORTANCE

HAM arose out of a need to have a better financial mechanism for road development. The BOT model ran into roadblocks with private investors not always coming forward to invest. First of all, the private investor had to fully arrange its finances – be it through equity contribution or debt. NPA, getting the investment by banks as a non-performing asset, riddled banks were becoming wary of lending to these projects. Also, if the compensation structure didn't involve a fixed compensation (such as annuity), developers had to take on the entire risk of low passenger traffic. In the past, many assumptions on traffic had gone awry, affecting returns. Now, they are unwilling to commit large sums of money in such models.

HAM is a good trade-off, spreading the risk between developers and the government. Here, the government finances 40% of the project cost – a type of viability-gap funding. This helps cut the overall debt and improves project returns. The annuity payment structure means that the developers aren't taking "traffic risk." From the government's perspective, it gets an opportunity to flag off road projects by investing a portion of the project cost. While it does take the traffic risk, it also earns better social returns by way of access and convenience to daily commuters.

### 1.3.2 WHY THE PROJECT AUTHORITY, DEVELOPER BUILDER AND PUBLIC SHOULD CARE

Infrastructure investments are key for economic growth, which may be linked for the nation's prosperity. So, it's important that these projects take off – HAM is a useful concept, thus rendering a big help for infrastructure development. Also, more and better roads mean smoother rides and less traffic congestion, who doesn't want that?

Additionally, once built, a high standard road generates high standard transport infrastructure, which in turn brings industrial and commercial growth in the regions the road passes through and in turn increases the worth of the land in the region.

The HAM projects are also being tested in urban infrastructure developments such as road and metro rail projects.

## 1.4  FEASIBILITY STUDY

A feasibility study involves both desk work and site visits, and addresses the key fundamentals of access and access constraints. Investigations and various surveys will also take place as to the current lawful use of the site and its current potential for generating traffic.

Initial assessments of predicted traffic flows are made based on the existing traffic. A judgment is then made as to the most appropriate form of junction to satisfactorily accommodate the desired level of traffic operation in relation to ease of access and impact on the network. The ability to physically accommodate the traffic at the selected junction on land, either within the development site or within the existing site, and the public highway, is critical to avoid. For example, limits of the public highway are noted along with any visible service utility apparatus or arrangements.

An initial view of the impact of the level of additional traffic is gained from the site visit where key junctions or highway links on the local road network can be identified. Depending on the nature of the development it may be appropriate to undertake a survey of existing traffic movements both on the highway network or generated by the existing land use. In addition, the access design, in respect to either partial control or full control, may require a spot speed survey of vehicles to be undertaken to determine key design features such as junction sight lines to be ascertained.

In the vicinity of the site, critical highway measurements are also taken, and a photographic record is usually made. The deliverability of the access and the initial view on the impact of the development traffic will form the nature and scale of the local development in relation to confirming that the "mix of development," dwelling numbers, or proposed floor space of retail/industrial/office/warehouse/leisure/healthcare uses are appropriate.

During the site visit a brief overview of the availability of facilities that assist access by sustainable modes will be undertaken. Where residential development is proposed, the availability of local facilities, shops, schools, doctors etc., will be briefly assessed as the National Planning Policy Framework seeks to link planning land use to sustainable transport facilities to ensure that sites should only be developed in areas that are well served by a variety of sustainable modes.

The base information given by a feasibility study report informs both the subsequent **scoping study** and the **transport assessment/transport statement**.

### 1.4.1  Preparation of Feasibility Studies Report (FSR)

What the feasibility studies report essentially contains is prepared in the following chapters:

- Executive Summary
- Chapter 1: Project Introduction

- Chapter 2: Socio Economic Profile
- Chapter 3: Traffic Survey, Analysis and Forecast
- Chapter 4: Engineering Surveys, Investigation Reports, Existing Level of Service
- Chapter 5: Improvement Proposal and Preliminary Design
- Chapter 6: Cost Estimate
- Chapter 7: Conclusion with Economic and Financial Analysis for Project Viability
- Chapter 8: Remarks of Project Development Mode

## 1.5 DETAIL PROJECT REPORT (DPR)

A detailed project report is an essential component of a highway project. It should be prepared carefully. Before finalizing the DPR, care should be given to carry out proper surveys, investigations, and designs. Sufficient details should be included to ensure proper appraisal, approval, and implementation of the project on time. Considering the importance of DPR preparation, a document intended for reference is detailed along with it. The guidelines provided in this document shall be adhered to strictly. In addition, the concessionaire can incorporate specific additional relevant details to supplement the base data. What the DPR essentially contains is prepared in the following chapters.

### 1.5.1 PREPARING THE DPR

The standard project report contains of the following volumes:

- Executive Summary
- Volume 1: Main Report including Project Profile
- Volume 2: Design Report for Roads, Bridges and Structures
- Volume 3: Materials Report
- Volume 4: Bill of Quantities
- Volume 5: Analysis of Rates
- Volume 6: Cost Estimation
- Volume 7: Technical Specifications
- Volume 8: Contract Agreement
- Volume 9: Construction Drawings
- Volume 10: Environmental Impact Assessment (EIA) Report
- Volume 11: Rehabilitation and Resettlement (R & R) Report

The main structure of the report typically has the chapters as detailed next:

#### 1.5.1.1 Chapter 1: Introduction
1.1 Project Background
1.2 Description of the Project Road
1.3 Objective of Consultancy Services
1.4 Scope of Consultancy Services

### 1.5.1.4   Chapter 4: Engineering Surveys and Investigation Report

### 1.5.1.5   Chapter 5: Improvement Proposals and Preliminary Design

### 1.5.1.6  Chapter 6: Cost Estimate

### 1.5.1.7  Chapter 7: Conclusion

The various reports are submitted as a draft report to the project authorities for study, upon receiving their comments the final reports are prepared.

In case the project is decided to be done using the design build finance and operate (DBFO) model using the BOT pattern, then the DPR is prepared by the concessionaire. The EPC is awarded the project for execution.

During the execution of the project the supervision consultant ensures the construction follows the technical specifications and the design drawings/reports. The contractor also prepares the complete set of "As-Built drawings" for various road, bridge and structural work for their submission to the concessionaire and the project authorities.

## BIBLIOGRAPHY

*Transportation Engineering & Planning* by Papacostas, P. D. Prevedouros, by Prentice Hall India.

*Manual for Road Investment Decision Mod Technical Papers, Consultants Role for the Preparation of Project for the Development of Rural Road Infrastructure in India* by H.S. Bhatia, S. Goswami, Premal D. Mehta and Mrs. Rachna Chopra, for "Seminar on Integrated development of Rural and Arterial Road Network for Socio-Economic Growth," Vol. I, 5–6 December, 2003, IRC.

*Theory and Applications of Economics in Highway and Transport Planning* by Dr. Vinay Maitri, Dr. P. K. Sarkar, by Standard Publishers Distributors.

*Traffic Engineering and Transport Planning* by Dr. L. R. Kadiyali, by Khanna Publishers.

*Road Development Plan Vision: 2021*, Ministry of Road Transport & Highways, Govt. of India.

*Manual on Economic Evaluation of Highway Projects in India*, IRC SP 30-1993, IRC.

# 2 Design Standards

## 2.1 GENERAL

Geometric design practices with federal/national highways agencies and other designing agencies are not exactly same in many countries. Significant variation exists in the laws of various countries and states, which defines the limits of size and weight of motor vehicles. Differences also exist in the financing ability of various authorities and agencies. These along with other policy issues influences the engineers' decisions, who accordingly modify the adoption of fully uniform design standards. Differences in local conditions such as terrain type, weather conditions, and available construction materials, also affect the design practices and standards to some extent.

## 2.2 HIGHWAY DETAIL ENGINEERING

In this chapter, the book describes the design criteria, standards, design procedure, highway alignment, elements of cross sections, and other related topics. Preparing a comprehensive highway design needs proper establishment of a travel lane configuration, alignment location, and various dimensions related to the highway cross section. A three-dimensional formation of the highway is determined through calculations of horizontal and vertical alignments of the highway centerline by considering a variety of operational considerations. This chapter deals with the design of motor vehicle facilities. Specific design elements are considered with respect to the adopted design methodology.

### 2.2.1 DESIGN CRITERIA

Various federal, national, and state highway agencies prepare their own versions of specifications, and standard drawings, they mostly follow the national guidelines in preparing such specifications. In such preparations, the AASHTO standards provide useful information and guidelines. The formulation of requirements for the design of highway facilities serves to ensure that uniform highways are designed and constructed on a state wide basis.

### 2.2.2 TERRAIN CLASSIFICATION

The terrain classification system adopted for the project road is as follows:

**TABLE 2.1**
**Terrain Classification**

| Terrain | Cross Slope (%) |
|---------|-----------------|
| Plain | 0–10 |
| Rolling | 10–25 |
| Mountainous | 25–60 |
| Steep | >60 |

*Source:* Table 1, Geometric Design Standards for Rural (Non-Urban) Highways, IRC 73-1980

### 2.2.3 LANE REQUIREMENT

From the capacity analysis, it is concluded that provision of whether $1 \times 2$, $2 \times 2$, $2 \times 3$, or $2 \times 4$ lane road configuration would be sufficient to cater for the projected traffic volume at the desired level of service (say, LOS-B) during the project analysis period, for approximately 30 years.

### 2.2.4 LANE WIDTH

Lane width has a significant influence on the safety and comfort of the traveling traffic. The capacity of a roadway is markedly affected by the lane width. In general, safety increases with wider lanes up to a width of about 3.75 m. The commonly adopted lane width is 3.5 m for a design speed of 100 km/h.

### 2.2.5 SHOULDERS

Shoulders are a critical element of the roadway cross section. Shoulders provide recovery area for errant vehicles, a refuge for stopped or disabled vehicles, and access for emergency and maintenance vehicles. Shoulders can also provide an opportunity to improve sight distance through large cut sections. The shoulder widths on the outer side are proposed as given in the tables 2.2 and 2.3.

**TABLE 2.2**
**Width of Shoulders in Plain & Rolling Terrain**

| Type of Section | Width of the Shoulder (m) | | |
|-----------------|-------|---------|-------|
| | Paved | Earthen | Total |
| Open country with isolated built-up areas | 1.5 | 2.0 | 3.5 |
| Built-up area | 2.0 | | 2.0 |
| Approaches to grade-separated structures | 2.0 | | 2.0 |
| Approaches to bridges | 1.5 | 2.0 | 3.5 |

**TABLE 2.3**
**Width of Shoulders in Mountainous & Steep Terrain (Hilly Area)**

| Type of Section | Width of the Shoulder Including Drain and Crash Barrier as Applicable (m) | |
|---|---|---|
| Open country with isolated built-up areas | 1.5 (on hill side) 2.0 (on valley side) | Earthen shoulder |
| Built-up area and approaches to grade-separated structures and bridges | 1.5 (on hill side) 2.0 (on valley side) | Raised footpath along with provision of adequate drainage along and across the footpath |

**TABLE 2.4**
**Width of Median**

| Type of Section | Minimum Width of the Median (m) | | Remarks |
|---|---|---|---|
| | Plain & Rolling Terrain | Mountainous & Steep Terrain | |
| Open country with isolated built-up areas | 4.5 | 2.0 | Absolute minimum |
| Built-up area | 2.0 | 2.0 | is 1.0 m as crash |
| Approaches to grade-separated structures | 4.5 | 2.0 | barrier |

### 2.2.6 MEDIANS

Medians on divided highways serve a variety of important purposes related to safety, traffic operations, access control, aesthetics, physical separation of opposing traffic flows, storage area for right-turning vehicles, provision of pedestrian refuge space, control of access by restricting right-turns and U-turns to specific median openings, provision of physical space for traffic control devices and bridge piers, and provision of physical space for landscaping to enhance highway aesthetics. The general guidelines for the minimum width of median for various locations are proposed as shown in table 2.4.

### 2.2.7 SIDE SLOPES

As per general standard, slope of 1 V: 2 H has been adopted for earthen embankment up to 3 m height. Higher embankments have been designed for site specific condition with slope stabilization measures such as gabions/retaining structures. For cut section, slope of 1 V: 1 H has been adopted for cutting up to 2 m and up to 1 V:0.25 H for greater height depending upon rock characteristics and stability.

### 2.2.8 RIGHT OF WAY (ROW)

As per general guidelines, proposed right of way (ROW) is in range of 60 m, while 45 m in built up areas is considered desirable. Existing ROW generally varies from

20 m to 30 m in most of the two-lane roads. As far as possible the improvement schemes should be proposed by being within the proposed ROW except at isolated locations like major junctions, rest areas, toll plazas, and wayside amenities, etc., where additional land requirements will be proposed.

### 2.2.9 Cross Sectional Elements

Four types of cross sections are proposed. In straight reaches, a camber or crossfall of 2.5% uniformly from the inner edge to the outer edge of the bituminous pavement surface (including any paved shoulder) and 3.0% in the granular shoulder are provided.

For minor crossroads and service/slip road, only 5.5 m width (intermediate lane) may be considered as against 7.0 m width stipulated by general guidelines. This is in view of comparatively less local traffic envisaged in semi-urban areas. Typical cross sections drawings (TCS) for two/four/six/eight-lane roads are to be prepared for their inclusion in the project drawings.

### 2.2.10 Super Elevation

Super elevation is provided for all the horizontal curves with a radius less than 2000 m, in order to counteract the effect of centrifugal force (e.g. when the design speed is 100 k/mph, the super elevation is required at curves of radius less than 2000 m).

The super elevation "e" has been calculated from the formula:

$$e = \left[ V^2/127 * R \right] - f$$

Where V is the design speed, at say 100 km/h and R is the radius of the curve in meters (say 1000 m), and f is the coefficient of lateral friction, may be taken as equal to 0.15.

### 2.2.11 Plain and Rolling Terrain

Super elevation is provided for the horizontal curves in order to counteract the effect of centrifugal force. The maximum super elevation is generally limited to 5% to reduce roll over.

### 2.2.12 Hilly Terrain

Super elevation is provided for all the horizontal curves with a radius less than 450 m in order to counteract the effect of centrifugal force. The maximum super elevation is limited to 10% to reduce roll over.

### 2.2.13 Standards for Interchange Elements

Lengths of speed change lanes for interchanges recommended are given Table 5.10. Maximum vertical gradient of 3% generally would be adopted in a design.

**TABLE 2.5**
**Design Parameters for Plain/Rolling Terrain**

| S. No. | Description | Details |
|---|---|---|
| 1 | Design speed | 100/80 kmph |
| 2 | Lane width | 3.5 m |
| 3 | Camber (pavement & paved shoulder) | 2.50% |
| 4 | Camber (gravel shoulder) | 3.00% |
| 5 | Shy distance (edge strip) on footpath/separator side | 0.25 m |
| 6 | Maximum super-elevation | 5.00% |
| 7 | Minimum stopping sight distance (SSD) | 180/120 m |
| 8 | Minimum intermediate sight distance (ISD) | 360/240 m |
| 9 | Minimum radius of horizontal curve | 400/260 m |
| 10 | Minimum radius of horizontal curve without super elevation | 1800/1200 m |
| 11 | Minimum vertical gradient | 0.30%/0.0% at structures |
| 12 | Ruling maximum vertical gradient | 3.30% |
| 13 | Maximum grade change not requiring vertical curve | 0.50% |
| 14 | Minimum "K" Value for Summit Curves for 100/80 kmph | 74/33 |
|  | Minimum "K" Value for Valley Curves for 100/80 kmph | 42/26 |
| 15 | Vertical clearance over NH/SH | 5.5 m |
| 16 | Vertical clearance over rail | 6.75 m |
|  | Horizontal Curve Parameters | |
|  | (transition curves and super-elevation for maximum up to 100 kmph design speed) | |

| | Radius of Horizontal Curve (m) | Min. Transition Length (m) | Super-elevation |
|---|---|---|---|
|  | 400 | 110 | 5.00% |
|  | 500 | 90 | 5.00% |
| 17 | 600 | 75 | 5.00% |
|  | 700 | 70 | 5.00% |
|  | 800 | 70 | 5.00% |
|  | 900 | 70 | 5.00% |
|  | 1000 | 60 | 4.50% |
|  | 1200 | 50 | 3.80% |
|  | 1500 | 40 | 3.00% |
|  | 1800 | 35 | 2.50% |
|  | >1900 | NR | Normal camber |

## 2.2.14 MEDIAN OPENINGS

Median openings shall be provided for four/six/eight-lane divided carriageways as per standard guidelines and international practice. In general, median openings are provided at approximately 2.0 km intervals. However, median openings will be limited to authorized intersections with public roads and will not be provided for individual business needs. Where the median openings are provided at junctions, storage lanes have been considered.

## 2.2.15 UNDERPASSES/CATTLE CROSSINGS

In built-up areas, where pedestrian traffic and local traffic crossing is warranted across the main carriageway, the vertical profile would be designed to accommodate

**TABLE 2.6**
**Design Parameters for Hilly Terrain**

| S. No. | Description | Details |
|---|---|---|
| 1 | Design speed | 50/40 kmph |
| 2 | Right of way for 4-lane road | 45 m/30 m |
| | Right of way for two 2-lane roads | 24 m/18 m |
| 3 | Lane width | 3.5 m |
| 4 | Shoulder width for 4-lane road | 1.5 and 2.0 m |
| | Shoulder width for 2-lane road | $2 \times 0.9$ m |
| 5 | Roadway width for 4-lane road | 20 m |
| | Roadway width for 2-lane road | 8.8 m |
| 6 | Camber (pavement and paved shoulder) | 2.50% |
| 7 | Camber (gravel shoulder) | 3.00% |
| 8 | Maximum super-elevation | 10.0% |
| 9 | Minimum stopping sight distance (SSD) | 60/45 m |
| 10 | Minimum intermediate sight distance (ISD) | 120/90 m |
| 11 | Minimum radius of horizontal curve | 80/50 m |
| 12 | Minimum radius of horizontal curve without transition | 450/180 m |
| 13 | Minimun vertical gradient | 0.30% |
| 14 | Ruling maximum vertical gradient | 5.00% |
| 15 | Design service volume for 2 lanes | 5000/7000 PCU/day |
| 16 | Maximum grade change not requiring vertical curve | 1.00% |
| 17 | Minimum length of vertical curve | 30 m |
| 18 | Vertical clearance over NH/SH | 5.5 m |
| 19 | Vertical clearance over rail | 6.75 m |
| 20 | Horizontal Curve Parameters | |
| | (transition curves and super-elevation for 50 kph design speed) | |

| Radius of Horizontal Curve (m) | Min. Transition Length (m) | Super-elevation |
|---|---|---|
| 50 | NA | 10.00% |
| 70 | NA | 10.00% |
| 90 | 45 | 10.00% |
| 100 | 45 | 10.00% |
| 150 | 30 | 7.41% |
| 170 | 25 | 6.54% |
| 200 | 20 | 5.56% |
| 300 | 15 | 3.70% |
| 400 | 15 | 2.78% |
| >450 | NR | Normal camber |

an underpass connecting the service roads on both sides. At some locations where frequent cattle crossings are observed, then cattle crossings with less clear height will also be considered. Some of the box/slab culverts can be used as cattle crossings.

## 2.2.16 Standards for Interchanges

Where a requirement of providing interchanges arises, the proposal referring to relevant standards may be proposed for approval of the authority.

**TABLE 2.7**
**Details of Speed Change Lanes**

| | | | | Speed Change Lane | |
| | Design | | Stopping Sight | Acceleration | Deceleration |
| Description | Speed (kph) | Radius (m) | Distance (m) | Lane (m) | Lane (m) |
|---|---|---|---|---|---|
| Ramp | 80 | 230 | 130 | 300 | 130 |
| Loop | 60 | 130 | 80 | 400 | 150 |

## 2.2.17 STANDARDS FOR AT-GRADE INTERSECTIONS

Where requirements to provide either new or improvements to existing intersections arise, the proposal referring to relevant standards may be proposed for approval of the authority.

## 2.2.18 SUBSURFACE DRAINAGE

Adequate drainage is a primary requirement for maintaining the structural condition and functional effect of a good pavement structure including subgrade. Pavements must be protected from any ingress of water. Otherwise over a period of time it many weaken the subgrade by saturating it and cause distress in the pavement structure. The GSB layer shall extend through the full formation width and shall act as the drainage layer for effective subsurface drainage.

## 2.2.19 SURFACE DRAINAGE

The surface drainage shall be effected by providing a proper camber, say of 2.5%, in the pavement and 3.0% in the granular shoulder on either side in straight alignment reaches. In the horizontal curve portions where super elevations are introduced, the outer carriageway slopes toward the central median and so the collected water is to be discharged through concrete pipes embedded underneath, in the manner indicated in the relevant drawings. A minimum longitudinal gradient of 0.3% is considered adequate in most of the conditions to take care of surface drainage. In horizontal curves, median drainage system will be provided.

## 2.2.20 DESIGN STANDARDS FOR BRIDGES/STRUCTURES

The cross drainage structures shall be classified as culverts, minor bridges, and major bridges, depending on the length of the structure as per relevant standards. Structures up to 6 m in length fall into the category of culverts, more than 6 m and less than 60 m in length as minor bridges, and beyond this as major bridges. Widening of existing culverts and bridges will be for two-lanes with paved shoulders as per standard guidelines. All new structures are to be constructed for two-lane carriageway with paved shoulders as per standard guidelines.

In stretches, where a four-lane configuration is to be adopted, widening of existing culverts and bridges will be for two-lanes and hence all new structures are to be constructed for two-lane carriageway with an overall width of 12 m, as per standard guidelines, and manual for four-lane road.

The design standards and loading to be considered for culverts, bridges, underpasses, flyovers, and rail over bridges (ROBs), shall be as per latest standards and codes. Where the said codes are found wanting or are silent, other codes at national or international level may be followed in consultation with the project authority.

- The standard codes will be the basis of bridge designs, underpasses and flyover/ROBs. For items not covered by the codes, provisions of special publications and specification for roads and bridges published by the individual country shall be followed.
- Grades of concrete for superstructures will be as per technical specifications and standards. The minimum grade should be M40 for PSC and M30 for RCC respectively.
- For substructures and foundations, the concrete grade will not be lower than M30 except for well steining and bottom plug, where M25 grade of concrete may be used. For PCC substructures minimum grade of concrete of M20 should be adopted.
- For all new two-lane structures, three-lane live load will be considered as per relevant standard.
- Locations of new minor bridges will generally be guided by the alignment of the highway. But, for major bridges, the bridge location and its alignment shall override the highway requirement in that portion.
- On economic considerations and for ensuring good riding quality, wherever possible, for the new bridges the layout of the existing bridges having a number of small spans will be modified by decreasing the number of spans, maintaining the piers parallel and in line with those of the existing structure.
- The deck will have 2.5% unidirectional camber/crossfall and the wearing course will be of uniform thickness of 12 mm Mastic and 50 mm BC.
- In general, it has been observed during the preliminary study that existing bridges with open type foundations have not suffered any distress even after more than 30 years of service and accordingly open type foundations are proposed to be adopted for new structures at these locations.
- Pile/open foundations will be adopted for flyovers and ROB structures, depending on the properties of the strata based on sub-soil investigation reports.

## 2.2.21 WIDTH OF NEW BRIDGES

Schemes of two-lane and four-lane configurations are to be proposed to be adopted in different stretches as per the schedule proposed by the DPR consultant.

All new bridges to be constructed for two-lane carriageways with paved shoulders as per with standard overall width, which may vary case to case.

## 2.2.22 Minor Bridges

Minor bridges are of total lengths more than 6 meters and less than 60 meters.

- In case of realignment or reconstruction of minor bridges, an overall width of the new carriageway has been proposed as 13.0 m (new two-lane with paved shoulders).
- In case of widening of existing minor bridges, widened width has been proposed as 12.0 m (two-lane with paved shoulders).

## 2.2.23 Major Bridges

Major bridges are of total lengths more than 60 meters.

- In case of realignment, width of new carriageway has been proposed as 13.0 m (New two-lane, with paved shoulders).
- In case, widening of existing major bridge is not feasible, an additional carriageway of 12.0 m width (additional two-lane carriageway) may be proposed. Deck configuration may be proposed as: 12.0 m = 7.5 m (carriageway) + 1.5 m (paved shoulder) + 2*0.5 m (crash barrier on either side) + 1.5 m (footpath) + 0.5 m (railing)
- In case of widening of existing major bridges, having solid slab and open foundations, widened width has been proposed as 12.0 m (2 lane with paved shoulders).

All new bridges with four-lane configurations, may be proposed to be constructed with 12.0 m overall width.

## 2.2.24 Rail Over Bridges (ROBs)

All new ROBs may be proposed with span arrangement as $1 \times 11.5 + 1 \times 24 + 1 \times 11.5$.

- A new ROB with a two-lane configuration, parallel to an existing ROB, may be proposed with a deck width as $1 \times 12.0$ m.
- A new ROB with a four-lane configuration, may be proposed with a deck width as $2 \times 12.0$ m.

## 2.2.25 Underpasses

Vehicular underpasses (VUPs) with openings of $3 \times 15$ m $\times 5.5$ m in urban areas and $2 \times 15$ m $\times 5.5$ m in non-urban areas may be proposed. Pedestrian underpasses (PUPs) with 7 m width $\times 3.5$ m vertical clearance may be proposed for the crossing of roads with less importance. Deck width may be proposed as 12.0 m. Deck configuration may be proposed as mention below:

$$12\,m = 7.0\,m\,(carriageway) + 2 \times 2.0\,m\,(paved\ shoulder)$$
$$+ 2 \times 0.5\,m\,(crash\ barrier\ on\ either\ side)$$

### 2.2.26  FLYOVERS

Flyovers with a 30 m span may be proposed. Vertical clearance above the crossroad may be 5.5 m. Deck width may be proposed as 12.0 m. Deck configuration may be proposed as mention below:

$$12\,m = 7.0\,m\,(carriageway) + 2 \times 2.0\,m\,(paved\ shoulder)$$
$$+ 2 \times 0.5\,m\,(crash\ barrier\ on\ either\ side)$$

### 2.2.27  PLANNING FOR NEW BRIDGES

In general, the following aspects are taken into account while planning for the new bridges and structures:

- Proper positioning of bridge and geometrics of approaches.
- Linear waterways and minimum vertical clearances as per stream hydrology.
- Satisfactory geological conditions.
- Aligning the piers with those of the existing structure to avoid cross currents and obstruction to flow.
- Minimum distance from the existing structure consistent with construction requirements and hydraulic consideration.
- Modular approach in design for both superstructure and substructures.
- Minimum number of spans consistent with road deck levels and minimum vertical clearance above design HFL.

### 2.2.28  PLANNING FOR NEW CULVERTS

For culverts, following guidelines will be followed:

- For culverts in new carriageway, minimum span and vent height will be kept equal to that of those in the existing carriageway; raising of deck level according to highway alignment will be made wherever required.
- Weak and non-functional culverts are to be dismantled and new culverts are to be constructed with carriageway and median matching with highway plan and profile.
- For central widening to three-lane, existing PCC abutments will be widened on both sides of the existing culverts. Existing slab shall be widened with specified camber to be cast for the full length.

- Culverts in service road locations to be extended up to the roadside longitudinal drain.
- In cases where vent height is very small (<500 mm), i.e., difference between road formation level and adjacent ground level is very less and there is no waterlogged area in close vicinity, culvert location should be reviewed again.
- In new alignments and bypasses, sufficient numbers of balancing box culverts are to be provided wherever alignment crosses through flat agricultural fields and lies in close vicinity to high embankments of railways and flood bunds.
- In case of culverts whose bed and floor have been scoured off severely and considerable afflux is observed, the same will be replaced with new culverts having adequate vents or with a minor bridge, based on adequate hydrological studies.
- Culverts will be designed for loading as per relevant standard provisions.
- Culverts shall be constructed for full formation width of the roadway.
- For RCC pipe culverts, expansion chambers shall be provided at median or between main carriageway and service road for proper maintenance.
- All cross drainage RCC pipe culverts with less than 900 mm diameter should be replaced with new 1.2 m (minimum) diameter RCC pipe culverts.
- All new RCC pipe culverts may be of minimum 1.2 m diameter.

## 2.2.29 DESIGN LOADING

The bridges shall be designed to sustain safely the most critical combination of various loads, forces, and stresses that can co-exist as per the provisions of the relevant standard. The allowable stresses, and the permissible increase in stresses, for various load combinations shall be adopted as per the relevant design standard.

## 2.2.30 CARRIAGEWAY LIVE LOAD

Structures carrying the proposed project road with a carriageway width of 11 m may be designed for three-lanes of specified wheeled loading or one-lane of tracked loading plus one-lane of wheeled loading, whichever produces the strongest effect.

Structures carrying the proposed project road with a carriageway width of 7.5 m shall be designed for two-lanes of wheeled loading or one-lane of tracked loading whichever produces the strongest effect.

Structures carrying crossroads of a two-lane carriageway shall be designed for two-lanes of wheeled loading or one-lane of tracked loading whichever produces the strongest effect.

## 2.2.31 TRACTIVE AND BRAKING FORCE

The tractive and braking forces shall be considered as per the provisions of clause of the relevant design standard.

### 2.2.32 Footpath Live Load

The intensity of the footpath loading may be considered as 500 kg/m² or as per clause of the relevant design standard.

### 2.2.33 Wind Forces

The effect of wind as per relevant standard shall be considered for the design of the various components of the bridge.

### 2.2.34 Seismic Forces

Seismic zones through which the project road passes are to be considered to obtain seismic forces in accordance with the clause of the relevant design standard. Vertical seismic forces shall be considered, and seismic restrainers are to be provided for preventing the dislodging of the superstructure under seismic conditions.

### 2.2.35 Buoyancy Effects

The following buoyancy effects are proposed to be considered wherever applicable for the design of various components of the bridges:

- For foundations                        100%
- For substructure below water level     15%

### 2.2.36 Codes and Publications

The following codes and publications (latest editions) shall be used for the design of the approach road and bridge components:

- Specifications for road and bridge works.
- Standard for general features of design.
- Standard for loads and stresses.
- Standard for prestressed concrete road bridges.
- Standard for reinforced concrete design.
- Standard for substructure and foundations.
- Standard for metallic bearings (Part I).
- Standard for elastomeric bearings (Part II).
- Standard for pot cum PTFE bearings (Part III).
- Standard manual of specifications & standards for four-laning of highways through public private partnership.

### 2.2.37 Deck Levels of Structures

The deck levels of the structures carrying the project road should be proposed based on the following parameters:

- Vertical clearance required above the crossroads.
- Vertical profile of the proposed project road.
- Vertical clearance required above the high flood level.
- Opening found adequate as per stream hydrology studies.

The computer applications for the "feasibility study" is described in chapter one and detail engineering in respect to various items of highway design, drawings, and estimation, are presented in Chapters 2–15 of this book, referring to the same chapters in the "The Guide for Computer Application Tutorials" part of the book available for download from the website of the book "www.techsoftglobal.com/download.php".

## 2.3 ROADWAY CAPACITY AND LEVELS OF SERVICE (LOS)

### 2.3.1 LEVEL OF SERVICE (LOS)

The LOS characterizes the operating conditions on the roadway in terms of traffic performance measures related to speed and travel time, freedom to maneuver, traffic interruptions, and comfort and convenience. The LOS range from LOS A (least congested) to LOS F (most congested). The Highways Capacity Manual (HCM) provides the following LOS definitions (Table 2.8).

### 2.3.2 DESIGN VEHICLE

The selection of the appropriate design vehicle is a key element in good intersection design practice. For most major intersections along the project road, it is common practice to accommodate the minimum turning path of a large semi-truck trailer (WB-18 m). As per AASHTO (U.S. Practice) design guide the minimum turning radius of a tractor semi-trailer truck (WB-18) is 18.2 m.

### 2.3.3 CAPACITY ANALYSIS

The capacity figures used for determining the desired carriageway width in differing terrain, with respect to traffic volume and composition, will be as per the relevant standard. The capacity for different carriageway widths derived from the above mentioned source is given in the table 2.9.

These capacity values are based on a design hour traffic flow of 8–10% and directional distribution of 67/33%. The capacity for the urban road for different lane configurations may be calculated by using the above base hourly flows and applying actual design hour factor and directional split of the road.

The observed traffic volume when related with capacity reveals the volume capacity (V/C) ratio of road sections. Various sections may vary with different carriageway widths and paved shoulder, the V/C ratio has been worked out by considering the average pavement width for each of the homogeneous sections.

Capacity analysis is carried out to identify the present and future level of services at various sections of project road. LOS-B may be recommended for highways in

## TABLE 2.8
## Level of Service

### Recommended Design Service Volume
### PCUs per Hour

| | | Total Design Service Volumes for 'LOS – A' in Different Categories of Road | | | | Available Tentative Design Speed at LOS | | | | | |
| --- | --- | --- | --- | --- | --- | --- | --- | --- | --- | --- | --- |
| S. No. | Configuration | Arterial | Sub-Arterial | Collector | Rural | A | B | C | D | E | F |
| 1 | 2-Lane one way | 2,400 | 1,900 | 1,400 | | 50 | 40 | 20 | 10 | 5 | — |
| 2 | 2-Lane two way | 1,500 | 1,200 | 900 | | 40 | 30 | 20 | 10 | 5 | — |
| 3 | 3-Lane one way | 3,600 | 1,900 | 2,200 | | 60 | 50 | 40 | 30 | 5 | — |
| 4 | 4-Lane – undivided – two way | 3,000 | 2,400 | 1,800 | | 50 | 40 | 20 | 10 | 5 | — |
| 5 | 4-Lane – divided | 3,600 | 2,900 | — | | 60 | 30 | 20 | 10 | 5 | — |
| 6 | 6-Lane – undivided – two way | 4,800 | 3,800 | — | | 60 | 30 | 20 | 10 | 5 | — |
| 7 | 6-Lane – divided | 5,400 | 4,300 | — | | 60 | 30 | 20 | 10 | 5 | — |
| 8 | 8-Lane – divided | 7,200 | — | — | | 60 | 30 | 20 | 10 | 5 | — |
| 9 | 2-Lane – undivided – two way (Rural) | | | | 2,500 | 60 | 50 | 40 | 30 | 10 | — |
| 10 | 4-Lane – divided (Rural) | | | | 2,500 | 80 | 60 | 40 | 30 | 10 | — |
| 11 | 6-Lane – divided (Rural) | | | | 4,300 | 100 | 80 | 60 | 40 | 20 | — |
| 12 | 8-Lane – divided (Rural) | | | | 6,300 | 100 | 80 | 60 | 40 | 20 | — |

LOS – A: Free flow
LOS – B: Reasonable free flow
LOS – C: Stable flow
LOS – D: Approaching unstable flow
LOS – E: Unstable flow
LOS – F : Forced/Breakdown flow

## TABLE 2.9
## Hourly Capacities for Different Lane Configurations

| Lane Configuration | Capacity (PCUs per hour) |
| --- | --- |
| 2 lane | 2500 |
| 4 lane | 4300 |
| 6 lane | 6300 |

**TABLE 2.10**
**Design Service Volumes**

| Carriageway Width | Terrain | Curvature (Degree/km) | Design Service Volume (PCU/Day) |
|---|---|---|---|
| Two lane | Plain | Low (0–50)<br>High (above 50) | 15,000<br>12,500 |
| Four lane | Plain | | 35,000 (earth shoulder)<br>40,000 (paved shoulder) |

non-urban sections. Thus, it will be identified whether LOS-B is being maintained during the designed period of the project. The design service volumes are commonly considered are provided in table 2.10.

These values of design service volume have been kept in view while considering improvement proposals for the project road.

## 2.4 DESIGN SPEED

### 2.4.1 DRIVING SPEED

The speed of driving is adopted by the drivers depending upon the road layout and the prevailing conditions, which is also influenced by the performance of the vehicles to some extent. The main factors which influence the driving speed are visibility, curvature, road width, surface condition, conflicts causing interruptions, and speed limit regulations. A road is never designed with features adequate for the fastest drivers, rather it is essential to ensure safe driving at a consistent speed appropriate for the type of road which is adopted by majority of the road users in good conditions and moderate traffic.

The design speed is the link which ensures safety of the users at all features on the road for the design parameters like stopping distances, horizontal and vertical alignments and cross-sectional elements used for the engineering design of the road.

### 2.4.2 SELECTING OF DESIGN SPEED

In the case of designing any new road, the factors that influence the selection of the design speed are road class, urban or rural location, existing development, terrain type, and economics. In the case of designing local roads, the objectives of the town planner should be taken into consideration, which might desire to maintain a low traffic speed to maintain a calm environment.

In the case of upgrading existing roads, the generally adopted traffic speed is measured and normally the 85th percentile of the speed is considered as the design speed. The various features are then designed as the part of the new upgraded road with that design speed.

### 2.4.3 EFFECT OF TERRAIN

Mountainous terrain often necessitates the selection of low design speeds because of the presence of significant lengths of gradients and topographical constraints.

## TABLE 2.11
### Design Speed by Road Class

| | Design Speed (km/hr) | | | | |
| | Rural Section | | Urban Section | | Absolute Minimum for |
| Road Class | Minimum | Minimum | Maximum | Maximum | Mountainous Terrain |
| --- | --- | --- | --- | --- | --- |
| Local road | 60 | 40 | 60 | 30 | 30 |
| Collector | 80 | 60 | 80 | 50 | 50 |
| Secondary Arterial | 90 | 60 | 90 | 60 | 50 |
| Primary Arterial | 100 | 80 | 100 | 80 | 60 |
| Expressway | 120 | 100 | 120 | 80 | 80 |
| Freeway | 140 | 120 | 120 | 100 | 80 |

*Source:* Table 3.1, Design Manual for Roads & Bridges, Ministry of Public Works and Kuwait Ministry, State of Kuwait

The minimum design speeds as given in Table 2.11 and may be adopted in such circumstances. In rolling terrain an intermediate value may be appropriate to adopt.

### 2.4.4 POSTED SPEED RELATED TO DESIGN SPEED

Speed limits applied to roads tend to restrain fast moving vehicle drivers. The posted speeds are displayed on the roads and are to be strictly followed by the drivers. Still, the design engineer must provide a margin of safety for the vehicles driven at a speed higher than the speed limit. The various posted speeds appropriate for given design speeds are indicated in Table 2.12.

## TABLE 2.12
### Recommended Posted Speed

| Design Speed (km/hr) | Maximum Posted Speed (km/hr) | Minimum Posted Speed (km/hr) |
| --- | --- | --- |
| 30 | 25 | |
| 40 | 25 | |
| 50 | 45 | |
| 60 | 60 | |
| 70 | 60 | |
| 80 | 80 | 40 |
| 90 | 80 | 40 |
| 100 | 100 | 60 |
| 110 | 100 | 60 |
| 120 | 120 | 80 |

*Source:* Table 3.2, Design Manual for Roads & Bridges, Ministry of Public Works and Kuwait Ministry, State of Kuwait

## 2.4.5 EXISTING ROADS

It is to be remembered that the Table 2.12 must not be used in reverse, which means that the design speed of an existing road must not be determined from the posted speed. The design speed may be obtained from the 85th percentile speed available from the survey of actual speed distribution of vehicles using the road. An appropriate posted speed may be determined from the selected design speed after the improvement of the existing road.

The design engineer has to take care to avoid discontinuity in standards while a new road leads to an existing road. Where the existing road has a lower design speed compared to the new or improved road, the transition should not be abrupt at the interface zone.

In all cases the posted speeds need to be properly displayed by the traffic signage in accordance with the countries standard for the road signage.

## 2.4.6 LOCATIONS OF CHANGING DESIGN SPEED

Even along the new or improved road, the design speed may change, which is most likely at the interface between urban and rural conditions. The driver should not face an abrupt downward change in the speed provision standards.

At places where there is a change from higher to lower design speed, it is desirable to provide design speeds above the minimum standards for sight distances, horizontal and vertical curvature over the initial length of the section at a lower design speed.

## 2.4.7 INTERCHANGES

The ramps connecting two different roadways within a grade separated interchange should normally have a lower design speed than the main line. The indicative values are shown in Table 2.13.

**TABLE 2.13**
**Minimum Design Speed for Connecting Roadways**

| Mainline Design Speed (km/hr) | Minimum Design Speed for Connecting Roadways (km/hr) | | |
|---|---|---|---|
| | Maximum Posted Speed (km/hr) | Minimum Posted Speed (km/hr) | Loops |
| 60 | 60 | 50 | 40 |
| 70 | 60 | 50 | 40 |
| 80 | 70 | 50 | 40 |
| 90 | 80 | 60 | 50 |
| 100 | 90 | 70 | 50 |
| 110 | 100 | 80 | 50 |
| 120 | 110 | 90 | 50 |

*Source:* Table 3.3, Design Manual for Roads & Bridges, Ministry of Public Works and Kuwait Ministry, State of Kuwait

### 2.4.8  Reduction Below Standards

In certain circumstances the alignment design may not suit the prescribed standards and may be uneconomic. In such cases it may be required to reduce the standard of the road locally. For example, at a given design speed the values of a circular curve radius, lengths of the transition curves, or the sight distance, may not fit the site condition and may be reduced for the particular situation.

The design speed for the length of route under consideration must be maintained throughout and not locally reduced. If the reduction from the values prescribed for the design speed is to be adopted at the site of a particularly difficult section, prior approval must be obtained from the project authority. This will restrict the frequent reduction from the prescribed standards by any one and at any location.

## 2.5  SIGHT DISTANCE

### 2.5.1  General

While driving, the driver is always attentive to both the forward and backward views. But the term sight distance means the forward view only. For safe maneuvers of the vehicle the driver must have adequate forward visibility. In the following three situations the forward visibility is particularly important:

- Stopping before reaching a stationary obstruction.
- Overtaking on an undivided single carriageway road.
- Making a decision where a choice of actions presents itself.

The corresponding distances for these three situations are:

- Stopping site distance (SSD).
- Safe passing site distance (SPSD).
- Decision sight distance (DSD).

The sight distance is always measured in a straight line between points on the centerline of each lane. On horizontal curves the most critical lane is the lane which is nearest to the center of the curve.

In order to ensure the required sight distance on horizontal curves, the objects on the inside of a curve may need to be set back further from the inner edge of the carriageway compared to their normal positions on the straight section of the road.

### 2.5.2  Driver's Eye Height and Object Height

The area between the driver's eye height and the object height is known as the visibility envelope for the sight distance.

The visibility envelope is to be checked in both horizontal and vertical planes, between two points in the center of the lane nearest to the center of the curve. On divided or dual carriageway roads this is to be checked on the innermost lanes of both the carriageways.

For design purposes the minimum and maximum eye height are commonly taken as 1.05 m and 2.4 m respectively.

The minimum and maximum height of stationary object is commonly taken as 0.15 m and 2.0 m respectively.

For passing purposes in overtaking situations on undivided single carriageway roads, the object heights are taken as the same as the driver's eye height, which range therefore from 1.05 m and 2.4 m.

### 2.5.3 STOPPING SIGHT DISTANCE (SSD)

Figure 2.1 describes the visibility envelope for SSD as:

- Driver's eye height          1.05 m–2.40 m
- Object height                0.15 m–2.00 m

SSD is composed of two elements:

- Perception and reaction distance.
- Braking distance.

The various values of SSDs for design purposes are given in Table 2.14. The vertical gradients of the road affect the braking distance and therefore the braking distances are longer on down-grades, while braking distances are shorter on upgrades. But these don't change the SSDs. The "level" values may be used on all upgrades.

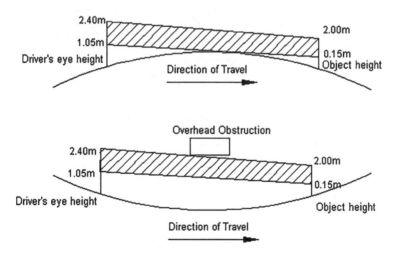

**FIGURE 2.1**   Visibility Envelopes for Stopping Sight Distance.
*Source:* Figure 4.1, Design Manual for Roads & Bridges, Ministry of Public Works and Kuwait Ministry, State of Kuwait

**TABLE 2.14**
**Stopping Sight Distances for Design**

| Design Speed | Stopping Sight Distances (m) | | | |
|---|---|---|---|---|
| (km/hr) | Level | 2% Downgrade | 4% Downgrade | 6% Downgrade |
| 30 | 35 | 35 | 35 | 35 |
| 40 | 50 | 50 | 50 | 50 |
| 50 | 65 | 65 | 70 | 70 |
| 60 | 85 | 85 | 90 | 95 |
| 70 | 105 | 110 | 115 | 120 |
| 80 | 130 | 135 | 140 | 145 |
| 90 | 160 | 160 | 170 | 175 |
| 100 | 185 | 190 | 200 | 210 |
| 110 | 220 | 225 | 235 | 245 |
| 120 | 250 | 260 | 270 | 285 |

*Source:* Table 4.1, Design Manual for Roads & Bridges, Ministry of Public Works and Kuwait Ministry, State of Kuwait

### 2.5.4 Safe Passing Sight Distance (SPSD) or Overtaking Sight Distance (OSD)

SPSD applies to undivided single carriageway two-lane roads having two-way traffic. On these roads, while a vehicle is performing an undertaking or overtaking maneuver, it moves into the lane used by vehicles traveling in the opposite direction.

Figure 2.2 describes the visibility envelope for SPSD or OSD as:

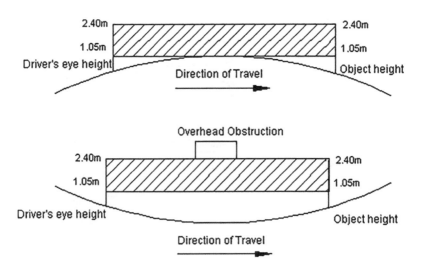

**FIGURE 2.2**   Visibility Envelopes for Safe Passing Sight Distance.
*Source:* Figure 4.2, Design Manual for Roads & Bridges, Ministry of Public Works and Kuwait Ministry, State of Kuwait

- Driver's eye height          1.05 m–2.40 m
- Object height                1.05 m–2.40 m

SPSD or OSD is composed of the following elements:

- The initial maneuver.
- The occupation of the left lane.
- The clearance length.
- The opposing vehicle distances.

As vertical gradients of the road do not have any effect on SPSD or OSD, there are no specified adjustments to be adopted. However, there may be a desirability of increasing the visibility beyond the minimum standard in case passing or overtaking is to be accommodated on a section of road with significant grades. The values of SPSD or OSD may be adopted from Table 2.15.

## 2.5.5  DECISION SIGHT DISTANCE (DSD)

At the junctions, intersections, and some other certain points on the road, the driver may have to decide to select a route or to stop. This needs adequate visibility to allow the decision to be made in suitable time. DSD is longer than the SSD as the correct course of action may be to stop and sometimes vehicles move a significant distance when maneuvering without any reduction in the speed.

The visibility envelope for DSD is the same as it is for SSD and is shown in Figure 2.1:

- Driver's eye height          1.05 m–2.40 m
- Object height                0.15 m–2.00 m

**TABLE 2.15**
**Safe Passing Sight Distances for Design**

| Design Speed (km/hr) | Safe Passing Sight Distance (m) |
|---|---|
| 30 | 200 |
| 40 | 270 |
| 50 | 345 |
| 60 | 410 |
| 70 | 485 |
| 80 | 540 |
| 90 | 615 |
| 100 | 670 |

*Source:* Table 4.2, Design Manual for Roads & Bridges, Ministry of Public Works and Kuwait Ministry, State of Kuwait

**TABLE 2.16**
**Decision Sight Distances for Design**

| | Decision Sight Distance | | | |
| | To Stop Control | | All Other Situations | |
| Design Speed (km/hr) | Rural | Urban | Rural | Urban |
|---|---|---|---|---|
| 50 | 70 | 155 | 145 | 195 |
| 60 | 95 | 195 | 170 | 235 |
| 70 | 115 | 235 | 200 | 275 |
| 80 | 140 | 280 | 230 | 315 |
| 90 | 170 | 325 | 270 | 360 |
| 100 | 200 | 370 | 315 | 400 |
| 110 | 235 | 420 | 330 | 430 |
| 120 | 265 | 470 | 360 | 470 |

*Source:* Table 4.3, Design Manual for Roads & Bridges, Ministry of Public Works and Kuwait Ministry, State of Kuwait

DSD is composed of three elements:

* Detection and recognition phase.
* Decision and response phase.
* Maneuver.

DSD is measured from the vehicle location to the point which demands decision, for example the top sign or the start of a bend of a ramp or loop. The values for varying design speeds are given in Table 2.16. There is no adjustment to be done for vertical gradients of the road.

DSD should be provided where any of the following situations occur:

* Unusual or unexpected maneuvers required at interchanges or intersections.
* Significant visual distractions such as traffic control devices and illuminated advertisements.
* Changes in the road cross section, for example a lane drop.

Typical examples of circumstances where the DSD is to be considered are:

* A rural road leading directly to a STOP signage or signal.
* An urban road leading directly to a STOP signage or signal.
* An off ramp leading to an abrupt change in the direction.
* At approach to a lane drop.
* A complexed weaving section having more than two entries or exits.

Commonly it is not always possible to provide the full DSD and in such circumstances it is advisable to increase the provisions of warning signs.

## 2.5.6  MAINTAINING SIGHT DISTANCES

On existing roads, sight distances may be measured directly on the ground by observing from the relevant eye height to the target at object height along the centerline of each lane.

For new or improved roads, the sight distances should be checked at the design stage by direct measurement on the plan and elevation drawings, either on prints or on CAD display. It is to be checked that no substantial objects obstruct the sightlines, including traffic signs, barriers, bridge parapets, etc. However, isolated slim objects such as street light posts, signage posts, and individual tree trunks can be ignored.

SSD should be maintained throughout the length of the route under design or improvement, and this may have a constraining influence on various elements in the geometric design of the road. The situations for adopting DSD are already discussed in previous section.

On horizontal curves, the various visibility obstructions which are located on the inside of the curve are to be adequately set back from the edge of the carriageway. Appropriate setbacks are to be provided to the faces of barriers and the bridge parapets located on the inside of the curve. This may require widening of the verge or median. In cut sections the side slopes may obstruct the visibility and are to be checked in a three-dimensional perspective, using drive through simulation at different driving speeds.

On crest vertical curves, the minimum values of curvature set out for different speeds are adequate to cater for SSD, but the designer should always check the DSD wherever applicable.

On sag vertical curves, the upper bound of the sight distance envelope should be checked against any overhead obstructions like a rail or road over bridge, or signage truss, etc.

## 2.5.7  PROVISION OF SAFE PASSING SITE DISTANCE

On a two-lane undivided single carriageway road it is not always possible to provide passing provisions. But frustration and dangerous maneuvers may take place in the case of non-availability of a passing facility on longer sections to allow vehicles to pass each other safely. At least half the total length of the route should permit safe passing. If this is not achieved, then provision of an auxiliary lane may be considered following AASHTO guidelines.

## BIBLIOGRAPHY

*A Policy on Geometric Design of Highways and Streets 2001*, AASHTO.
*Overseas Road Note 6, A Guide to Geometric Design*, Overseas Unit, Transport and Road Research Laboratory, Department of Transport, Crowthorne, Berkshire, United Kingdom.
*A Guide on Geometric Design of Roads*, Arahan Teknik (Jalan) 8/86, JKR, Malaysia.
*A Guide to the Design of At Grade Intersections*, JKR 201101-0001-87, Malaysia.
*A Guide to the Design of Interchanges*, Arahan Teknik (Jalan) 12/87, JKR, Malaysia.

*Manual on Traffic Control Devices Road Marking & Delineation*, Arahan Teknik (Jalan) 2D/85, JKR, Malaysia.

*Guidelines for Presentation of Engineering Drawings*. Arahan Teknik (Jalan) 6/85, Pindaan 1/88, JKR, Malaysia.

*Design Standards, Interurban Toll Expressway System of Malaysia*, Government of Malaysia, Malaysia Highway Authority, Ministry of Works and Utilities.

*Design Manual for Roads & Bridges*, Ministry of Public Works and Kuwait Ministry, State of Kuwait.

*Geometric Design Manual for Dubai Roads*, Dubai Municipality, Roads Department.

*Geometric Design Standards for Rural (Non-Urban) Highways*, IRC 73-1980, IRC.

*Vertical Curves for Highways*, IRC SP 23-1993, IRC.

*Guidelines for the Design of Interchanges in Urban Areas*, IRC 92-1985, IRC.

*Lateral and Vertical Clearances at Underpasses for Vehicular Traffic*, IRC 54-1974, IRC.

*Guidelines for the Design of High Embankments*, IRC 75-1979, IRC.

*Type Design for Intersection on National Highways*, Ministry of Surface Transport, Govt. of India, 1995.

*Rural Roads Manual, Indian Roads Congress*, Special Publication 20.

*Manual of Standards and Specifications, for Two-Laning of State Highways on BOT Basis*, IRC SP 73-2007, IRC.

*Manual of Specifications & Standards for Four-Laning of Highways Through Public Private Partnership*, IRC SP 84-2009, IRC.

*Geometric Design Standards for Rural (Non-Urban) Highways*, IRC 73-1980, IRC.

# 3 Introduction to Computer Applications

## 3.1 GENERAL

The following are various tasks related to highway designs done with computer applications:

- Topographical survey or satellite data processing.
- New highway design with uniform cross section by the design of horizontal alignment, vertical alignment, cross section elements, earthwork estimation, and the design and construction drawings.
- Design of at-grade intersections.
- Design of multi-level grade separated interchanges.
- Design of flexible and rigid pavement.
- Design of highway drainage.

To download the installation setup of the highway design software, HEADS Pro, and the tutorials with videos, which contains the tutorials in two folders, one is "book tutorials" for the readers of this book as a learner. The other one is "HEADS Pro tutorials" which has various tutorial videos, tutorials, and examples for users at basic, professional, and advanced levels. Refer to "book tutorials" of the website at: "www. techsoftglobal.com/download.php",

- First, the software HEADS Pro is to be downloaded and installed on the user's computer.
- Next, the tutorial videos are to be downloaded from the website to the user's computer and to be watched to start with, always.
- Finally, all the data and guides in the folder "book tutorials" are to be used in conjunction with the study of the relevant chapters of the book.

## 3.2 TOPOGRAPHICAL MAPS OR AERIAL SURVEY DATA PROCESSING

Computer applications are very effective and helpful at every stage of a highway development project, starting from the conceptual phase, to the preliminary design, to the detailed engineering, and to construction. The collecting and processing of ground survey data is the first step in computer-aided highway design. The process includes the creating of a ground model, digital mapping to create the survey base plan, triangulation to create the Digital Terrain Model (DTM) and thus the ground contours, and finally the ground long and cross sections.

The collecting of ground data comes either from obtaining the field survey data, most commonly by using the total station instrument, or by obtaining the ground elevation data with the latest available satellite-based technology, which also helps in identifying the best possible route of the road, its realignment, and the widening pattern with respect to available space and land use.

**FIGURE 3.1**    Satellite Imagery and Map of the Terrain

## 3.3   PROCESSING OF TOPOGRAPHIC SURVEY DATA BY TOTAL STATION INSTRUMENT

Trimble GPS
Receiver R8 with
Communications

Trimble GPS
Receiver R10 with
Advanced

Trimble R8 Model
4 trimble gps
receiver

**FIGURE 3.2**    Various GPS Instrument Models

The project corridor may be identified by site reconnaissance survey and by using topographic maps, along with satellite imagery from Google Earth.

The ground elevation data may be obtained either through a total station survey or by downloading the data from shuttle radar topography mission (SRTM) on the internet. Users may procure the software program Global Mapper to download ground topography data from internet. Most highway design software programs use Total Station Survey Data for processing topographical details most effectively and economically in various highway design and construction projects.

The next job is to prepare the survey base plan or ground model from the ground survey data with Total Station Survey. By selecting the drawing feature symbol from the CAD Block library, the various texts obtained from the surveyors are also placed correctly in the base plan drawing. The drawing is made in a CAD layered system and is compatible to AutoCAD and other popular CAD programs.

Finally, the Digital Terrain Model (DTM) is created by using the total station survey data by triangulation and subsequently processed for contouring.

## 3.4   PROCESSING OF CROSS SECTION SURVEY DATA BY AUTOLEVEL INSTRUMENT

**FIGURE 3.3**   Various Autolevel Digital Level Instrument Models

The Autolevel data contains chainages at a regular/constant interval, the distances on the cross section on either side of the center point 0.0, left side in "-ve" & right side is "+ve" and the elevations (Z) at all distances on either side of the center point. The data does not have easting (X) and northing (Y) at these points where (Z) is available. So, to get easting (X) and northing (Y) at these points, an alignment is to be defined, which passes through the center point at "0.0" of the cross sections at every chainage. The alignment must have the chainages at the same intervals as those in the Autolevel cross section survey. Referring to this alignment, the program will obtain the X & Y coordinates of each point on the cross section, this helps in making triangulation, the Digital Terrain Model (DTM), and finally the contours of the ground. Only the survey base plan drawing by digital mapping will not be created from the ground model, which needs total station data for reading various features like houses, drains, electric poles, and boundary walls, etc., on the ground.

## 3.5   PROCESSING OF BEARING LINE SURVEY DATA

**FIGURE 3.4**   Various Total Station and GPS Instrument Models

The bearing line method of a survey is very widely used in the design of hill roads. Every hill road commonly has a hill on one side and a valley on the other side. In a bearing line survey, the traverse passes through the foothill side, which changes its side with respect to the road to stick to the foothill always.

## 3.6 PROCESSING OF GROUND ELEVATION DATA BY SATELLITE DOWNLOADED FROM INTERNET

By defining tentative alignment by drawing straights by "My Path" in Google Earth to define the project corridor and subsequently downloading ground elevation data by using Global Mapper, minimizes the exercise to transform the surveyor's coordinates in transverse Mercator (TM) into Global Coordinates in UTM (Universal Transverse Mercator)/GPS. The alignment file may be saved as a KML/KMZ file. This file may be opened in the future in Google Earth to mark various features along the corridor, for example, the road alignment with existing "landuse", etc. The area may also be identified or verified by a site reconnaissance survey, using topographic maps, or by procuring satellite images from the survey department.

Next, the KML/KMZ file may be opened with Global Mapper and saved as a DXF file. Also, the ground elevation data may be downloaded with Global Mapper from SRTM (Shuttle Radar Topography Mission). The DXF file of the alignment is opened in HEADS Viewer and the alignment geometrics are designed as the horizontal alignment of the project highway.

The downloaded ground elevation data is to be opened with Microsoft Excel for saving in the desired format. Next the ground data is opened with HEADS Pro to create the ground model, and next the Digital Terrain Model (DTM) along with contours, are to be generated for the entire area under the study. This enables the engineers to develop the survey base plan, Digital Terrain Model (DTM) by Delauny triangulation, ground contours at user given primary and secondary intervals, 3D surface with rendering, and ground long and cross sections. This enables the users to prepare survey base plan drawings by drawing the ground model over Google Image, by using the CAD facilities.

## 3.7 THE CONTOURS ARE SUPERIMPOSED ON THE SATELLITE IMAGERY

The ground survey work also includes obtaining traverse coordinates with closing error correction by Bowditch, transit and close link methods, EDM survey, coordinate transformation from WGS84 to Lambert conformal conic projection (long lat to east north) and the reverse with Spheroid Everest (1956). This allows for installing the survey reference pillars along the proposed highway route and for their reference during the construction.

HEADS Pro software will be used for the design of any highway widening with reference to an existing road with multiple cross sections and multiple alignments, where the road cross sections may change with different configurations and with the change in the widening pattern from left to right, to concentric along the route from origin to destination. The design has special treatment on hill stretches. Provision for tunnels is also possible with very special design techniques.

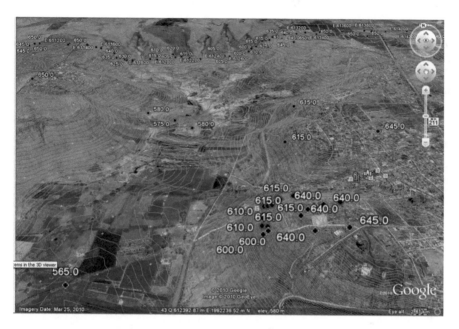

**FIGURE 3.5**   Contour Are Superimposed on the Satellite Imagery

## 3.8   GROUND MODELING

The survey data whether by Total Station Survey or by satellite data, must be made available in the desired format of five columns for: serial number, easting (meters), northing (meters), elevation (meters) and feature code and saved as a text file (for example SURVEY.TXT) which can be opened in "Notepad." It has five columns of data. When a data record with a "Serial No. 0" is met, then it discontinues the last feature and reads a new feature in the ground model. After creating the ground model, it is drawn to create the survey base map and is described in the relevant section of this book.

## 3.9   TRIANGULATION AND CONTOURS

The model name, string label and increments for primary, secondary and text are set with default names and values. If the ground is flat in nature, then a user has to reduce the primary contour increment as 0.5 or 0.1 to get contours for the relatively flat ground. The secondary contour increment is always five times the increment of the primary contours. Texts for contour elevations are displayed on the secondary contours.

## 3.10  TUTORIAL VIDEOS

The various tutorial videos are helpful to understand and master the above processes. By watching the tutorial videos and by using the file survey.txt, as given in the Book Tutorials, the process can be done by using any project survey data.

The various tutorial data are available in the two primary folders 'Book Tutorials' and 'Software Tutorials', which exist in the website of this book, as mentioned below:

1.  Website: http://www.roadbridgedesign.com, for downloading as follows:
    There is menu item 'Book Tutorials', under this menu item, select 'Computer Aided Highway Engineering', which contains 'User's Guide for Hands On Practice', and all other items of this book.
2.  Website: http://www.techsoftglobal.com, for downloading as follows:
    There is menu item 'Software Tutorials', under this menu item, select 'HEADS Pro', which contains various items used in this book, and are to be downloaded. Software 'HEADS Pro R24 Installation Setup' is to be run to install HEADS Pro in the computer.

    Chapter wise various data may be processed by following the step-by-step guide in the 'User's Guide for Hands On Practice', for respective chapters of this book.

## 3.11  NEW HIGHWAY OR LOW-COST RURAL ROAD DESIGN WITH UNIFORM SINGLE OR DUAL CARRIAGEWAY

These are highways or roads that have uniform single or dual carriageway cross sections throughout their alignment. These types of highways or roads are designed based on the design of the horizontal alignment and the vertical profile of the road's centerline. The various cross section elements are created by defining the horizontal and vertical offset distances with respect to either the centerline or any other reference lines in the cross section.

## 3.12  NEW HIGHWAY, EXPRESSWAY, OR FREEWAY DESIGN WITH MULTIPLE SECTIONS

These are highways or roads with different at-grade and elevated cross sections applied on various sections from chainage station to chainage station along the alignment of the highways or roads. These types of highways or roads are also designed based on the design of the horizontal alignment and the vertical profile of the road's centerline. The various cross section elements are designed by defining the horizontal and vertical offset distances with respect to either the centerline or any other reference lines in the cross section.

## 3.13 HIGHWAY, EXPRESSWAY, OR FREEWAY WIDENING WITH MULTIPLE SECTIONS

These are highways or roads constructed for widening from two/four lane configurations to four/six/eight lane configurations, with different at-grade and elevated cross sections applied on various sections from chainage station to chainage station along the alignment of the highways or roads. These types of highways or roads are also designed based on the design of the horizontal alignment and the vertical profile of the centerline of the carriageway on either side of the central verge or median. Here the alignment for the median centerline is drawn based on the desired widening pattern, either on the left or right side, or the concentric and horizontal alignment for the centerline are designed. Next, by setting horizontal offset distances with respect to the alignment of the median centerline, the centerlines for left and right carriageways are designed. Next, the vertical profile is separately designed for the centerlines of the left and right carriageways. Normally a pavement overlay is considered over the existing road carriageway and a full pavement structure is considered on the new construction side for widening the existing road. Most commonly the carriageways on either side are at different elevations. This requires drains in the median or central verge on curved sections to ensure the drainage of storm water outside the super elevated sections.

## 3.14 HILL ROAD DESIGN

Hill road designs are done either by using Total Station, Auto/Digital Level, or Bearing Line data. Most commonly hill roads are designed for a single carriageway configuration. In the case of widening from a single carriageway to a dual carriageway configuration, the widening is commonly done by cutting in to the hill side.

During hill road design, some special provisions are made in the cross sections of the retaining wall on the valley side, gabion walls on the hill side, varying slopes from flatter to steeper with the depth of the cut, etc. Special care is to be taken in the design of the horizontal alignment and the vertical profile by balancing the natural topography with ensuring adequate visibility.

Traffic management during construction is also quite difficult if an alternative route is not available. The excavated earth and debris are transported by dumpers and tippers which usually block the road during the loading operations. The excavators also operate on the road. So, sections may be undertaken for construction in small lengths. This will prevent the accumulation of excessive queuing traffic from either end.

## 3.15 DESIGN OF AT-GRADE INTERSECTIONS

The at-grade intersection is a provision in the road network where two or more roads meet or cross at-grade or at the same vertical level. There are three basic types of at-grade intersections, these are major or minor intersections, roundabout or rotary, and U-turns. Each of these types may be signalized or non-signalized.

Intersections are widely identified as the primary locations of accidents on all roads. Adequate safety is therefore important in the design of an intersection. These include visibility, driver perception, signing and road marking, traffic control, and pedestrian access.

## 3.16   DESIGN OF MULTI-LEVEL GRADE SEPARATED INTERCHANGES

Fully grade separated interchanges are used to vertically separate some or all of the conflicting streams of traffic by using one or more bridges or underpasses. Conflicts are eliminated, leaving only merging and diverging movements, which occur at ramp terminals.

## 3.17   DESIGN OF FLEXIBLE AND RIGID PAVEMENT

The pavement design features for design of flexible rigid pavement, rigid pavement and pavement overlay by using Benkelman beam rebound deflection data by the AASHTO (American Association of State Highway and Transportation Officials) method. The analysis for a layered pavement system is essential to study the improvements in the horizontal strain, vertical stress and vertical deflection of the pavement under applied loading by introducing fiber glass geogrids, jute geotextiles, etc., in the pavement structure composed of bound layer (bituminous), unbound layer (granular GSB or WMM) and subgrade.

## 3.18   DESIGN OF HIGHWAY DRAINAGE

The drainage design for long and cross sections of side drains is based on Manning's equation. Users may try various sections as desired for their adequacy to ensure effective drainage of storm water for a given duration.

## 3.19   MODES OF DATA PROCESSING

A user has to select the working folder at the start, for example, a working folder may be created as "Work" on the desktop, next, various menu items will be enabled for working. HEADS Pro has three modes of processing: **New/Open Project Workspace, Open a Project Data File** and **Open a Text Data File**.

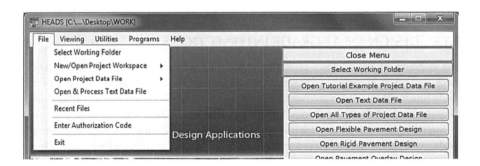

### 3.19.1   NEW/OPEN PROJECT WORKSPACE

This is how to begin producing deliverables for any project using HEADS Pro. Here the process is structured separately for each individual project. Most projects start with the total station survey data obtained from field survey. By processing the survey data for ground modeling, triangulation, and contour, the design process begins. The design process creates all the design reports and drawings for construction. A text file for design input data is also produced as "Project Data File.txt" by writing all the design data files in proper sequence.

#### 3.19.1.1   Selecting the Processing Mode as New/Open Project Workspace

In **Workspace** mode, from the HEADS Pro main screen, select **File** and menu item **New/Open Project Workspace**, to open the workspace with data panel and CAD window. Here the design for the road geometrics is done in a CAD environment. Here user doesn't have to write the data files, rather the data will be written by HEADS Pro upon inputting the required parameters for the design.

A **New Design** has the facility for CAD based interactive design for horizontal and vertical alignments. By opening HEADS Viewer and next, by selecting menu option **Design** in HEADS CAD Viewer, the horizontal and vertical alignments may be designed interactively as **New HIP** and **New VIP** methods. During the design of highways, hill roads, and tunnels, the main job is the geometric design of the horizontal alignment, vertical profile, and cross sections of the highway/hill road/tunnel. The set of all input data will be saved as **Project Data File** in the specific <Project Folder> located inside the **Working Folder** as selected by the user at the start of each session. *This process utilizes a large amount of system memory, so systems with insufficient memory, less that 4 GB, may slow down the process.*

There are various items featured under **Project Workspace**, the following, are the most frequent designs that are created here:

- New highway or low-cost rural road design with uniform single or dual carriageway.
- New highway, expressway, or freeway design with multiple sections.
- Highway, expressway, or freeway widening with multiple sections.
- Hill road design.
- Design of at-grade intersections.
- Design of multi-level grade separated interchanges.
- Design of flexible and rigid pavements.
- Design of highway drainage.

Under the **File** menu, the first item is the **Select Working Folder**, the user has to select the working folder, for example, folder "Work" on the desktop, as "Desktop\Work." Next, the user has to select the **File** menu item **New/Open Project Workspace**, then select the desired item to open the design workspace for processing the design in various steps. The process of the design for each item is explained step-by-step in subsequent chapters from four thru fifteen, in this guide.

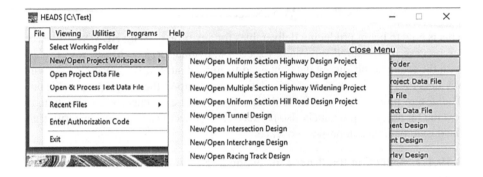

### 3.19.2 NEW/OPEN A PROJECT DATA FILE

This option is not for the design of the project. Having the file Project Data File.TXT created as output by the previous process of design by using Workspace is in order to process the survey data and all other data written in sequence in the project data file in batch mode, after any modification that is manually done by the user, by following the acceptable data format. In a fresh folder, the user has to copy the survey data file and the project data file as SURVEY.TXT and Project Data File.TXT, select the project data file to open for processing, a list will come up to process all the data in sequence as written in the project data file, the user has to click the button **Proceed** to start the batch processing.

In **Project Data File** mode, from the HEADS Pro main screen, select **File** menu and select the item **Open a Project Data File** to select a project data file as created in **Workspace** mode. Next the **Process List** is displayed, the various items in the **Process List** are in the same sequence of various major option data in the project data file. Upon clicking on button **Proceed**, the various items will be batch processed.

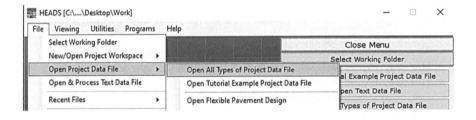

### 3.19.3 NEW/OPEN A TEXT DATA FILE

There are various types of data, for example DGM, MODELEDIT, HALIGN, VALIGN, OFFSET, PROOFFSET, XSEC, INTERFACE, VOLUME, PLAN, PROFILE, SECTION, etc., as described in the user manual. These major option data may be written in NOTEPAD as a text data file and may be saved in a fresh folder as a text file, say, 'Input Data.txt'. To process the data in this file, the user has to select this text data file and select the relevant menu item from the top menu in the main screen of HEADS Pro.

For example: Geometrics >> 200 Halignment, next the process dialog box comes up, and the user has to click on button **Proceed**. When the process is over, a message comes up informing the user about the report file created, the user has to click on **OK** and then **Finish**. The above process takes the input for 200 HALIGN data written anywhere in the selected text data file and creates outputs as a report and the model in model files. In case there are more than one 200 HALIGN in the selected text data file, then every time the topmost 200 HALIGN data will be processed and others below this data will not be processed. So, the user may have to cut and paste the desired data to the top for processing and saving the text data file every time. The same method is applied to the other data in the same text data file, for example, 300 Valignment, 400 Offset, etc.

There is a difference when processing a text data file using **Open a Text Data File**, where only one data set is processed at a time upon running the relevant process selected by the user from the menu option and **Open a Project Data File**, where all the data are processed available in the project data text file by automatically selecting the process by itself. In **Text Data File** mode, from HEADS Pro main screen, select **File** menu item **Select & Process Text Data File** to select a text data file for a major option. Next, process the selected major option data by clicking the menu item specific to the major option data in the selected text data file, in the same way of old classic mode. Here the user can process individual major option data for various components of DGM, geometrics, and drawings, like Create DGM/GEN file, HALIGN, VALIGN, OFFSET, XSEC, INTERFACE, PLAN, Profile and Section etc.

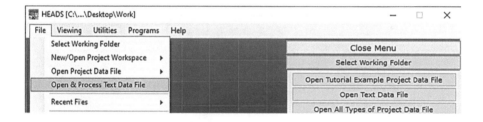

## BIBLIOGRAPHY

*HEADS Pro Design Manual.*
*HEADS Pro Example Manual.*
*HEADS Pro User's Manual.*
*HEADS Pro Quick Reference Guide.*
*HEADS Pro Installation Guide.*

The above items are available for download from the website http://www.techsoftglobal.com/download.php by selecting menu item 'Downloads', then further selecting 'HEADS Pro Release 24' and 'Book Tutorials', and from the website http://www.roadbridgedesign.com, by selecting menu item 'Book Tutorials', then further selecting 'Software Tutorials' and 'HEADS Pro'.

# 4  Topographical Survey and Data Collection

## 4.1  GENERAL

The basic objective of the topographic survey would be to capture the essential ground features along the alignment to consider improvements and for working out improvements, rehabilitation, and upgrading costs. The detailed topographic surveys should normally be taken up after the completion of reconnaissance surveys.

## 4.2  TOPOGRAPHICAL SURVEYS AND INVESTIGATIONS

### 4.2.1  RECONNAISSANCE AND ALIGNMENT

1. The consultants should make an in-depth study of the available land width or Right of Way (ROW) topographic maps, satellite imageries and air photographs of the project area, and other available relevant information collected by them concerning the existing alignment. The consultant must arrange the required maps and the information needed by him from the potential sources.
2. The detailed ground reconnaissance may be taken up immediately after the study of maps and other data. The primary tasks to be accomplished during the reconnaissance surveys include:
   - Topographical features of the area.
   - Typical physical features along the existing alignment within and outside ROW i.e., land use pattern.
   - Possible alignment alternatives, vis-à-vis, scheme for the construction of additional lanes parallel to the existing road.
   - Realignment requirements including the provision of bypasses, ROBs/ Flyovers, and viaduct for pedestrian crossings.
   - Preliminary identification of improvement requirements including treatments and measures needed for the crossroads.
   - Traffic pattern and preliminary identification of traffic homogenous links.
   - Sections through congested areas.
   - Inventory of major aspects including land width, terrain, pavement type, carriageway type, bridges, and structures (type, size, and location), intersections (type, crossroad category, location) urban areas (location, extent), geologically, sensitive areas, environmental features.
   - Critical areas requiring detailed investigations.
   - Requirements for carrying out supplementary investigations.
   - Soil (textural classifications) and drainage conditions.
   - Type and extent of existing utility services along the alignment (within ROW).

The data derived from the reconnaissance surveys are normally utilized for planning and programming the detailed surveys and investigations. All field studies including the traffic surveys should be taken up based on information derived from the reconnaissance surveys.

1. The data and information obtained from the reconnaissance surveys should be documented. The data analysis and the recommendations concerning alignment and the field studies should be included in the Inception Report. The data obtained from the reconnaissance surveys should form the core of the database which would be supplemented and augmented using the data obtained from detailed field studies and investigations.

2. The data obtained from the reconnaissance surveys should be compiled in the tabular as well as graphical chart form indicating the major physical features and the proposed widening scheme for NHAI's comments. The data and the charts should also accompany the rationale for the selection of traffic survey stations.

## 4.2.2 TOPOGRAPHIC SURVEYS

1. The carrying out of topographic surveys will be one of the most important and crucial field tasks under the project. The detailed field surveys shall be carried out using high precision instruments i.e., GPS, Total stations. The data from the topographic surveys shall be available in (x, y, z) format for use in a sophisticated DTM.

2. The detailed field surveys would essentially include the following activities:
   • Topographic Surveys along with the Existing ROW: Running a continuous open traverse along the existing road and realignments, wherever required, and fixation of all cardinal points such as horizontal intersection points (HIP's), center points and transit points, etc. and properly referencing the same with a pair of reference pillars fixed on either side of the centerline at safe places within the ROW.
   • Collection of details for all features such as structures (bridges, culverts, etc.) utilities, existing roads, electric and telephone installations (both O/H as well as underground), huts, buildings, fencing, and trees (with girth greater than 0.3m) oil and gas lines, etc. falling within the extent of the survey.

3. The width of the survey corridor will generally be as given under:
   • The width of the survey corridor should consider the layout of the existing alignment including the extent of embankment and cut slopes and the general ground profile. While carrying out the field surveys, the widening scheme of highway or bridges (i.e., right, left, or symmetrical to the centerline of the existing carriageway) should be taken into consideration so that the topographic surveys cover sufficient width beyond the centerline of the proposed divided carriageway. Normally the surveys should extend a minimum of 30m beyond either side of the centerline of the proposed divided carriageway or land boundary whichever is more.

- In case the reconnaissance survey reveals the need for bypassing the congested locations, the traverse lines would be run along the possible alignments to identify and select the most suitable alignment for the bypass. The detailed topographic surveys should be carried out along the bypass alignment approved by the authority. At locations where grade-separated interchanges could be the obvious choice, the survey area will be suitably increased. Fieldnotes of the survey should be maintained which would also provide information about traffic, soil, and drainage, etc.
- The width of the surveyed corridor will be widened appropriately where developments and/or encroachments have resulted in a requirement for adjustment in the alignment, or where it is felt that the existing alignment can be improved upon through minor adjustments.
- Where existing roads cross the alignments, the survey will extend a minimum of 100m on either side of the road centerline and will be of sufficient width to allow improvements, including at grade intersection to be designed.
4. The surveyed alignment shall be transferred on to the ground as under:
Reference Pillar and Benchmark/Reference pillar of size 15cm × 15cm × 60cm shall be cast in RCC of grade M15 with a nail fixed in the center of the top surface. The reference pillar shall be embedded in concrete up to a depth of 45 cm with CC M10 (5 cm wide all around). The balance 15 cm above ground shall be painted yellow. The spacing shall be 250m apart, in case the Benchmark pillar coincides with the reference pillar, only one of the two need to be provided.
- Establishing Benchmarks at a site connected to GTS Benchmarks at an interval of 250m on Benchmark pillar made of RCC as mentioned above with RL and BM No. marked on it with red paint.

### 4.2.2.1 Longitudinal and Cross Sections

1. The topographic surveys for longitudinal and cross sections shall cover the following:
- Longitudinal section levels along the final centerline at every 25m interval, at the locations of curve points, small streams, and intersections, and at the locations of elevation change.
- Cross sections at every 50m interval in the full extent of a survey covering enough spot levels on the existing carriageway and adjacent ground for profile correction course and earthwork calculations. Cross sections shall be taken at the closer interval at curves.
- Longitudinal section for crossroads for length adequate for design and quantity estimation purposes. Longitudinal and cross sections for major and minor streams as per recommendations contained in the relevant standard.

### 4.2.2.2 Details of Utility Services and Other Physical Features

1. The design consultants shall collect details of all important physical features along the alignment. These features affect the project proposals and should

normally include buildings and structures, monuments, burial grounds, crema-
tion grounds, places of worship, railway lines, water mains, sewers, gas/oil
pipes, crossings, trees, plantations, utility services such as electric, and telephone lines (O/H & U/G) and poles, optical fiber cables (OFC), etc. The survey would cover the entire ROW of the road on the adequate allowance for
possible shifting of the central lines at some of the locations of the
intersection.

2. The information collected during reconnaissance and field surveys shall be
   shown on a strip Plan so that the proposed improvements can be appreciated
   and the extent of land acquisition with L. A schedule, utility removals of each
   type, etc. assessed and suitable actions can be initiated. A separate strip Plan
   for each of the services involved shall be prepared for submission to the concerned agency.

## 4.3   TRAVERSE SURVEY AND ELECTRONIC DISTANCE MEASUREMENT (EDM) APPLICATIONS

### General Theodolite Concepts

There are three important lines or axes in a theodolite, namely the line of sight,
the horizontal axis, and the vertical axis. The line of sight has to be perpendicular to the horizontal axis (trunnion axis), and their point of intersection has to lie on the vertical axis (Figure 4.1). The line of sight then
coincides with the line of collimation which, in a correctly adjusted theodolite, describes a vertical plane when rotated about the horizontal axis. In use,
the vertical axis has to be centered as accurately as possible over the station
at which angles are being measured and that axis has to be truly vertical.

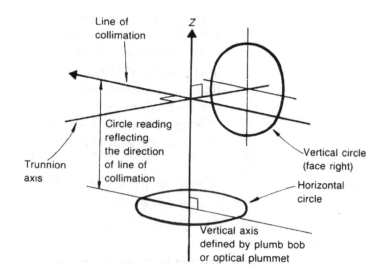

**FIGURE 4.1**   Concept of Theodolite.
*Source:* Figure 3.1, Text Book "Solving Problems in Surveying" by A. Bannister, Raymond
Baker, ELBS with Longman

### Errors in Horizontal Circle Reading

Errors in horizontal circle readings due to certain maladjustments of the theodolite are given below, h being the altitude of the signal observed.

| Maladjustment | Error |
|---|---|
| • Line of collimation making an angle (90-c) with the trunnion axis, i.e., vertical hair of diaphragm displaced laterally. | c sec h |
| • Trunnion axis inclined at (90-i) to the vertical axis, i.e., tilted from the horizontal. | i tan h |

### Errors in Vertical Circle Reading

It can be taken that errors in the vertical circle readings due to (a) and (b) are negligible and that, in each case, the mean of horizontal circle observations taken face left and face right will be free from error. This is not the case when the vertical axis is not truly vertical. The error in the horizontal circle reading is of the form given for (b) above but I is now variable, depending upon the pointing direction of the telescope. Its maximum value occurs when the trunnion axis lies in that plane containing the vertical axis of the instrument and the true vertical.

### Traverse

Employing a traverse survey, a framework of stations or control points can be established, their positions being determined by measuring the distances between the stations and the angles subtended at the various stations by their adjacent stations. For accurate work, the theodolite is used to measure the angles, with distances measured, although stadia tacheometry would only be used in low-order work. Modern EDM instruments can be mounted on or incorporated within theodolites and this allows the two to be used in combination efficiently, particularly with data recorders.

### Coordinates

Normally, plane rectangular coordinates are used to identify the stations on a traverse. A specific point is defined by its perpendicular distances from each of two coordinate axes which are based on north-south and east-west directions. The former is the reference axis, and it can be:

- True north.
- Magnetic north.
- National grid north.
- A chosen arbitrary direction, which could be one of the traverse lines if so wished.

The intersection of the axes gives the origin for the survey and usually, it is to the south and west to ensure that all points have positive coordinates.

### Easting/northing

The coordinates given by the perpendicular distances from the two main axes are termed:

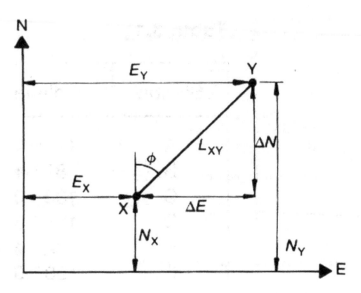

**FIGURE 4.2    Concept of Easting/Northing.**
*Source:* Figure 3.2, Text Book "Solving Problems in Surveying" by A. Bannister, Raymond Baker, ELBS with Longman

$$\text{eastings } \left(\text{distance from the north-south axis}\right)$$
$$\text{northings } \left(\text{distance from the east-west axis}\right)$$

As indicated in Figure 4.2 for the points X and Y. Relative positions are given by coordinate differences

$$\Delta E = E_Y - E_X$$

and

$$\Delta N = N_Y - N_X$$

**Bearings**

The position of the point may also be referenced by stating length XY and bearing $\phi$ of line XY, and these are referred to as polar coordinates. The bearing $\phi$ is a termed the whole circle bearing (WCB) of XY. It is measured clockwise from 0° to 360° at X between the north-south reference direction and the direction of Y from X. In Figure 4.3 the WCB of YZ is $\theta$, and the WCB of ZY is ($\theta$ - 180°). Similarly, for YX is Figure 2.2 the bearing is ($\phi$ + 180°) and, in general, bearing of line 1–2 = bearing of line 2-1 ± 180°, 1 and 2 being points within the system.

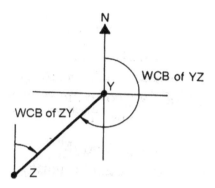

**FIGURE 4.3    Concept of Bearings.**
*Source:* Figure 3.3, Text Book "Solving Problems in Surveying" by A. Bannister, Raymond Baker, ELBS with Longman

**Example 4.1:    Calculation of Bearings**

The included angles given in Table 4.1 are recorded at stations forming a closed traverse survey around the perimeter of a field.

To determine the amount of angular error in the survey and adjust the values of the included angles. If the WCB of the line BC is 45° we must calculate the WCBs of the traverse lines and the corresponding values in the centesimal system.

True north refers to the north geographical pole. The true or geographical meridian through a point is the trace of the plane through the north and south poles and the point in question.

Magnetic north does not coincide with geographical north, the magnetic meridian is the direction revealed by a freely floating magnetic needle. The angle between it and the true meridian is termed declination.

The WCB of a line has been defined previously as the angle, lying between 0° and 360°, between the direction of north and the direction of the line, measured clockwise.

**Solution:** We have to determine the angular error and apply corrections.

**TABLE 4.1**

| Station | Included Angle |
|---------|----------------|
| A | 92° 47' 40" |
| B | 158° 06' 40" |
| C | 122° 42' 20" |
| D | 87° 16' 40" |
| E | 133 ° 08' 20" |
| F | 125 ° 55' 20" |

Figure 4.4 shows the traverse survey, the orientation of line BC being 45° from the meridian. This form of the traverse is known as a closed loop traverse since it begins and ends at the same point. The sum of the internal angles of a polygon is (2n – 4) right angles, where n is the number of angles.

Thus, the sum of the six angles of this example must be eight right angles, or 720° 00' 00", whereas by measurement it is 719° 56'00" (Table 4.2). The total error is, therefore – **4' 00"** or –240", and hence a total correction of + 240" must be applied.

The corrections can be applied equally to each angle on the assumption that conditions were constant at the time of measurement and that the angles had been measured with the same accuracy. Hence a correction of (+ 240" /6) = + 40" is given to each angle in this example. Next, we calculate WCBs. It is usual to proceed in an anti-clockwise manner around the traverse when internal angles have been measured.

To determine the WCB of the line to the forward station it is necessary to add the WCB of the previous line, i.e., that from the back station to the internal angle at the station, and then to add or deduct 180° depending upon whether that sum is less or greater than 180°. For instance, at A we require the WCB of AF knowing that of BA.

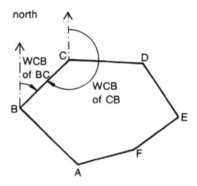

**FIGURE 4.4    Calculation of Bearings**.
*Source:* Figure 3.4, Text Book "Solving Problems in Surveying" by A. Bannister, Raymond Baker, ELBS with Longman

**TABLE 4.2**

| Angle | Observed Value | Correction | Adjusted Value |
|-------|----------------|------------|----------------|
| A | 122° 42' 20" | +40" | 122° 43'00" |
| B | 87° 16' 40" | +40" | 87° 17'20" |
| C | 133° 08' 20" | +40" | 133° 09' 00" |
| D | 125° 55' 20" | +40" | 125° 56' 00" |
| E | 92° 47' 40" | +40" | 92° 48' 20" |
| F | 158° 05' 40" | +40" | 158° 06' 20" |
| Total | 719° 56' 00" | Total | 720° 00' 00" |

In this example, we are given the bearing of BC as 45°, but to move around the traverse in the anti-clockwise direction B, A, F, E, D, C, B we need the bearing of CB instead.

Now WCB of BC = 45°

Therefore, a WCB of CB = 45° + 180° = **225°**. We can now proceed, using the quoted rule, to determine the WCBs of the six traverse lines as follows:

| | |
|---|---|
| WCB of CB | = 225° 00′ 00″ |
| Add angle B | = 87° 17′ 20″ |
| | ------------------ |
| | 312° 17′ 20″ (Exceeds 180°) |
| Deduct 180° | = 180° 00′ 00″ |
| | ------------------ |
| Therefore WCB of BA | = **132° 17′ 20″** |
| Add angle A | = 122° 43′ 00″ |
| | ------------------ |
| | 255° 00′ 20″ (Exceeds 180°) |
| Deduct 180° | = 180° 00′ 00″ |
| | ------------------ |
| Therefore WCB of AF | = **75° 00′ 20″** |
| Add angle F | = 158° 06′ 20″ |
| | ------------------ |
| | 233° 06′ 40″ (Exceeds 180°) |
| Deduct 180° | = 180° 00′ 00″ |
| | ------------------ |
| Therefore WCB of FE | = **53° 06′ 40″** |
| Add angle E | = 92° 48′ 20″ |
| | ------------------ |
| | 145° 55′ 00″ (Less than 180°) |
| Add 180° | = 180° 00′ 00″ |
| | ------------------ |
| Therefore WCB of ED | = **325° 55′ 00″** |
| Add angle D | = 125° 56′ 00″ |
| | ------------------ |
| | 451° 51′ 00″ (Exceeds 180°) |
| Deduct 180° | = 180° 00′ 00″ |
| | ------------------ |
| Therefore WCB of DC | = **271° 51′ 00″** |
| Add angle C | = 133° 09′ 00″ |
| | ------------------ |
| | 405° 00′ 00″ |
| Deduct 180° | = 180° 00′ 00″ |
| | ------------------ |
| Therefore WCB of CB | = 225° 00′ 00″ (Checked and matched at the start value) |

In the centesimal system the major graduations of the instrument range from zero to 400 gon, as against zero to 360° in the sexagesimal system. Subdivision in the centesimal system is carried out in steps of ten and reading may be made to 0.0001 gons. It will be evident that:

$$1 \text{degree is equivalent to } \frac{10}{9} \text{gon}$$

$$1 \text{minute is equivalent to } \frac{1}{54} \text{gon}$$

$$1 \text{second is equivalent to } \frac{1}{3240} \text{gon}$$

Hence in the case of line FE, 53° 06′ 40″ in the sexagesimal system is equivalent to

$$53 \times \frac{10}{9} = 58.8889$$

$$6 \times \frac{1}{54} = 0.1111$$

$$40 \times \frac{1}{3240} = 0.0123$$

Therefore, 53° 06′ 40″ = 58.8889 + 0.1111 + 0.0123 = 59.0123 gon

Similar calculations for whole circle bearings for all straights are shown in Table 4.3.

**TABLE 4.3**

| | Whole Circle Bearing | |
|---|---|---|
| Line | Sexagesimal System | Centesimal System |
| AF | 75° 00′ 20″ = (75×10/9)+(6/54)+(20/3240) | = 83.4506 gon |
| FE | 53° 06′ 40″= (53×10/9)+(6/54)+(40/3240) | = 59.0123 gon |
| ED | 325° 55′ 00″= (325x10/9)+(55/54)+(0/3240) | = 362.1296 gon |
| DC | 271° 51′ 00″= (271x10/9)+(51/54)+(0/3240) | = 302.0555 gon |
| CB | 225° 00′ 00″= (225×10/9)+(0/54)+(0/3240) | = 250.0000 gon |
| BA | 132° 17′ 20″= (132x10/9)+(17/54)+(20/3240) | = 146.9876 gon |

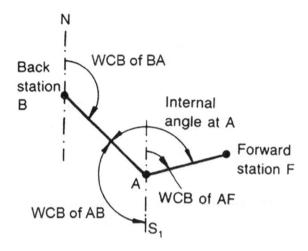

**FIGURE 4.5**  Whole Circle Bearing.
*Source:* Figure 3.5, Text Book "Solving Problems in Surveying" by A. Bannister, Raymond Baker, ELBS with Longman

Therefore, from Figure 4.5, WCB of AF = BA + A − 180°

## 4.3.1  CORRECTING A TRAVERSE BY BOWDITCH'S METHOD

**Example 4.2:   A survey was carried out on a closed loop traverse with six sides. With the traverse labeled anti-clockwise as shown in Figure 4.6 the data in Table 4.4 were obtained.**

The coordinates of point A are 1000 mE, 1000 mN, and the whole circle bearing of line A - F is 166° 45' 52" (156° 35' 52").

After adjustment by Bowditch's method, we must determine the coordinates of the other five traverse stations.

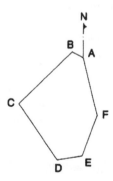

**FIGURE 4.6**  Example Traverse Correction by Bowditch's Method.
*Source:* Figure 3.6, Text Book "Solving Problems in Surveying" by A. Bannister, Raymond Baker, ELBS with Longman

**TABLE 4.4**

| Station | Internal angle | | Length |
|---|---|---|---|
| A | 120° 15' 48" | AB | 14.503 |
| B | 115° 15' 20" | BC | 85.213 |
| C | 94° 35' 25" | CD | 76.384 |
| D | 126° 21' 12" | DE | 27.296 |
| E | 122° 45' 17" | EF | 28.855 |
| F | 140° 47' 10" | FA | 87.613 |

**TABLE 4.5**

| Angle | Observed value | Correction | Adjusted value |
|---|---|---|---|
| A | 120° 15' 48" | –2" | 120° 15' 46" |
| B | 115° 15' 20" | –2" | 115° 15' 18" |
| C | 94° 35' 25" | –2" | 94° 35' 23" |
| D | 126° 21' 12" | –2" | 126° 21' 10" |
| E | 122° 45' 17" | –2" | 122° 45' 15" |
| F | 140° 47' 10" | –2" | 140° 47' 08" |
| Total | 720° 00' 12" | | 720° 00' 00" |

**Solution:** The first stage in the process is to determine the angular error and apply corrections as discussed in example 4.1.

Table 4.5 shows the tabulated angular data which sums to 720° 00' 12",

$$\text{Expected Closure} = (2n - 4) \times 90°$$
$$= 720°$$
$$\text{or} = 720°00'12" - 720°$$
$$= +12"$$

Thus, assuming that the angles have been measured with equal accuracy, a correction of – 2" should be applied to each as shown in Table 4.5.

The next stage is to calculate the whole circle bearings. In example 1 a method of calculating whole circle bearings was presented adding or subtracting 180° to the bearing of the back station plus the internal angle at the station.

The same set of bearings can be obtained by approaching the problem differently. The reader is advised to try both methods and adopt the one that is found to be easiest; the two methods cannot be mixed in any one calculation. Consider Figure 4.7.

$$WCB_{BA} = WCB_{AB} \pm 180° \,(\text{to lie in range } 0° < WCB_{BA} < 360°)$$

$$WCB_{BC} = WCB_{BA} + B \,(\text{subtract } 360° \text{ if greater than } 360°)$$

$$WCB_{CB} = WCB_{BC} \pm 180° \text{ etc}$$

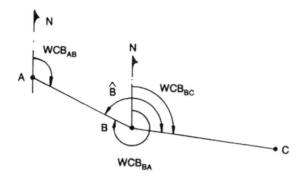

**FIGURE 4.7** Example Traverse Correction.
*Source:* Figure 3.7, Text Book "Solving Problems in Surveying" by A. Bannister, Raymond Baker, ELBS with Longman

Readers can consider themselves to be at each of the stations in turn logically proceeding around the traverse in an anti-clockwise direction.

| | |
|---|---|
| At A WCB$_{AF}$ | $=156°\ 35'\ 52''$ |
| add A | $=120°\ 15'\ 46''$ |
| | --------------- |
| WCB$_{AB}$ | $=276°\ 51'\ 38''$ (exceeds 180°) |
| | $-180°\ 00'\ 00''$ |
| | --------------- |
| At B WCB$_{BA}$ | $=96°\ 51'\ 38''$ |
| add B | $=115°\ 15'\ 18''$ |
| | ----------------- |
| WCBBC | $=212°\ 06'\ 56''$ (exceeds 180°) |
| | $-180°\ 00'\ 00''$ |
| | --------------- |
| At C WCB$_{CB}$ | $=32°\ 06'\ 56''$ |
| add C | $=94°\ 35'\ 23''$ |
| | -------------------- |
| WCB$_{CD}$ | $=126°\ 42'\ 19''$ (less than 180°) |
| | $+180°\ 00'\ 00''$ |
| | -------------------- |
| At D WCB$_{DC}$ | $=306°\ 42'\ 19''$ |
| add D | $=126°\ 21'\ 10''$ |
| | -------------------- |
| | $433°\ 03'\ 29''$ (exceeds 360°) |
| | $-360°\ 00'\ 00''$ |
| | ----------------- |
| WCB$_{DE}$ | $=73°\ 03'\ 29''$ (less than 180°) |
| | $+180°\ 00'\ 00''$ |
| | -------------------- |
| At E WCB$_{ED}$ | $=253°\ 03'\ 29''$ |
| add E | $=122°\ 45'\ 15''$ |
| | -------------------- |
| | $375°\ 48'\ 44''$ (exceeds 360°) |
| | $-360°\ 00'\ 00''$ |
| | -------------------- |

| $WCB_{EF}$ | 15° 48′ 44″ (less than 180°) |
| | +180° 00′ 00″ |
| | -------------------- |
| At F $WCB_{FE}$ | = 195° 48′ 44″ |
| add F | = 140° 47′ 08″ |
| | -------------------- |
| $WCB_{FA}$ | = 336° 35′ 52″ (exceeds 180°) |
| | -180° 00′ 00″ |
| | -------------------- |
| $WCB_{AF}$ | = 156° 35′ 52″ (Checked and |
| | matched at the start value) |

The next stage is to determine the easting and northing differences. In the introduction, it was pointed out that:

a. Easting difference $\Delta E = E_Y - E_X$
b. Northing difference $\Delta N = N_Y - N_X$

for points X and Y, which can be the stations at each end of a line as in Figure 4.8. It will be seen that

$$\Delta E = l_{XY} \sin(WCB)$$

and

$$\Delta N = l_{XY} \cos(WCB)$$

The signs of $\Delta E$ and $\Delta N$ automatically follow the trigonometrical terms. Since the whole circle bearing lies between 180° and 270° both sin (WCB) and cos

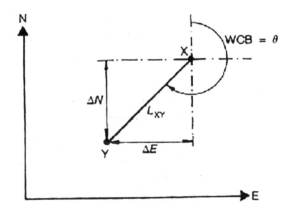

**FIGURE 4.8**   Example of Traverse Correction.
*Source:* Figure 3.8, Text Book "Solving Problems in Surveying" by A. Bannister, Raymond Baker, ELBS with Longman

**TABLE 4.6**

| Line | WCB | Length (M) | ΔE (m) | ΔN (m) |
|------|-----|-----------|--------|--------|
| AB | 276° 51' 38" | 14.503 | −14.399 | 1.732 |
| BC | 212° 06' 56" | 85.213 | −45.301 | −72.173 |
| CD | 126° 42' 19" | 76.384 | 61.238 | −45.654 |
| DE | 73° 03' 29" | 27.296 | 26.111 | 7.954 |
| EF | 15° 48' 44" | 28.855 | 7.862 | 27.763 |
| FA | 336° 35' 52" | 87.613 | −34.798 | 80.405 |
| **Totals** | | **319.864** | **+0.7128** | **+0.0259** |

(WCB) are negative. This causes ΔN and ΔE to be negative in respect of the positive directions of N and E in Figure 4.8. The resultant easting differences and northing differences can now be computed for the traverse lines, as in Table 4.6.

Next, the closing error is determined. The algebraic sums of the easting differences and northing differences should be zero because the traverse starts and ends at A, but in fact, we have total errors in so far as this traverse is concerned of

$$\Delta E = 0.7128 m$$

and

$$\Delta N = 0.0259 m$$

$$\text{Closing error} = \sqrt{\left(\Delta E^2 + \Delta N^2\right)}$$
$$= \sqrt{\left(0.7128^2 + 0.0259^2\right)}$$
$$= 0.713 m$$

Expressed fractionally in terms of the total length of the traverse the linear error is 0.713m in 319.864m or 1 in 449.

The error in easting and northing distance should now be corrected. There are several methods of carrying out this task; Bowditch's method is commonly adopted in civil engineering surveys since it has some theoretical background and is relatively simple to apply. The method assumes that the error in the bearing of a line caused by inaccurate angular measurement produces a displacement at one end of a line (C) relative to the other end (B) which is equal and perpendicular to the displacement along that line due to an error in linear measurement, which is taken to the proportional to $\sqrt{L_{BC}}$. Figure 4.9 shows that the method causes C to be displaced to C', and consequently the bearing of BC changes.

For individual lines, Bowditch's method states:

$$\text{Correction to easting difference } \Delta E_{BC} = \frac{dE \times \text{length of line BC}}{\text{Total length of the traverse}}$$

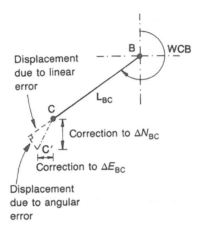

**FIGURE 4.9**   Example of Traverse Correction.
*Source:* Figure 3.9, Text Book "Solving Problems in Surveying" by A. Bannister, Raymond Baker, ELBS with Longman

$$\text{Correction to northing difference } \Delta N_{BC} = \frac{dN \times \text{length of line BC}}{\text{Total length of the traverse}}$$

in which dE and dN are the total corrections required for the easting differences and northing differences, respectively.

In this example dE = -0.067m and dN = + 0.07 m. Hence the corrections for line BC are:

$$\text{Correction to easting difference} = -0.7128 \times \frac{85.213}{319.864}$$

$$= -0.189 \text{m}$$

$$\text{Correction to northing difference} = +0.0259 \times \frac{85.771}{319.864}$$

$$= +0.007 \text{m}$$

Note also that in Figure 4.9,

$$\text{The change in bearing of the line} = \tan^{-1} \frac{\text{correction to } \Delta E_{BC}}{\text{correction to } \Delta N_{BC}} = \tan^{-1} dN$$

This is the bearing of the closing error and the correction in bearing applies throughout the traverse for all lines. The corrections can be tabulated as in Table 4.7.

$$\textbf{Correction to } \Delta E = (14.503, 85.213, 76.384, 27.296, 28.855, 87.613)$$
$$\times (-0.7128 // 319.864)$$
$$= -0.032, -0.189, -0.170, -0.061, -0.064, -0.195$$

## TABLE 4.7

| Line | ΔE | Correction to ΔE | Corrected ΔE | ΔN | Correction to ΔN | Corrected ΔN |
|---|---|---|---|---|---|---|
| AB | −4.399 | −0.032 | −14.431 | +1.732 | +0.001 | +1.733 |
| BC | −5.302 | −0.189 | −45.491 | −72.173 | +0.006 | −72.167 |
| CD | 61.238 | −0.170 | 61.068 | −45.655 | +0.006 | −45.649 |
| DE | 26.111 | −0.061 | 26.050 | +7.954 | +0.002 | +7.956 |
| EF | 7.862 | −0.064 | 7.798 | +27.763 | +0.002 | +27.765 |
| FA | −34.798 | −0.195 | −34.993 | +80.405 | +0.007 | +80.412 |
| Totals | +0.713 | −0.711 | 0.0 | +0.026 | +0.024 | 0.05≈0.0 |

## TABLE 4.8

| Station | Co-ordinates E (m) | N (m) |
|---|---|---|
| A | 1000.00 | 1000.00 |
| (line AB) | −14.431 | +1.733 |
| B | = 985.569 | = 1001.733 |
| (line BC) | −45.491 | −72.167 |
| C | = 940.078 | = 929.566 |
| (line CD) | 61.068 | −45.649 |
| D | = 1001.146 | = 883.917 |
| (line DE) | 26.050 | +7.956 |
| E | = 1027.196 | = 891.873 |
| (line EF) | 7.798 | +27.765 |
| F | = 1034.994 | = 919.638 |
| (line FA) | −34.993 | +80.412 |
| A | = 1000.0 | = 1000.0 |

$$\text{Correction to } \Delta N = (14.503,85.213,76.384,27.296,28.855,87.613)$$
$$\times(+0.0259//319.864)$$
$$= +0.001,+0.006,+0.006,+0.002,+0.002,+0.007$$

The final stage is to determine the coordinates of the stations by applying the corrected difference ΔE and ΔN to the previous station coordinates, i.e., for B. Easting coordinate of B = easting coordinate of A + $\Delta E_{AB}$. Northing coordinate of B = northing coordinate of A + $\Delta N_{AB}$. Hence, we obtain the values given in Table 4.8.

**The computer-aided processing of traverse survey data with corrections of closing errors by Bowditch's method is being described in section 4.1 of Chapter 4 in the "The Guide for Computer Application Tutorials" part of the book available for download from the website of the book.** To download the book tutorials and other items, visit the websites at: www.roadbridgedesign.com and www.techsoftglobal.com/download.php

## 4.3.2 Correcting a Traverse by the Transit Rule

**Example 4.3: A closed-loop traverse survey ABCDEA, shown in Figure 4.10, gave the information in Table 4.9.**

We must determine the closing error and hence, after adjustment by the transit rule, determine the coordinates of the traverse stations given that the coordinates of A are 1200.00 mE, 1200.00 mN.

The previous example covered the first stage in traverse computation in that the measured angles were assessed for error, and then duly corrected. This was then followed by the determination of the whole circle bearings of the lines. This example allows us to follow the remainder of the procedure applied to the transit rule.

This method has no theoretical background, but it is such that if a line has no easting difference it will not be given an easting correction. This is not so with the Bowditch approach. The rule states

$$\text{Correction to } \Delta E_{DE} = \frac{dE \times \Delta E_{DE}}{\Sigma \Delta E}$$

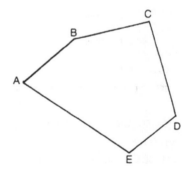

**FIGURE 4.10  Example Traverse Correction by Transit Rule.**
*Source:* Figure 3.10, Text Book "Solving Problems in Surveying" by A. Bannister, Raymond Baker, ELBS with Longman

### TABLE 4.9

| Line | Length (m) | Whole circle bearing |
|------|-----------|----------------------|
| AB | 586.54 | 45° 11' 14" |
| BC | 1441.66 | 72° 04' 05" |
| CD | 994.24 | 161° 50' 55" |
| DE | 1046.68 | 228° 44' 20" |
| EA | 1523.74 | 300° 41' 40" |

## TABLE 4.10

| Line | Length (m) | WCB | ΔE (m) | ΔN (m) |
|------|-----------|-----|--------|--------|
| AB | 586.54 | 45° 11' 14" | +416.09 | +413.40 |
| BC | 1441.66 | 72° 04' 55" | +1371.53 | +443.74 |
| CD | 994.24 | 161° 50' 55" | +309.61 | −944.55 |
| DE | 1046.68 | 228° 44' 20" | −786.87 | −690.27 |
| EA | 1523.74 | 300° 41' 40" | −1310.27 | +777.80 |
| **Totals** | | | **+0.09** | **−0.12** |
| **Sum** | | | **4194.37** | **3269.76** |

$$\text{Correction to } \Delta N_{DE} = \frac{dN \times \Delta N_{DE}}{\Sigma \Delta N}$$

**Solution:** We first calculate the values of $\Delta E$ and $\Delta N$ in the same manner as in example 4.2. The data are tabulated in Table 4.10.

$$45°11'14'' = 45.187°, \sin(45.187) \times 586.54$$
$$= +416.09, \cos(45.187) \times 586.54 = +413.40$$

$$72°04'55'' = 72.082°, \sin(72.082) \times 1441.66$$
$$= +1371.53, \cos(72.082) \times 1441.66 = +443.74$$

$$161°50'55'' = 161.849°, \sin(161.849) \times 994.24$$
$$= +309.61, \cos(161.849) \times 994.24 = -944.55$$

$$228°44'20'' = 228.739°, \sin(228.739) \times 1046.68$$
$$= -786.87, \cos(228.739) \times 1046.68 = -690.27$$

$$300°41'40'' = 300.694°, \sin(300.694) \times 1523.74$$
$$= -1310.27, \cos(300.694) \times 1523.74 = +777.80$$

In the example $\Sigma \, \Delta E = 4194.37$m and $dE = +0.09$
also $\Sigma \, \Delta N = 3269.76$m and $dN = -0.12$.
Magnitudes of the differences are considered, and the signs are ignored.
Thus, for line AB we have

$$\text{Correction to } \Delta E_{AB} = 416.09 \times (+0.09 / /4194.37) = +0.04$$
$$\text{Correction to } \Delta N_{AB} = 413.40 \times (-0.12 / /3269.76) = -0.01$$

For the complete traverse, we obtain the values in Table 4.11.

To determine the coordinates of the stations we apply the corrected difference $\Delta E$ and $\Delta N$ to the previous station coordinates in the same manner as example 2, and hence we obtain the values in Table 4.12.

## TABLE 4.11

| Line | $\Delta E$ | Correction to $\Delta E$ | Corrected $\Delta E$ | $\Delta N$ | Correction to $\Delta N$ | Corrected $\Delta N$ |
|------|-----------|--------------------------|----------------------|-----------|--------------------------|----------------------|
| AB | +416.09 | +0.04 | 416.13 | +413.40 | −0.01 | +413.39 |
| BC | +1371.53 | +0.15 | 1371.68 | +443.74 | −0.01 | +443.73 |
| CD | +309.61 | +0.03 | 309.64 | −944.55 | −0.03 | −944.58 |
| DE | −786.87 | −0.09 | −786.96 | −690.27 | −0.02 | −690.29 |
| EA | −1310.27 | −0.15 | −1310.42 | +777.80 | −0.02 | +777.78 |
| Totals | +0.09 | −0.02 | 0.00 | +0.12 | −0.039 | 0.0 |

## TABLE 4.12

| Station | Co-ordinates | |
|---------|-------------|---|
|  | E (m) | N (m) |
| A | **1000.0** | **1000.0** |
| (line AB) | 416.13 | +413.39 |
| B | **1416.13** | **1413.40** |
| (line BC) | 1371.68 | +443.73 |
| C | **2787.81** | **1857.13** |
| (line CD) | 309.64 | −944.58 |
| D | **3097.45** | **912.55** |
| (line DE) | −786.96 | −690.29 |
| E | **2310.49** | **222.26** |
| (line EA) | −1310.42 | +777.78 |
| A | **1000.0** | **1000.00** |

**Vector Misclosure of a Traverse**

**Example 4.4: A five-sided loop traverse (whose angles have an accepted misclosure) has been computed giving the coordinate differences in Table 4.13 for each leg.**

- We must determine the easting, northing, and vector misclosure of the traverse.
- The vector misclosure indicates a mistake of 1m in the length of one of the sides of the traverse. Find which side contains the mistake and, after eliminating its effect, recompute the easting, northing, and vector misclosures.

### TABLE 4.13

| Leg | $\Delta E$ (m) | $\Delta N$ (m) |
|-----|---------------|---------------|
| AB | −41.44 | −58.32 |
| BC | +66.93 | −32.97 |
| CD | +48.31 | +45.70 |
| DE | −21.86 | +69.70 |
| EA | −51.04 | −24.57 |

**TABLE 4.14**

| Leg | E | N |
|---|---|---|
| AB | −41.44 | −58.32 |
| BC | +66.93 | −32.97 |
| CD | +48.31 | +45.70 |
| DE | −21.86 | +69.70 |
| EA | −51.04 | −24.57 |
| Misclosures | +0.90 | −0.46 |

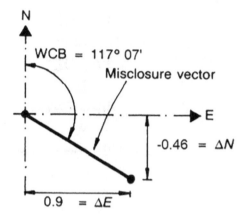

**FIGURE 4.11** Vector Misclosure of the Traverse.
*Source:* Figure 3.11, Textbook "Solving Problems in Surveying" by A. Bannister, Raymond Baker, ELBS with Longman

**Solution:**
Calculate the vector misclosure of the traverse. The easting misclosure is +0.90m and the northing misclosure is −0.46 m, as indicated in Figure 4.11.

$$\text{Whence vector misclosure} = \sqrt{\left((+0.90)^2 + (-0.46)^2\right)}$$

$$= 1.01\text{m}$$

$$\text{with a bearing of } \tan^{-1}\frac{0.90}{-0.46} = 117.07°\text{say}$$

Now to find the side that is in error. We are given that the angles as measured produced an acceptable misclosure, and accordingly the magnitude of the misclosure cannot be attributed to their measurement. In such a situation we must search for a side in the traverse which has the same bearing, approximately, like that of the closing error. Bearings have not been given in this example, but the data reveals that line BC is the only one whose easting difference and northing difference bear some proportional relationship with the corresponding differences of the closing error, i.e., +66.93m for ΔE and −32.97m for ΔN against the closing errors +0.90m and –0.46 m, respectively. Thus, we can assume that BC is in error by the amount stated.

$$\text{The apparent length of BC} = \sqrt{\left((66.93)^2 + (-32.97)^2\right)}$$

$$=74.61\text{m}$$

$$\text{Also, for BC,} \frac{\Delta E}{\Delta N} = \frac{66.93}{-32.97} = -2.03$$

$$\text{and the closing error } \frac{\Delta E}{\Delta N} = \frac{0.90}{-0.46} = 1.95$$

If we accept that the length of BC to be taken as 73.61m (given that 74.61 has been subjected to an error of 1 m) and recalculate the magnitudes of the misclosures.

$$\text{Corrected } \Delta E \text{ for BC} = \frac{+73.61}{74.61} \times 66.93 = +66.03\text{m}$$

$$\text{Corrected } \Delta N \text{ for BC} = \frac{-73.61}{74.61} \times 32.97 = -32.53\text{m}$$

Hence, the closing errors are revised as,
$\Delta E = -(0.90-(66.93-66.03)) = 0.0$ and $\Delta N = -(0.49-(32.97-32.53)) = -0.05$

$$\text{The amended closing errors, } \Delta E = +0.0\text{m} \quad \Delta N = -0.05\text{m}$$
$$\text{magnitude} = \sqrt{\left((0.0)^2 + (0.05)^2\right)} = 0.05\text{m}$$
$$\text{bearing} = \tan^{-1}\frac{0.0}{-0.05} = 180°.$$

**The computer-aided processing of traverse survey data with corrections of closing errors via Transit method is being described in Section 4.3 of Chapter 4 in the "The Guide for Computer Application Tutorials" part of the book available for download from the website of the book.**
To download the book tutorials and other items visit the websites at:
www.roadbridgedesign.com and www.techsoftglobal.com/download.php

## 4.3.3 CLOSED LINK TRAVERSE

Rather than starting and finishing at one point, in an extended form, this traverse is still closed, since it runs between two points whose coordinates are fixed and two lines whose bearings are fixed. It is known as a closed link traverse and it can be readily adjusted.

The measured angles are shown in Figure 4.12. Keeping the bearings XA and EY and the coordinates of A and E fixed, the adjusted coordinates for B, C, and D are determined, by using an equal shift angular adjustment and Bowditch linear adjustment.

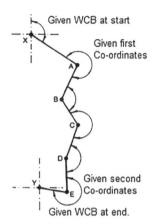

**FIGURE 4.12**    Correction of Traverse by Closed Link Method.
*Source:* Figure 3.12, Text Book "Solving Problems in Surveying" by A. Bannister, Raymond Baker, ELBS with Longman

The whole circle bearing, starting with the bearing of XA is obtained at the end to match with the given bearing of EY, if not then there will be some angular error to balance.

**The computer-aided processing of traverse survey data with corrections of closing errors by closed link method is being described in Section 4.4 of Chapter 4 in the "The Guide for Computer Application Tutorials" part of the book, available for download from the website of the book.**

To download the book tutorials and other items, visit the websites at:

www.roadbridgedesign.com and www.techsoftglobal.com/download.php

### 4.3.4    ELECTRONIC DISTANCE MEASUREMENT (EDM)

**EDM in Distance Measurement**

Slant distance L between the EDM and the reflector located at respective stations has been measured and reduced horizontal length, l, required. To obtain l we require the vertical angle $\theta$ which the measured length makes with the horizontal. This incorporates measured angle $\alpha$ together with two corrections $\beta$ and $\gamma$, defined as "eye and object" corrections, due to the differences in height above the ground stations of the measuring devices and their targets (Figure 4.13).

It is known that the measured slope length $L$ between two stations can be reduced to its chord length at the mean height of the stations. Accordingly, the equivalent chord length at mean sea level can be computed and a correction then added to determine the spheroidal distance between the stations. Assuming that the slope distance was established by EDM, the relevant corrections are $L^3/43\ R^2$ and $L^3/33\ R^2$ for microwave and infra-red systems, respectively. These include corrections for the curvature of the path of the signal and are of particular importance for long lines. Note that the theoretical difference between spheroidal distance and chord length is Rc − 2R sin c/2 = $d^3/24\ R^2$ (Figure 4.14).

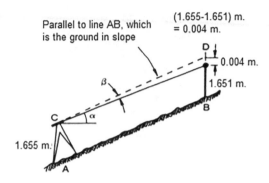

**FIGURE 4.13**  Distance Measurement with EDM.
*Source:* Figure 2.8, Text Book "Solving Problems in Surveying" by A. Bannister, Raymond Baker, ELBS with Longman

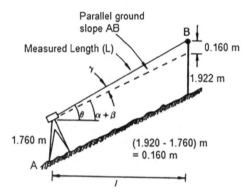

**FIGURE 4.14**  Example for Distance Measurement with EDM.
*Source:* Figure 2.9, Text Book "Solving Problems in Surveying" by A. Bannister, Raymond Baker, ELBS with Longman

To calculate the horizontal length $l$, the first step is to determine the corrections $\beta$ and $\gamma$. If the height of the instruments and their respective "targets" above the ground stations were equal, the corrections would be zero since the lines of sight would be parallel to AB, and measured angle $\alpha$ would have given the required slope angle.

**Example 4.5:  The slope distance measured by EDM and corrected (L) = 115.155 m.**

Heights of EDM and the reflector above ground level = 1.760 m. and 1.920 m.

Heights of theodolite and its target above ground level = 1.655 m. and 1.651 m.

The vertical angle was measured by theodolite and corrected ($\alpha$) = +4° 35' 12".

Relative accuracy of reduced horizontal distance (dl) = 1/100, 000
Standard error for reduced horizontal distance (dθ) = +/- 2 mm.= 0.002 m.
Both EDM instrument and theodolite were situated at the lower station.
From Figure 4.13.

$$\beta = \frac{1.655 - 1.651}{115.155} \text{radian}$$
$$= \frac{0.004 \times (180 \times 60 \times 60 / 3.1416)}{115.155}$$
$$= 7.2''$$

From Figure 4.14

$$\gamma = \frac{1.920 - 1.760}{115.155} \text{radian}$$
$$= \frac{0.160 \times (180 \times 60 \times 60 / 3.1416)}{115.155}$$
$$= 286.6''$$
$$= 4'46.6''$$

Therefore

$$\text{slope angle } \theta = \alpha + \beta + \gamma$$
$$= +4°35'12'' + 7.2'' + 4'46.6''$$
$$= 4°40'5.8''$$

Therefore,

$$\text{horizontal length } l = L\cos\theta$$
$$= 115.155 \times \cos 4°40'5.8''$$
$$= 115.155 \times \cos\left(\left(\left(5.8/60\right) + 40\right)/60\right) + 4\right)$$
$$= 115.155 \times \cos\left(4.668\right)$$
$$= 114.773$$

Note the assumption that length CD is equal to the measured length of
115.155m when calculating β and γ.

In this case β and γ are positive corrections, first because the height of the
target at B was less than the theodolite height at A, thereby causing the line of
sight to be depressed with respect to CD: thus α was measured low. Second,
because the EDM height was less than the reflector height the line of sight was
elevated with respect to the parallel to AB.

Next, we find the precision of the slope angle.
Since $l = L \cos \theta$
Then $dl = - L \sin \theta \, d\theta$
(an increase dθ causes a decrease in $l$).

$$\text{Relative accuracy} \frac{dl}{l} = \frac{1}{100000} = \frac{L\sin\theta d\theta}{L\sin\theta} (\text{neglecting the negative } sign)$$
$$= \tan\theta d\theta = \tan 4°40'5.8'' \times d\theta$$
$$= \tan(4.668) \times d\theta$$
$$= 0.082 \times d\theta.$$

$$\text{Therefore, } d\theta = 0.082/1,00,000$$
$$= 0.00000082 \text{ radian}$$
$$= 0.00000082 \times 180 \times 60 \times 60/3.1416$$
$$= 0.169 \text{seconds.}$$

$$\text{If } dl = \pm 2mm$$
$$0.002 = L\sin\theta d\theta$$
$$= 115.155 \times \sin 4°40'5.8'' \times d\theta.$$

$$d\theta = 0.002/(115.155 \times \sin(4.668))$$

$$\text{Therefore } \quad d\theta = 0.0002 \text{ radian}$$
$$= 0.0002 \times 180 \times 60 \times 60/3.1416$$
$$= \pm 41.25 \text{ seconds.}$$
$$\text{For } \quad dl = \pm 2mm.$$

**The computer-aided processing of traverse survey data with corrections of closing errors by EDM is being described in Section 4.5 of Chapter 4 in the "The Guide for Computer Application Tutorials" part of the book, available for download from the website of the book.**

To download the book tutorials and other items, visit the websites at: www.roadbridgedesign.com and www.techsoftglobal.com/download.php

### Survey data either by Total Station or Satellite

Computer applications are very effectively helpful in every stage of highway development projects starting from the Conceptual phase to preliminary design, to Detail engineering to construction. The collecting and processing of ground survey data is the first step in computer-aided highway design. The process includes creating the ground model, Digital mapping to create the survey base Plan, triangulation to create the DTM and thus the ground contours and finally, the ground long and cross sections.

The collecting of ground data uses either obtaining the field survey data most commonly by using the Total Station instrument or obtaining the ground elevation data with the latest available satellite-based technology which also helps in identifying the best possible route of the road, its realignment, and the widening scheme.

## Ground Modeling

The survey data whether by Total Station survey or by satellite data, but the data must be made available in the desired format of five columns for Serial Number, Easting (meters), Northing (meters), Elevation (meters), and Feature Code and saved as a text file (for example SURVEY.TXT) which can be opened in "Notepad." If it has five columns of data, when a data record with Serial No. "0" is met then it discontinues the last feature and reads a new feature in the ground model. After creating the ground model, it is drawn to create the survey base map and is described in the relevant section of this book.

## 4.4 TOTAL STATION SURVEY

| Serial No: | Easting (Meters) | Northing (Meters) | Elevation (Meters) | Feature Code |
|---|---|---|---|---|
| 1 | 372233.83400 | 524236.04200 | 580.46100 | AV |
| 2 | 372208.47200 | 524235.86300 | 580.08500 | AV |
| 3 | 372181.75000 | 524236.53000 | 579.60500 | AV |
| 4 | 372156.01300 | 524238.11100 | 579.17200 | AV |
| 5 | 372130.14900 | 524240.05700 | 578.86400 | AV |
| 6 | 372104.71500 | 524242.51400 | 578.48700 | AV |
| 7 | 372080.13900 | 524245.40600 | 577.72400 | AV |
| 0 | 0.00000 | 0.00000 | 577.72400 | AV |
| 8 | 372233.72900 | 524242.52300 | 580.24600 | AV |
| 9 | 372208.48700 | 524242.26900 | 579.85500 | AV |
| 10 | 372182.41900 | 524242.85700 | 579.42300 | AV |
| 11 | 372156.85200 | 524244.33400 | 579.06200 | AV |
| 12 | 372131.12600 | 524246.37700 | 578.84000 | AV |
| 13 | 372105.74300 | 524248.78900 | 578.35300 | AV |
| 14 | 372081.05400 | 524251.80900 | 577.65300 | AV |
| 0 | 0.00000 | 0.00000 | 577.65300 | AV |
| 15 | 370913.03200 | 524404.60200 | 576.79500 | AV |
| 16 | 370897.74300 | 524405.41400 | 576.61100 | AV |
| 17 | 370891.85000 | 524405.83800 | 576.52600 | AV |

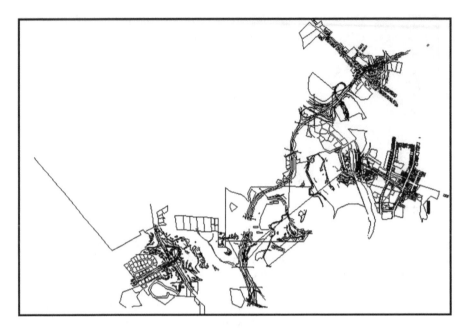

**FIGURE 4.15**    Survey Base Plan (Ground Model).
*Source:* Author of this Book

### Processing of Ground Elevation Data by Satellite Downloaded from Internet

By defining tentative alignment by drawing straights by "My Path" in Google Earth to define the project corridor and subsequently downloading ground elevation data by using Global Mapper minimizes the exercise to transform the Surveyors coordinates in TM into Global coordinates in UTM/GPS. The alignment file may be saved as KML/KMZ file. This file may be opened in the future in Google Earth to mark various features along the corridor, for example, the road alignment with existing "Land use" etc. The area may also be identified or verified by site reconnaissance survey or by using topographic maps or by procuring satellite imageries from the survey department.

Next, the KML/KMZ file may be opened with Global Mapper and saved as a DXF file. Also, the ground elevation data may be downloaded with Global Mapper from SRTM (Shuttle Radar Topography Mission). The DXF file of the alignment is opened in HEADS Viewer and the alignment geometrics are design as Horizontal Alignment of the project highway.

The downloaded ground elevation data is to be opened with software MS-Excel for saving in the desired format. Next, the ground data is opened with the software HEADS Pro to create the ground model, and next the DTM along with contours are to be generated for the entire area under the study. This enables the engineers to Develop the survey base Plan, DTM by Delauny triangulation, Ground contours at user given primary & secondary intervals, 3D Surface with rendering, Ground Long and Cross Sections. The Digital mapping facility by drawing ground model enables the users to prepare survey base Plan drawings over Google Pictures using the CAD engine of software HEADS Pro.

**Triangulation and Contours**

The model name, String Label, and increments for primary, secondary, and text are set with default names & values. If the ground is flat, then the user must reduce the primary contour increment to 0.5 or 0.1 to get contours for the relatively flat ground. The secondary contour increment is always five times of increment of the primary contours. Texts for contour elevations are displayed on the secondary contours (Figure 4.15).

## 4.5 CROSS SECTION SURVEY BY AUTOLEVEL

In case of "cross section" survey done by Autolevel the elevations of various points on either both or same side of the Traverse at each chainage station are written in text files following the format as described below:

Autolevel data: interval 50

| | | |
|---|---|---|
| 0 | 0.000 | 99.815 |
| | -1.500 | 99.840 |
| | -3.000 | 99.860 |
| | -4.500 | 99.485 |
| | -6.000 | 99.655 |
| | 1.500 | 99.685 |
| | 3.000 | 99.615 |
| | 4.500 | 99.635 |
| | 6.000 | 99.525 |
| | 7.500 | 99.530 |
| 25 | 0.000 | 99.870 |
| 50 | 0.000 | 99.880 |
| | -1.500 | 99.780 |
| | -3.000 | 99.940 |
| | -4.500 | 99.840 |
| | 1.500 | 99.975 |
| | 3.000 | 99.740 |
| | 4.500 | 100.170 |
| | 5.000 | 99.880 |
| 75 | 0.000 | 99.780 |
| 100 | 0.000 | 100.010 |
| | -1.500 | 99.895 |
| | -3.000 | 99.870 |
| | -4.500 | 99.840 |
| | -6.000 | 99.825 |
| | -7.500 | 99.820 |
| | 1.500 | 100.060 |
| | 3.000 | 100.145 |
| | 4.500 | 100.145 |

The above data starts with the line "interval 50", which means that the cross sections are taken at 50 meters interval and Long sections are taken at 25metres intervals. There are three columns in the data, the first column is for the chainage, which is 0, 25, 50, 75, 100. The data in the second column is for the distances and the data in the third column is for ground elevations or Reduced Levels. The cross section at each chainage is taken as ground elevations at different distances measured from the centerline on either side. At the centerline, the distance is "0.0," The distances on the left side are with the "-" sign and the distances without the "-" sign are on the right side of the centerline. Cross sections are taken at chainages 0.0, 50.0, 100.0 meters, and the Long sections are taken on the centerline only and at chainages 0.0, 25.0, 50.0, 75.0, and 100.0 meters.

## 4.6 GEOGRAPHIC COORDINATE SYSTEM TRANSFORMATIONS

The geographic coordinate system transformation is a class of coordinate transformations. The diagram on coordinate transformations describes how this class fits into the total schema of coordinate transformations (Figure 4.16).

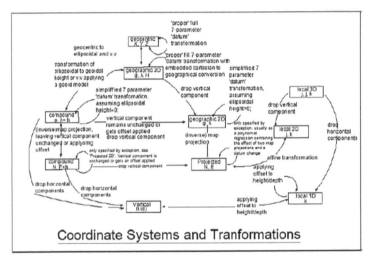

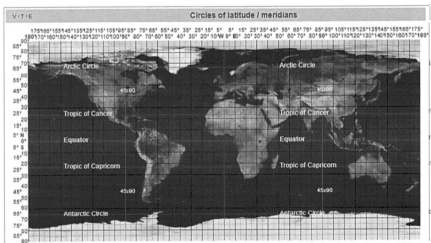

**FIGURE 4.16** Coordinate System Transformations
*Source:* Wikipedia

### 4.6.1  GEOGRAPHIC COORDINATE SYSTEM TRANSFORMATION METHODS

It has been noted earlier that it is frequently required to convert coordinates derived in one geographic coordinate system to values expressed in another. For example, land and marine seismic surveys are nowadays most conveniently positioned by GPS satellite in the WGS 84 geographic coordinate system, whereas the national geodetic system in use for the country concerned will probably be a much earlier coordinate system. It may therefore be necessary to convert the observed WGS 84 data to the national geodetic system to avoid discrepancies. This form of Epicenter coordinate transformation has to be between source and target geographical coordinate systems with the coordinates of both expressed in geographical terms. It is not executed by a transformation between projected coordinates.

The transformations may be most readily achieved by first expressing the observed or source geographical coordinates in terms of three-dimensional XYZ cartesian values (geocentric coordinates) instead of the normal angular expressions of latitude and longitude. However, to derive the three true cartesian coordinates of a point on the earth's surface one must recognize that as well as having a latitude and longitude the point will also have an elevation above the ellipsoid. While XYZ cartesian coordinates may be derived from a mere latitude and longitude by assuming that the point lies on the ellipsoid's surface, in fact very few earth points do. Therefore, the height of the point must be taken into account and the height required is the height above the ellipsoid. It is this height that is delivered by GPS about the WGS84 ellipsoid. Heights above other ellipsoids are not generally immediately available. Instead, the height that is generally available is the height above the national vertical datum, which normally means sea level at a particular coastal point measured over a particular period. Levels will normally have been derived by conventional surveying methods and values will relate to the geoid surface. Hence to derive heights above the ellipsoid it is necessary to know the height of the geoid above the ellipsoid or vice versa and for most areas of the world this is not known with great precision. There exist various mathematical models of the geoid which have been derived for individual countries or parts of the world or the entire world and as satellite and terrestrial gravity data accumulate, they are being steadily refined. The best available geoidal data should be used to convert surveyed heights (surveyed other than by GPS) to ellipsoidal heights for use in Epicenter's geodetic coordinate transformation. If the true ellipsoidal height of a point cannot be obtained the assumption that it is zero will often allow a geodetic transformation without introducing significant error in the horizontal coordinates.

In the early days of satellite surveying when relationships between datums were not well defined and the data itself was not very precise, it was usual to apply merely a three-parameter dX, dY, dZ shift to the XYZ coordinate set in one datum to derive those in the second datum. This assumed, generally erroneously, that the axial directions of the two ellipsoids involved were parallel. For localized work in a particular country or territory, the consequent errors introduced by this assumption were small and generally less than the observation accuracy of the data. Nevertheless, as knowledge and data have accumulated and surveying methods have become more accurate, it has become evident that a three-parameter transformation is neither appropriate for worldwide use, nor indeed for widespread national use if one is seeking the maximum possible accuracy from the satellite surveying and a single set of transformation parameters.

For petroleum exploration purposes it may well be that a three-parameter transformation within a particular area of interest is quite adequate but it should not be assumed that the same transformation parameters are appropriate for use with different data in an adjoining area of interest or for another purpose.

The simplest transformation to implement involves applying shifts to the three geocentric coordinates. **Molodenski** developed a transformation that applies the geocentric shifts directly to geographical coordinates. Both of these methods assume that the axes of the source and target systems are parallel to each other. As indicated above this assumption may not be true and consequently, these transformation methods result in only moderate accuracy, especially if applied over large areas.

Improved accuracy can be obtained by applying a **Helmert 7-parameter transformation** to geocentric coordinates. However, there are two opposing sign conventions for the three rotation parameters. In Epicenter, these are two different transformation methods, - either a **position vector transformation** or a **coordinate frame** transformation. (The position vector transformation is also called the Bursa-Wolf transformation). The signs of the rotation parameters must be consistent with the rotation convention being applied.

It is also possible to interpolate geographical or grid shifts for points based on known shifts for several control points in a specifically defined area. One such application is the coordinate transformation introduced to enable the conversion of coordinates expressed in the North American Datum of 1927 (including the Clarke 1866 ellipsoid) to coordinates expressed in the new North American Datum of 1983 which takes the GRS 1980 ellipsoid. Because the North American survey control network was built by conventional terrestrial survey observations and suffers from the inevitable instrumental and adjustment shortcomings of the time, the old network, based on the non-geocentric Clarke 1866 ellipsoid and a single datum point at Meades Ranch in Kansas, is not wholly consistent when compared with data which can be more readily and accurately secured nowadays with the advantages of satellite technology, modern instrumentation, and electronic computational techniques. Hence to convert between coordinates on the old system to values in the new datum, it is not appropriate to merely apply the type of orthogonal transformation represented by the Molodenski or Helmert transformations described above. Different points in different parts of the North American continent need to undergo different positional or coordinate shifts according to their position within the continental network. This is known in Epicenter as the NADCON transformation method.

The conversion is achieved by **Bilinear gridded interpolation** for the new NAD 83 Latitudes and Longitudes using US Coast & Geodetic Survey NADCON control point grids. Bilinear interpolation is also used in Canada for the same purpose using the National Transformation (NT) application and grids. As with the US, longitude differences are applied to longitudes that are positive west. The NAD27 and NAD83 geographical coordinate systems documented by EPSG use the positive east longitude convention. The Canadian gridded file format has recently been adopted by Australia.

Alternatively, a polynomial expression with listed coefficients for both latitude and longitude may be used as the transformation method. A transformation, applicable to offshore Norway to effect the transformation between coordinates expressed in the imperfect European Datum 1950 (ED50) and the newer ED87 uses this approach. Statens Kartwerk, the Norwegian Survey authority, publishes a document that lists 15 coefficients for each of separate latitude and longitude polynomial conversion formulas, involving up to fourth-order expressions of latitude and longitude expressed in degrees.

If and when other countries decide to convert to using Geocentric Datums to suit satellite positional data acquisition methods, coordinate transformations similar to NADCON or the Norwegian example may well be employed to facilitate the conversion between the old terrestrial survey derived coordinates and the new geocentric datum values.

Note that it is especially important to ensure that the signs of the parameters used in the transformations are correct in respect of the transformation is executed. Preferably one should always express transformations in terms of "From" & "To", thus avoiding the confusion which may result from interpreting a dash as a minus sign or vice versa.

## 4.6.2 Geographic Coordinate System Transformation Formulas

### 4.6.2.1 Geographic/Geocentric Conversions

Latitude, $\varphi$, and Longitude, $\lambda$, in terms of geographic coordinate system may be expressed in terms of a geocentric (earth-centered) Cartesian Coordinate system (X, Y, Z) with the Z-axis corresponding with the polar axis positive northwards, the X-axis through the intersection of the Greenwich meridian and equator, and the Y-axis through the intersection of the equator with longitude 90°E.

Geocentric coordinate systems are conventionally taken to be defined with the X-axis through the intersection of the Greenwich meridian and equator. This requires that the equivalent geographic coordinate system is based on the Greenwich meridian. In the application of the formulas below, geographic coordinate systems based on a non-Greenwich prime meridian should first be transformed to their Greenwich equivalent.

If the earth's spheroidal semi-major axis is **a**, semi-minor axis **b**, and inverse flattening **1/f**, then

$$X = (v + h)\cos\varphi\cos\lambda$$
$$Y = (v + h)\cos\varphi\sin\lambda$$
$$Z = ((1 - e^2)v + h)\sin\varphi$$

where is $\nu$ the prime vertical radius of curvature at the latitude $\varphi$ and is equal to

$$\upsilon = \frac{a}{\sqrt{1 - e^2 \sin^2 \varphi}}$$

$\varphi$ and $\lambda$ are respectively the latitude and longitude (related to the prime meridian) of the point

h is the height above the ellipsoid, (see note below), and

e is the eccentricity of the ellipsoid where

$$e^2 = \left(a^2 - b^2\right)/a^2 = 2f - f^2$$

(*Note:* *that* h *is the height above the ellipsoid. This is the height value that is delivered by GPS satellite observations but is not the gravity-related height value that is normally used for national mapping and leveling operations. The gravity-related height* (H) *is usually the height above mean sea level or an alternative level reference for the country. If one starts with a gravity-related height* H, *it will be necessary to convert it to an ellipsoid height* h *before using the above transformation formulas.* h *= H + N, where N is the geoid height above the ellipsoid at the point and is sometimes negative. The geoid is a gravity surface approximating mean sea level. For the WGS84 ellipsoid the value of N, representing the height of the geoid relative to the ellipsoid, can vary between values of -100m in the Sri Lanka area to +60m in the North Atlantic. Geoid heights of points above the nationally used ellipsoid may not be readily available.*)

For the reverse conversion, cartesian coordinates in the geocentric coordinate system may then be converted to the geographical coordinates in terms of the geographic coordinate system by:

$$\varphi_B = \arctan\left(\frac{Z_B + e^2 \cdot v \cdot \sin\varphi}{\sqrt{X_B^2 + Y_B^2}}\right) \quad \text{by iteration}$$

$$\lambda_B = \arctan\left(Y_B/X_B\right)$$

$$h_B = X_B \sec\lambda \sec\varphi - \upsilon$$

where $\lambda_B$ is relative to Greenwich. If the geographic system has a non-Greenwich prime meridian, the Greenwich value of the local prime meridian should be applied to longitude, see the example below.

### 4.6.2.2   Offsets

Several transformation methods utilizing offsets in coordinate values are recognized. These include longitude rotations, geographical coordinate offsets, and vertical offsets.

Mathematically, if the origin of a one-dimensional coordinate system is shifted along the positive axis and placed at a point with ordinate A (where A > 0), then the transformation formula is:

$$X_{new} = X_{old} - A$$

However, it is common practice in coordinate system transformations to apply the shift as an addition, with the sign of the shift parameter value having been suitably reversed to compensate for the practice. Since 1999 this practice has been adopted by EPSG. Hence transformations allow calculation of coordinates in the target system by <u>adding</u> a correction parameter to the coordinate values of the point in the source system:

$$X_t = X_s - A$$

where $X_s$ and $X_t$ are the values of the coordinates in the source and target coordinate systems and A is the value of the transformation parameter to transform source coordinate system coordinate to target coordinate system coordinate.

### 4.6.2.3 Geocentric Translations

If we assume that the minor axes of the ellipsoids are parallel and that the prime meridian is Greenwich, then shifts **dX, dY, dZ** in the sense from source geocentric coordinate system to target geocentric coordinate system may then be applied as

$$X_t = X_s + dX$$
$$Y_t = Y_s + dY$$
$$Z_t = Z_s + dZ$$

**Example 4.5.1:   This example combines the geographic/geocentric transformation with a geocentric transformation.**

Consider a North Sea point with coordinates derived by GPS satellite in the WGS84 geographical coordinate system, with coordinates of:

| | | |
|---|---|---|
| latitude $\varphi_s$ | = | 53 deg 48 min 33.82 sec N |
| longitude $\lambda_s$ | = | 2 deg 07 min 46.38 sec E |
| ellipsoidal height $h_s$ | = | 73.0 m |

whose coordinates are required in terms of the ED50 geographical coordinate system which takes the International 1924 ellipsoid. The three-parameter geocentric translations transformation <u>from</u> WGS84 <u>to</u> ED50 for this North Sea area is given as dX = +84.87m, dY = +96.49m, dZ = +116.95m. The WGS84 geographical coordinates first convert to the following geocentric values using the formulas mentioned above:

$$X_s = 3\ 771\ 793.97\ m$$
$$Y_s = 140\ 253.34\ m$$
$$Z_s = 5\ 124\ 304.35\ m$$

Applying the quoted geocentric translations to these, we obtain new geocentric values now related to ED50:

$$X_t = 3\,771\,793.97 + 84.87 = 3\,771\,878.84 \text{ m}$$
$$Y_t = 140\,253.34 + 96.49 = 140\,349.83 \text{ m}$$
$$Z_t = 5\,124\,304.35 + 116.95 = 5\,124\,421.30 \text{ m}$$

Using the reverse conversion above, these convert to ED50 values on the International 1924 ellipsoid as:

latitude $\varphi_t$ = 53 deg 48 min 36.565 sec N
longitude $\lambda_t$ = 2 deg 07 min 51.477 sec E
ellipsoidal height $h_t$ = 28.02 m

Note that the derived height is referred to the International 1924 ellipsoidal surface and will need a further correction for the height of the geoid at this point to relate it to mean sea level.

### 4.6.2.4   Abridged Molodenski Transformation

As an alternative to the above computation of the new latitude, longitude, and height above ellipsoid in discrete steps, the changes in these coordinates may be derived directly by formulas derived by Molodenski. Abridged versions of these formulas, which are quite satisfactory for three-parameter transformations, are as follows:

$$d\varphi'' = \left(-dX \sin\varphi\cos\lambda - dY \sin\varphi\sin\lambda + dZ \cos\varphi + \left[a\cdot df + f\cdot da\right]\sin 2\varphi\right)/\left(\rho\sin 1''\right)$$
$$d\lambda'' = \left(-dX \sin\lambda + dY \cos\lambda\right)/\left(\upsilon\cos\varphi\sin 1''\right)$$
$$dh = dX \cos\varphi\cos\lambda + dY \cos\varphi\sin\lambda + dZ \sin\varphi + \left(a\cdot df + f\cdot da\right)\sin^2\varphi - da$$

where the dX, dY, and dZ terms are as before, and r and $\upsilon$ are the meridian and prime vertical radii of curvature at the given latitude $\varphi$ on the first ellipsoid, da is the difference in the semi-major axes of the target and source ellipsoids ($a_t - a_s$) and df is the difference in the flattening of the two ellipsoids ($f_t - f_s$).

The formulas for $d\varphi$ and $d\lambda$ indicate changes in $\varphi$ and $\lambda$ in arc-seconds.

### 4.6.2.5   Helmert Transformation

Transformation of coordinates from one geographic coordinate system into another (also colloquially known as a "datum transformation") is usually carried out as an implicit concatenation of three transformations:

[geographical to geocentric >> geocentric to geocentric >> geocentric to geographic]

The middle part of the concatenated transformation, from geocentric to geocentric, is usually described as a simplified 7-parameter Helmert transformation, expressed in matrix form with seven parameters, in what is known as the "Bursa-Wolf" formula:

$$\begin{pmatrix} X_T \\ Y_T \\ Z_T \end{pmatrix} = M * \begin{pmatrix} 1 & -R_Z & +R_X \\ -R_Z & 1 & -R_X \\ +R_Y & +R_X & 1 \end{pmatrix} * \begin{pmatrix} X_S \\ Y_S \\ Z_S \end{pmatrix} + \begin{pmatrix} dX \\ dY \\ dZ \end{pmatrix}$$

The parameters are commonly referred to as defining the transformation "from source coordinate system to target coordinate system", whereby $(X_S, Y_S, Z_S)$ are the coordinates of the point in the source geocentric coordinate system and $(X_T, Y_T, Z_T)$ are the coordinates of the point in the target geocentric coordinate system. But that does not define the parameters uniquely; neither is the definition of the parameters implied in the formula, as is often believed. However, the following definition, which is consistent with the position vector transformation convention, is common E&P survey practice:

(dX, dY, dZ): Translation vector, to be added to the point's position vector in the source coordinate system to transform from source system to target system; also: the coordinates of the origin of source coordinate system in the target coordinate system $(R_X, R_Y, R_Z)$: Rotations to be applied to the point's vector. The sign convention is such that a positive rotation about an axis is defined as a clockwise rotation of the position vector when viewed from the origin of the Cartesian Coordinate System in the positive direction of that axis, e.g., a positive rotation about the Z-axis only from the source system to target system will result in a larger longitude value for the point in the target system. Although rotation angles may be quoted in any angular unit of measure, the formula given here requires the angles to be provided in radians.

The scale correction to be made to the position vector in the source coordinate system to obtain the correct scale in the target coordinate system. $M = (1 + dS*10^{-6})$, whereby dS is the scale correction expressed in parts per million.

**Example 4.5.2:** **This example combines the geographic/geocentric transformation with a position vector transformation.**

The transformation from WGS 72 to WGS 84 (EPSG transformation code 1238). Transformation parameters:

$$dX = 0.000 \text{ m}$$
$$dY = 0.000 \text{ m}$$
$$dZ = +4.5 \text{ m}$$
$$R_X = 0.000 \text{ seca}$$
$$R_Y = 0.000 \text{ seca}$$
$$R_Z = +0.554 \text{ seca}$$
$$dS = +0.219 \text{ ppm}$$

Input point coordinate system: WGS 72 (geographic 3D)

| | | |
|---|---|---|
| Latitude $\varphi_S$ | = | 55 deg 00 min 00 sec N |
| Longitude $\lambda_S$ | = | 4 deg 00 min 00 sec E |
| Ellipsoid height $h_S$ | = | 0 m |

Using the geographic to geocentric transformation method, this transforms to cartesian geocentric coordinates:

$$X_S = \ 3\ 657\ 660.66 \text{ m}$$
$$Y_S = \ 255\ 768.55 \text{ m}$$
$$Z_S = \ 5\ 201\ 382.11 \text{ m}$$

Application of the seven parameter position vector transformation (code 1238) results in

$$X_S = \ 3\ 657\ 666.78 \text{ m}$$
$$Y_S = \ 255\ 778.43 \text{ m}$$
$$Z_S = \ 5\ 201\ 387.75 \text{ m}$$

Using the reverse formulas for the geographic/geocentric transformation method given in Section 3.6.2.1 this converts into:

Latitude $\varphi_T$          =   55 deg 00 min 00.090 sec N
Longitude $\lambda_T$         =   4 deg 00 min 00.554 sec E
Ellipsoid height $h_T$   =   +3.22 m

Although being common practice particularly in the European E&P industry, the position vector transformation sign convention is not universally accepted. A variation on this formula is also used, particularly in the USA E&P industry. That formula is based on the same definition of translation and scale parameters, but a different definition of the rotation parameters. The associated convention is known as the "coordinate frame Rotation" convention. The formula is:

$$\begin{pmatrix} X_T \\ Y_T \\ Z_T \end{pmatrix} = M * \begin{pmatrix} 1 & +R_Z & -R_X \\ -R_Z & 1 & +R_X \\ +R_Y & -R_X & 1 \end{pmatrix} * \begin{pmatrix} X_S \\ Y_S \\ Z_S \end{pmatrix} + \begin{pmatrix} dX \\ dY \\ dZ \end{pmatrix}$$

and the parameters are defined as:

(dX, dY, dZ): Translation vector, to be added to the point's position vector in the source coordinate system to transform from source system to target system; also: the coordinates of the origin of source coordinate system in the target coordinate system $(R_X, R_Y, R_Z)$: Rotations to be applied to the coordinate frame. The sign convention is such that a positive rotation of the frame about an axis is defined as a clockwise rotation of the coordinate frame when viewed from the origin of the cartesian coordinate system in the positive direction of that axis, that is a positive rotation about the Z-axis only from source coordinate system to target coordinate system will result in a smaller longitude value for the point in the target coordinate system. Although rotation angles may be quoted in any angular unit of measure, the formula given here requires the angles to be provided in radians.

M: The scale factor to be applied to the position vector in the source coordinate system to obtain the correct scale of the target coordinate system: $M = (1 + dS*10^{-6})$, whereby dS is the scale correction expressed in parts per million.

In the absence of rotations, the two formulas are identical; the difference is solely in the rotations. The name of the second method reflects this.

Note that the same rotation that is defined as positive in the first method is consequently negative in the second and vice versa. It is therefore crucial that the convention underlying the definition of the rotation parameters is clearly understood and is communicated when exchanging datum transformation parameters, so that the parameters may be associated with the correct coordinate transformation method (algorithm).

The same example as for the position vector transformation can be calculated, however, the following transformation parameters must be applied to achieve the same input and output in terms of coordinate values:

$$dX = \quad 0.000 \text{ m}$$
$$dY = \quad 0.000 \text{ m}$$
$$dZ = \quad +4.5 \text{ m}$$
$$R_X = \quad 0.000 \text{ seca}$$
$$R_Y = \quad 0.000 \text{ seca}$$
$$R_Z = \quad -0.554 \text{ seca}$$
$$dS = \quad +0.219 \text{ ppm}$$

Please note that only the rotation has changed sign as compared to the position vector transformation.

## 4.7  COMPUTER-AIDED DESIGN OF PROCESSING FOR COORDINATE CONVERSION

The Parameters for Transforming Longitude/Latitude to Easting/Northing to be given to yield the desired transformations are described below:

1. Spheroid: Everest 1956
2. Lambert Conformal Conic Projection

```
// Spheroid Parameters
A = 6377301.243; // India 1956
F = 1.0/300.8017; // Everest
// False Northing & Easting
x_offset = 3000000.0;
y_offset = 1000000.0;
// Origin - Zone Specific
```

lat_offset = 19D 15m 14.9818s
long_offset = 79.0D 0m 0s;
// Standard Paralells Specific
PHI1 ∅1 = 17D 15m 0s;
PHI2 ∅2 = 21D 15m 0s

1. The program converts Co-ordinates from Longitude Latitude to Easting Northing Remember that the order is LONGITUDE and then LATITUDE, by using the above parameters which are different at various locations in the world.
2. The program converts Co-ordinates from Easting Northing to Longitude Latitude, remember that the order is EASTING and then NORTHING, by using the above parameters which are different at various locations in the world.

**The computer-aided processing of coordinate conversion is being described in Section 4.6 of Chapter 4 in the "The Guide for Computer Application Tutorials" part of the book available for download from the website of the book.**
To download the book tutorials and other items, visit the websites at:
www.roadbridgedesign.com and www.techsoftglobal.com/download.php

## 4.8  TRIANGULATION

For developing a geometric figure using a given set of points the triangulation algorithm is used to connect the points and find the convex hull. Most commonly the Delauney triangulation is used in the development of 3D geometric figures for which the complexity of the problem is more than the 2D case. The method is particularly suited when we do not introduce any constraints on the set of points to be connected. Additionally, Delaunay triangulation has properties of optimal Equiangularity and Uniqueness in 2D.

The development work for three-dimensional GIS has generated effective techniques for the visual display and analysis of three-dimensional rasters. The facility takes care of the indexing of vector polylines in three-dimensional space. Still, the interpolation of sparse point data sets into coherent and robust models in unseen structurally complex geoscientific domains remains a central and largely unsolved problem.

The method of triangulation-based interpolation extends into two to three dimensions, suggests hybrid implementation mechanisms to optimize performance, and discusses the requirements of geoscientific interpolation as they affect the triangulation procedure (Figure 4.17).

### 4.8.1  DELAUNAY TRIANGULATION AND ALGORITHM

The technical aspect of the problem is of connecting a set of points. Earlier a way was proposed to subdivide a given domain into a set of convex polygons. When two points "Pu" & "Pj" are given in the plane "T", the perpendicular to the segment PiPj

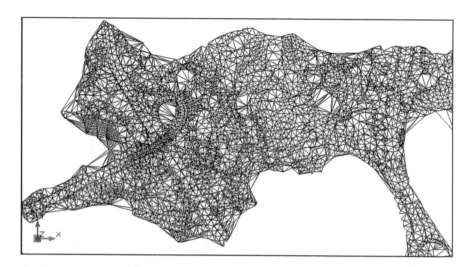

**FIGURE 4.17**   Triangulation

in the middle point divides the plane "T" into two regions "Vi" & "Vj". Region "Vi" contains all only the points closest to "Pi" than to "Pj". In case we have more points, we can extend this concept by defining that region "Vi" is assigned to "Pi", so that each point belonging to "Vi" is closest to "Pi" than to any other point.

The subdivision or region of the space is defined by a set of distinct points where each point is associated with that region of the space, then the region nearer to that point than any other point is known as "Dirichlet Tessellation".

This process applies to a closed domain, this generates a set of convex distinct polygons which are called "Voronoi" regions, this covers the entire domain. This definition while applied to three dimensions, then the "Voronoi" regions are convex polyhedrons. By connecting all the pairs of points sharing a border of a "Voronoi" region we obtain a triangulation of the convex space containing those points. This triangulation is known as "Delaunay triangulation".

We can obtain a triangulation for in three dimensions, if we connect all pairs of points sharing a common facet in the "Voronoi" region diagram, it develops a set of tetrahedra filling the entire domain.

### 4.8.1.1   THE DELAUNAY TRIANGULATION HAS THE FOLLOWING INTERESTING PROPERTIES

1. IN CIRCLE:
   For a triangle "T (Pi, Pj, Pk)" belonging to a "Delaynay Triangulation (DT)" of a set of points "P", then there will be no point of "P" which is inside the circle defined by "Pi. Pj, Pk". This property defines a mechanism to automatically build a "Delaunay triangulation" with a set of given points.
2. MAX MIN:
   A quadrilateral is associated with four points, the diagonal split it into two triangles. This is optimal in terms of maximizing the lesser of the internal angles.

This property ensures the shapes of the triangles as best, for the given set of four points in the quadrilateral.

### 4.8.1.2 Developing a Delaunay Triangulation

There are several algorithms available to build a Delaunay triangulation for a set of points. One of these algorithms is described here and is considered a highly effective method in terms of the quality of elements generated in three dimensions. This is an incremental algorithm, here the points are added one at a time into an existing triangulation. This algorithm applies to two- and three-dimensional points. The procedure is automatic and does not require any user intervention.

Let us take "Tn" as the "Delaunay Triangulation" of a set of points,

$$Vn = \{Pi|,I = 1|,2|,...|,n\},$$

By defining a simplex as any "n-dimensional" polygon and a convex hull as the domain to which these points belong, is formalized as "Rs" the radius circumscribed to each simplex "S" of "T", and as "Qs" the center of the "n" dimensional circumscribed sphere.

Next, if we insert a new point "Pn+1" in the convex hull of "Vn" and define B = {S |S Tn| d(Pn+1,Qs) < Rs} where d(p,Q) is the Euclidean distance between points "P" & "Q",

Now, "B" is not empty as Pn+1 lies in the convex hull of "Vn" and inside a simplex "S1" belonging to "Tn". So at least s1 belongs to B.

When "B" is removed from "T", the region "C" is formed and is simply connected and contains point "Pn+1", "Pn+1" is visible from all points which form the border of "C", now it is possible to generate a triangulation for a set of points Vn+1 = Vn {Pn+1}, this connects "Pn+1" with all the points that form the border of "C", this is the "Delaunay triangulation".

### 4.8.1.3 Degeneracies

From the definition of Delaunay triangulation, it is evident that problem arises in the procedure when certain degeneracies occur in the data:

1. When two points are coincident:
   The Delaunay triangulation for a set of points is applicable only when the points are distinct, the uniqueness of the points in the given set of points is a prerequisite for the applicability of the algorithm.
2. When three 2D points of a triangle are co-linear or four 3D points of a tetrahedron are co-planner: This means that it is not possible to get a valid center for the circumcircle for the triangle or the center of the inscribed sphere for the tetrahedra.
3. When four or more points are in the same circle.

The usual way to deal with such degeneracies are:

- Reject the point.
- Delay the point insertion.
- Shift its coordinates.

The choice of the best option often depends upon the application, for example, "Reject of the point" may be chosen if we have many points creating high point density.

When a set of points in two dimensions is given, the Delaunay triangulation is best applied, and it is unique. But in some cases, the triangulation is not unique as the ways of connecting points are different, and all results in a valid triangulation. This degeneracy is common for regular distribution of points, as in the case when four points in two dimensions lie on a circle and the Voronoi vertices are coincident.

#### 4.8.1.4 Data Structures

The data structures:

- Must be supporting efficient operations on the objects.
- Capable of handling queries on attributes.
- Capable of handling queries on the object's position within the domain.

The N-Tree data structure implemented in this work is a generic dynamic n-dimensional tree used to recursively divide the domain and provide efficient search operations on geometric objects such as points edges, triangles, and tetrahedra.

This data structure also provides an efficient way of verifying object uniqueness. So, when an initial set of points is taken for triangulation the points are all inserted in an N-Tree, a filtering action is done for duplicate points, points that are equal due to limited machine precision, points which are equal due to rounding off error which fall into the same partition. These points are eliminated to satisfy the basic requirement of Delaunay triangulation.

The N-Tree data structure is similar to the Octree data structure. Both of them are recursive space subdivision data structures. In the N-Tree there is no fixed resolution or data density for the subdivision which means that there is no limit for the number of objects present in the tree, at any point in time. The data structure itself is the model, so the Octree represents an object where the model and the data structure are synonymous. Whereas the N-Tree is more generic, this is a superset of the Octree, as its main purpose is to store more objects rather than represent them. This is true for basic geometric objects like points, edges, triangles, tetrahedra, etc. The way of representing an object, which is the domain in this case with full advantages of a dynamic data structure is done by Delaunay triangulation by storing a domain in an N-Tree as a set of geometric objects like triangles or tetrahedra. This helps the regeneration of all or parts of the model as desired. There is scope to generalize and interpolate features by taking full advantage of the N-Tree data structure.

#### 4.8.1.5 Shape Refinement

The triangulation of a set of points is an important method with many applications including the simulations of finite elements. Various algorithms are available for triangulating a set of points in two or three dimensions. Some of them consider optimizing the shape of the triangular elements.

Triangular meshes are required of bounded aspect ratio to keep ill conditioning in simulation and discretization errors to the minimum. The aspect ratio of triangles or tetrahedral is the ratio of radii of the inscribing to that of the circumscribing circles in case of triangles and spheres in the case of tetrahedral. In a two-dimensional case, the maximum-minimum angle property of the Delaunay triangulation ensures the best possible shape of the elements, but it does not hold in three dimensions. any algorithm is hardly available which triangulates the convex hull of a set of three-dimensional points with tetrahedral of guaranteed quality of aspect ratio.

The Bowyer-Watson algorithm is strong in two dimensions, it further divides the triangle or tetrahedral elements adding points so that the shape of the element obtained after split is better than that of the original element.

There are many reasons to build well shaped elements. If elements have very low aspect ratios, then the triangulation is not desirable because:

- In visual graphics, this results in irregularities in shading.
- In finite element analysis, this may lead to ill conditioned numerical problems.
- The centroid as a representative location becomes weak.

If our model is based on tetrahedral elements with vertices at the given data points, then each point has an attribute that can be associated with the tetrahedral by using the local piecewise interpolation method.

This is a simple way to calculate the average of the values of its vertices. Whichever interpolation method has used the shape of the tetrahedral must reflect the representative location of the attributes if we want good results from the interpolation process.

### 4.8.1.6  Complexity

The complexity of the algorithm depends upon the components which are dependent on the number of points and thus play an important part in the procedure. When a new point is inserted into an existing triangulation, we need to search the list of the triangles to find those elements whose circumcircle contains the new point.

Once the elements forming the void are obtained, then the time needed to reconnect the points can be considered constant. The time required to triangulate "N" points can be expressed as T = (Tk + T1k), where "Tk" is the time to find the first triangle to be deleted and "T1k" is the time to find all the others.

By maintaining a topological relationship among the elements to store the reference of all the neighbors of a triangle in the triangle then we can consider "T1k" as proportional to the number of elements in the void which is independent of "k". Excepting the first search, all other operations are local and may be carried out at a time independent of the number of points currently in the structure. So, we can estimate the time complexity of the algorithm as tentatively proportional to "Tk". The search time "Tk" will be proportional to "k" leading to the overall time complexity of "O(N²)". Using the "N-Tree" data structure the first search can be reduced to "O (logN)" by giving an overall time complexity for the triangulation algorithm of "O (NlogN)".

### 4.8.1.7  Environment for Implementation

The programming for the algorithm must be well suited to define both geometric objects and data structures. The advantages of the object-oriented programming language used are the following:

- Easy extendibility to higher dimensions.
- Straight forward structured coding allows experimenting with different solutions.

### 4.8.1.8  Various Modules in the Triangulation Process

- Basic geometric objects both in two and three dimensions like points, edge, triangle, tetrahedra, etc.
- The N-Tree data structure for two and three dimensions.
- Delaunay triangulation in two and three dimensions.
- Constrained Delaunay triangulation in two and three dimensions, there is a provision to specify a set of constraints like edges, boundary, etc. to be included in the triangulation process by maintaining all its properties.
- Enrichment procedures are used to split and improve the average shape of the elements.
- In the triangulation process, each sample point has one or more attributes attached to it.

Each separate module is organized as a library that can be easily linked to an application.

### 4.8.1.9  Application to Three-Dimensional Geoscientific Modeling

Most of the geoscientific applications need to collect data about spatial objects and domains such as features of solid earth, aquifers, oceans, atmosphere weather fronts which are enclosed in a three-dimensional space. The geometric representation of these domains requires defining each known location in an x,y,z coordinate system. For a qualitative representation of the object, the attributes will have to be linked to the geometric descriptions.

The approaches of three-dimensional representation and structuring of geo-objects can be categorized as raster, vector, and function based. The raster solutions are mostly around the voxel, as a basic unit that is not necessarily cubic. Many theories have shown the advantages of terrain modeling are based on TINs.

The modeling is more adaptive, and the handling of terrain data is more flexible compared to the traditional grid-based techniques. The TIN based modeling has been partly exploited in two dimensions but hardly any application is done in three dimensions.

Following are the reasons to describe a geo-model using a three-dimensional triangulation:

- The generation is objective and so fully automatic.
- Space is defined uniquely, and cells are spatially indexed.
- It is possible to adjust the Size of elements locally as a function of the complexity of the model.
- The model is easily editable manually.
- The topology is obtained from neighborhood relationships.
- To represent trends the vectors or surface constraints can be used for constrained Triangulation.
- As the use of triangular elements is the basis for rendering techniques so is perfect for visualization.
- Compared to the block models the accuracy and approximation are better.
- It is easy to calculate and efficient for the integral properties.
- It is easy to extract the three-dimensional triangulated surface, which is the boundary, from the three-dimensional solid representation of an object.
- It is easy to implement Spatial searches and Relation queries.
- It has good performance of Boolean operations.

The representation of relationships and interpolation of sparse geoscientific data is indeed a difficult task. So, although all the above triangulation algorithms work well for the random data sets, still, when triangulation points are obtained from solid models or boreholes they are not randomly positioned and usually form highly degenerate sets of points aligned on the "z-axis" which form sub-optimum element shapes. This is the reason for applying post-processing for shape refinement to the initial set of elements as described in Figures 4.3 and 4.4.

## 4.9 CONTOURS

A contour line (also isoline, isopleth, or isarithm) of a function of two variables is a curve along which the function has a constant value so that the curve joins points of equal value. It is a cross section of the three-dimensional graph of the function $f(x, y)$ parallel to the x, y plane. In cartography, a contour line (often just called a "contour") joins points of equal elevation (height) above a given level, such as mean sea level. A contour map is a map illustrated with contour lines, for example, a topographic map, which thus shows valleys and hills, and the steepness or gentleness of slopes. The contour interval of a contour map is the difference in elevation between successive contour lines (Figure 4.18).

More generally, a contour line for a function of two variables is a curve connecting points where the function has the same particular value.

The gradient of the function is always perpendicular to the contour lines. When the lines are close together the magnitude of the gradient is large: the variation is steep. A level set is a generalization of a contour line for functions of any number of variables.

Contour lines are curved, straight or a mixture of both lines on a map describing the intersection of a real or hypothetical surface with one or more horizontal planes. The configuration of these contours allows map readers to infer the relative gradient of a parameter and estimate that parameter at specific places. Contour lines may be

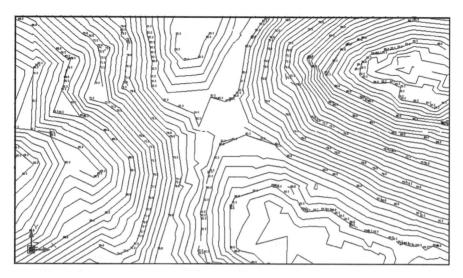

**FIGURE 4.18**   Contours with Elevations

either traced on a visible three-dimensional model of the surface, as when a photo-grammetrist viewing a stereo-model plots elevation contours, or interpolated from estimated surface elevations, as when a computer program threads contours through a network of observation points of area centroids. In the latter case, the method of interpolation affects the reliability of individual isolines and their portrayal of slope, pits, and peaks.

**The computer applications for the "Topographical Survey and Data collection" are explained in Chapter 4 in the "The Guide for Computer Application Tutorials", part of the book available for download from the website of the book.**

To download the book tutorials and other items, visit the websites at:

www.roadbridgedesign.com and www.techsoftglobal.com/download.php

## 4.10   COMPUTER-AIDED DESIGN APPLICATION IN SURVEY DATA PROCESSING

**This Design Project will be continued from Chapters 4 to 8**
   **By Using Total Station Data, the Following are the Essential Steps:**

- Formatting the data in standard GENIO or DGM formats.
- Processing the formatted data to create the ground model.
- If the Total Station Data is in TM format, then transform it into UTM format.
- Process ground model for creating a triangulation model.
- Process triangulation model for creating a contour model.
- Draw the ground model.
- Draw a triangulation model in 3D.
- Draw a contour model in 3D.

In this session, we shall see the processing of survey data design of highway with uniform cross section applied from start to end of the alignment. The design is based on the alignment of the centerline of the road passing through the center of the median or central verge of the dual carriageway cross section.

**The computer-aided processing of topographic survey data is being described in Section 4.7.1 and the processing of cross section survey data is being described in Section 4.7.2 of Chapter 4 in the "The Guide for Computer Application Tutorials" part of the book available for download from the website** www.roadbridgedesign.com.

To download the book tutorials and other items, visit the websites at:

www.roadbridgedesign.com and www.techsoftglobal.com/download.php

## BIBLIOGRAPHY

Surveying by A. Bannister, S. Raymond, ELBS with Longman

Solving Problems in Surveying by A. Bannister, Raymond Baker, ELBS with Longman

Fundamentals of Geographic Information Systems by Michael N. Demers, John Wiley & Sons. Inc.

Textbook on Remote Sensing in Natural Resources Monitoring and Management, by C. S. Agarwal, P. K. Garg, Wheeler Publishing

Principles of Highway Engineering and Traffic Analysis by Fred L. Mannering, Walter P. Kilareski, Scott S. Washburn John Wiley & Sons Inc.

Pavements and Surfacings for Highways and Airports, by Michel Sargious, Applied Science Publishers Ltd.

# 5 Design of Horizontal Alignment

## 5.1 GENERAL

The horizontal alignment of a road has to ensure that travel along the road is safe with adequate visibility and at a continuous uniform design speed.

## 5.2 DESIGN CONSIDERATIONS

The following factors influence the degree of horizontal curvature of a road:

- Safety
- Design speed
- Terrain topography, adjacent land use and obstruction
- Vertical alignment
- Maximum allowable superelevation
- Road classification
- Cost

A balanced blend of all these factors produces a good alignment. Poor design may produce a reduction in safety and capacity of the road. Some specific guidance is given in this chapter. Additionally, there are some more considerations which are also important for the design of a safe and economic design. These are essential in the design of high-speed travel routes and are listed below:

- In the case of a design of a new alignment, it is always preferable to use a curve of greater radius than the specified minimum value.
- Compound circular curves are composed of two or more circular arcs of different radii joined end-to-end in one direction. The radius and design speed should be so chosen that the section doesn't require any super elevation which will be difficult to fit under such circumstances. In that case, the compound curve can be avoided by using a simple curve if that is a better choice. Where compound curves are used, then for each two connecting curves, the radius of the flatter curve should not be more than 50% of the radius of the sharper curve. This consideration does not apply in locations of intersection, roundabouts, etc., having a lower travel speed.
- In reverse circular curves, two successive curves turn in opposite directions. The two curves should include an intervening transition section of sufficient length so that it can accommodate the reversal of superelevation between circular curves on high-speed road sections. The tangent section with a normal

crown between the curves will provide sufficient distance between the reverse curves to accommodate the superelevation runoff and the tangent runout for both the curves. In case it is needed and the superelevation is to be reversed without any intervening normal crown section, then the length between the reverse curves should be such that the superelevation runoff lengths abut, and it is done by providing an instantaneous level section across the pavement.

- Broken back curves consist of two curves in the same direction of turning, connected by a short tangent and should be avoided. These are unexpected by drivers and are not pleasing in appearance.
- Horizontal alignment should be consistent with other design features and the topography. The horizontal alignment must have coordination with the vertical alignment.
- On divided dual carriageway roads, consideration may be given to possibility of providing independent horizontal and vertical alignments for each carriageway. In computer-aided design, this is possible by using software programs with string modeling technology which is versatile for various design requirements and superior compared to software programs based on template technology.

## 5.3   MAXIMUM SUPERELEVATION

On straight sections of the roadway, transverse drainage is accomplished by providing crossfall at a standard rate, say of 2%. On undivided single carriageway roads, the surface normally slopes down outwards from both sides of the central crown and is known as normal crown, while on divided dual carriageway roads the surface of each carriageway normally slopes down outwards from the central median.

On horizontal curves the crossfall during turning maneuvers becomes more difficult for drivers on the outside of the curve and so for radii below a certain value it is essential to eliminate this adverse crossfall by making the whole width of road fall toward the inside edge of the curve, resulting a superelevation of 2%. On sharper curves of smaller radii, a higher superelevation value can be adopted to assist drivers traveling round the corner.

The maximum value of superelevation is governed by the speed of slow-moving vehicles which find it difficult and unexpected to exert a driving force against the direction of the curve. During a rainy spell after a long dry period, it can result in low surface friction factors and so, a high value of crossfall may be avoided. The normally adopted values of maximum superelevation are described in Table 5.1.

## 5.4   MINIMUM CURVATURE

The relationship between the speed of the vehicle, radius of the curve, the superelevation, and the side friction between the tire and the road surface is given below:

$$R = \frac{V^2}{127\left(e + fs\right)}$$

where

R = radius of curve (m)
V = vehicle speed (km/hr)
e = superelevation (%) divided by 100
fs = side friction factor

From observations, the side friction factor has been found to lie in the range of 0.35–0.50 on dry roads which may drop to 0.20 on wet road surfaces. To make the design safe even lower values are adopted. The commonly adopted values are given in Table 5.2, which vary from 0.17 at design speed 30 km/hr, to 0.113 at design speed 120 km/hr.

For a given design speed, the minimum curve radii can be obtained for a range of superelevation rates which are given in Table 5.3. On local roads in residential areas having design speeds of 50 km/hr or less, full superelevation should not be provided to prevent giving an impression to the drivers that a higher speed may be applied. The radii shown in columns (3)–(5) in Table 5.3 may be used but restricting the maximum superelevation to 2%.

**TABLE 5.1**
**Maximum Superelevation**

| Road Class | Maximum Superelevation (%) | |
|---|---|---|
| | Rural | Urban |
| Local Road | 4% | 2% |
| Secondary Road | 6% | 4% |
| Primary Road | 8% | 6% |
| Special Road | 8% | 8% |
| Loops (within interchanges) | 8% | 8% |

**TABLE 5.2**
**Side Friction Factors for Design**

| Design Speed | Side Friction Factor |
|---|---|
| 30 | 0.170 |
| 40 | 0.164 |
| 50 | 0.157 |
| 60 | 0.151 |
| 70 | 0.145 |
| 80 | 0.138 |
| 90 | 0.132 |
| 100 | 0.125 |
| 110 | 0.119 |
| 120 | 0.113 |

**TABLE 5.3**
**Minimum Horizontal Curvature**

| | Minimum Radii (m) | | | | |
| | Normal Crown (–2%) | Superelevation | | | |
| | (1) | (2) | (3) | (4) | (5) |
| Design Speed (km/hr) | | 2% | 4% | 6% | 8% |
| 30 | 50 | 40 | 35 | N/A | N/A |
| 40 | 90 | 70 | 60 | N/A | N/A |
| 50 | 145 | 110 | 100 | 90 | 80 |
| 60 | 220 | 165 | 150 | 135 | 125 |
| 70 | 310 | 235 | 210 | 190 | 175 |
| 80 | 430 | 320 | 285 | 255 | 230 |
| 90 | 570 | 420 | 375 | 335 | 305 |
| 100 | 750 | 545 | 480 | 430 | 395 |
| 110 | 965 | 685 | 600 | 535 | 500 |
| 120 | 1220 | 850 | 740 | 655 | 590 |

At intersections other than roundabouts, the normal crown or superelevation of the main road should be continued through the intersection and tying the longitudinal profile of the minor road to the cross-sectional profile of the main road. At roundabouts the considerations are different and will be discussed in the relevant chapter for the design of roundabouts.

## 5.5   CALCULATION OF SUPERELEVATION

For a given design speed and radius, the superelevation is calculated as follows (Ref. Table 5.3):

- For radii in column (1) and larger, provide a normal crown with 2% crossfall and no superelevation.
- For radii in the range between column (2) and column (1), provide crossfall of, say 2% toward the inside of the curve.
- For radii in the range between column (5) and column (2), provide superelevation calculated by using the formula:

$$e\% = \left( \frac{V^2}{127} - fs \right) \times 100$$

## 5.6   TRANSITION CURVES

### 5.6.1   Applications

When there is a change from a straight to a circular travel path, a good design provides the drivers with a natural transition of travel. Incorporation of transition curves

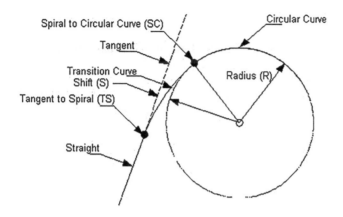

**FIGURE 5.1**  Typical Layout of Transition Curve.

also improves the appearance of the alignment and helps in the attainment of super-elevation before entering the circular curve.

The use of the Euler spiral or clothoid transition curves are more common compared to a cubic parabola in most countries. The degree of curvature of spiral or clothoid transition curves varies directly with the length of the curve.

Transition curves are not required with circular curves of radii equal to or greater than the values given in column (1) of Table 5.3. They are also not required on roads with design speeds 70 km/hr or less.

The layout of a typical transition curve joining a straight tangent and circular curve in the road alignment is shown in Figure 5.1.

### 5.6.2  LENGTH OF TRANSITION CURVE

As shown in Figure 5.1, the length of a transition curve is from TS to SC. The length depends on the radius of the circular curve to which it is joined and is defined by the following formula:

$$Ls = \frac{V^3}{(46.7 \times q \times R)}$$

where
  Ls = Length of spiral (m)
  V  = Design speed (km/hr)
  q  = Rate of change of centripetal acceleration (m/sec$^3$)
  R  = Radius of circular curve (m)

The value of q is commonly taken as 0.3 m/sec$^3$ from comfort considerations, however the value up to 0.6 m/sec$^3$ can be adopted if required by the design. The rousnded values of computed spiral lengths are given in Table 5.4, for the radii in column (4) of Table 5.3. The maximum length of a transition curve is limited to

**TABLE 5.4**
**Basic Spiral Lengths for Minimum Radii at 6% Superelevation**

| Design Speed (km/hr) | Minimum Radius at 6% Superelevation (m) | Length of Spiral | |
|---|---|---|---|
| | | Desirable ($q = 0.3$ m/sec³) | Minimum ($q = 0.6$ m/sec³) |
| 60 | 135 | 115 | 60 |
| 70 | 190 | 130 | 65 |
| 80 | 255 | 145 | 75 |
| 90 | 335 | 155 | 80 |
| 100 | 430 | 170 | 85 |
| 110 | 535 | 180 | 90 |
| 120 | 655 | 190 | 95 |

$\sqrt{(24\text{ R})}$ m, where R is the radius of the circular curve. The designer should check that the applied transition length is less than this value.

### 5.6.3 LENGTH OF SUPERELEVATION IN APPLICATIONS

The driver comfort and satisfactory appearance is usually achieved by ensuring that the profile of the carriageway edge does not have gradient more than 0.5% from the line about which the carriageway is rotated and by the smoothing of all changes in edge profile. The minimum length of superelevation is determined by the following formula:

$$L = 1.5 \times W \times \Delta e$$

where
　　L  = length required to accommodate the change in superelevation (m)
　　W = width of carriageway over which change in superelevation occurs
　　$\Delta e$= change in gradient (%)

The minimum lengths required to accommodate the change in superelevation are calculated for a range of superelevation rates and carriageway widths, these are given in Table 5.5. If longer lengths are used then the change of gradient will be flatter and

**TABLE 5.5**
**Minimum Lengths to Accommodate Change in Superelevation**

| Change in Gradient $\Delta e$ | Length Required to Accommodate Change in Superelevation (m) | | |
|---|---|---|---|
| | 1 Lane (3.7 m) | 2 Lanes (7.4 m) | 3 Lanes (11.1 m) |
| 2% | 12 | 23 | 34 |
| 4% | 23 | 45 | 67 |
| 6% | 34 | 67 | 100 |
| 8% | 45 | 89 | 134 |
| 10% | 56 | 112 | 167 |

less than 5%, in such case designer has to check for the adequacy of the sufficient gradient required for drainage.

### 5.6.4 APPLICATION OF SUPERELEVATION

The developing of superelevation is illustrated in Figure 5.2 by rotation of the cross section of the road about the edges and about the center of the road. In design, the

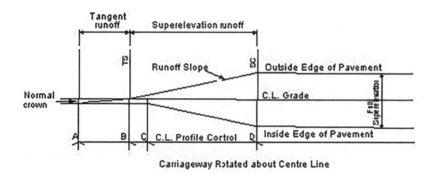

Carriageway Rotated about Centre Line

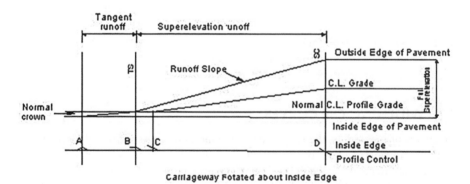

Carriageway Rotated about Inside Edge

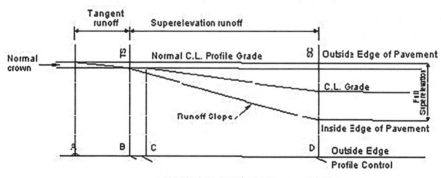

Carriageway Rotated about Outside Edge

**FIGURE 5.2**   Development of Superelevation.

most suitable method is to be adopted. On divided roads, the designer has to consider the topography, cut and fill, catchments, and median drainage. Adopting of different horizontal and vertical geometry for the two separate carriageways may also be considered. In reverse curves the superelevation is applied by calculation using the following relationship:

$$Y = 35\,x^2/L^2 - 2\,s\,x^3/L^3$$

where

    Y  = offset of channel with respect to the line, about which the carriageway is rotated, at a distance x (m)

    x  = distance from the start point of application (m)

    s  = maximum offset of channel superelevation with respect to the line, about which the carriageway is rotated (m)

    L  = length of application of superelevation (m)

If transition curves are considered, then superelevation or elimination of adverse camber shall generally be applied on or within the length of the transition curve. On roads without transition curves 1/2 to 2/3 of the superelevation is to be introduced on the approach straight and the remainder at the beginning of the circular curve.

    A  = Normal crown

    B  = Adverse camber removed

    C  = Superelevation at normal crossfall rate

    D  = Full superelevation

## 5.7 WIDENING ON CURVES

On curves of low radius, the rear wheels of vehicles do not exactly follow the track of the front wheels and so some extra widening becomes necessary (Table 5.6(A)).

Widening depends on vehicle geometry, particularly on wheelbase, lane width, and curve radius. Widening should be applied in both the directions of travel to have the lane width on the circular curve as given in Table 5.6(B). On divided roads it is only necessary to widen the outer lane on both the carriageways, all other lanes remaining with their normal width.

**TABLE 5.6(A)**
**Minimum Lane Width on Curves**

| Radius (m) | Lane Width (m) |
|---|---|
| >125–300 | 4.5 |
| >300–400 | 4.0 |
| More than 400 | Normal Width |

**TABLE 5.6(B)**
**Width of Roadways**

| Curve Radius (m) | Case 1: Single Lane Width (m) | Case 2: Single Lane Width with Space to Pass Stationery Vehicle (m) | Case 3: Double Lane Width (m) |
|---|---|---|---|
| 15 | 6.9 | 8.7 | 12.6 |
| 25 | 5.7 | 8.1 | 11.1 |
| 30 | 5.4 | 7.5 | 10.5 |
| 50 | 5.1 | 7.2 | 9.9 |
| 75 | 4.8 | 6.9 | 9.3 |
| 100 | 4.8 | 6.6 | 9.0 |
| 125 | 4.8 | 6.6 | 8.7 |

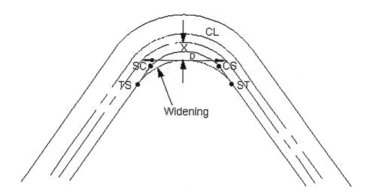

**FIGURE 5.3** Application of Carriageway Widening on Curves.

To provide the additional carriageway width by widening the inside of the curve is always good practice. In Figure 5.3 the widening is developed over the length of the transition, this provides full widening around the circular part of the curve.

## 5.8 LATER CLEARANCES

Signposts and lighting posts fall within the roadside verge area over which the visibility is maintained, but no structure which causes significant obstruction to visibility should exist in the verge area. In some situations when the curve radius is relatively small, the sight line envelope extends outside the right of way and gets obstructed by properties. In such cases it is important to maintain proper setback to obstructions to ensure proper sight distances.

Sight distances are measured between points on the centerline of the innermost lane of the carriageway, the setback is to be maintained from that line. The sight distances are given in Tables 5.7, 5.8, and 5.9.

**TABLE 5.7**
**Minimum Setback to Maintain Stopping Sight Distance (Level Road)**

| Radius (m) | Offset (m) from Centre Line of Nearest Lane, at Design Speed (km/hr) of | | | | | | |
|---|---|---|---|---|---|---|---|
|  | 60 | 70 | 80 | 90 | 100 | 110 | 120 |
| 100 |  |  |  |  |  |  |  |
| 200 | 4.6 | 7.0 |  |  |  |  |  |
| 300 | 3.0 | 4.6 | 7.1 |  |  |  |  |
| 400 | 2.3 | 3.5 | 5.3 | 8.1 | 10.8 |  |  |
| 500 |  | 2.8 | 4.2 | 6.4 | 8.6 | 12.3 |  |
| 750 |  | 1.8 | 2.8 | 4.3 | 5.7 | 8.1 | 10.5 |
| 1000 |  |  | 2.1 | 3.2 | 4.3 | 6.1 | 7.8 |
| 1250 |  |  | 1.7 | 2.6 | 3.4 | 4.8 | 6.3 |
| 1500 |  |  |  | 2.1 | 2.9 | 4.0 | 5.2 |
| 2000 |  |  |  |  | 2.1 | 3.0 | 3.9 |

**TABLE 5.8**
**Minimum Setback to Maintain Safe Passing Sight Distance (SPSD)**

| Radius (m) | Offset (m) from Centre Line of Nearest Lane, at Design Speed (km/hr) of | | | | |
|---|---|---|---|---|---|
|  | 60 | 70 | 80 | 90 | 100 |
| 100 |  |  |  |  |  |
| 200 |  |  |  |  |  |
| 300 | 81 | 124 | 169 |  |  |
| 400 | 57 | 82 | 105 | 144 | 182 |
| 500 | 44 | 63 | 79 | 106 | 129 |
| 750 | 29 | 40 | 51 | 66 | 79 |
| 1000 | 21 | 30 | 37 | 49 | 58 |
| 1250 | 17 | 24 | 30 | 39 | 46 |
| 1500 | 14 | 20 | 25 | 32 | 38 |
| 2000 | 11 | 15 | 19 | 24 | 29 |
| 2500 | 8.4 | 12 | 15 | 19 | 23 |
| 3000 | 7.0 | 9.8 | 12 | 16 | 19 |
| 4000 | 5.3 | 7.4 | 9.1 | 12 | 14 |
| 5000 | 4.2 | 5.9 | 7.3 | 9.5 | 11 |
| 6000 | 3.5 | 4.9 | 6.1 | 7.9 | 9.4 |
| 7000 | 3.0 | 4.2 | 5.2 | 6.8 | 8.0 |
| 8000 | 2.6 | 3.7 | 4.6 | 5.9 | 7.0 |
| 9000 | 2.3 | 3.3 | 4.1 | 5.3 | 6.2 |

**TABLE 5.9**
**Minimum Setback to Maintain Sight Distance (Urban/No Stop Situation)**

| Radius (m) | Offset (m) from Centre Line of Nearest Lane, at Design Speed (km/hr) of | | | | | | |
|---|---|---|---|---|---|---|---|
| | 60 | 70 | 80 | 90 | 100 | 110 | 120 |
| 100 | | | | | | | |
| 200 | 38.2 | 54.8 | | | | | |
| 300 | 24.0 | 33.4 | 44.7 | | | | |
| 400 | 17.6 | 24.4 | 32.3 | 42.8 | 53.6 | | |
| 500 | 14.0 | 19.3 | 25.5 | 33.5 | 41.7 | 48.6 | |
| 750 | 9.3 | 12.7 | 16.7 | 21.9 | 27.2 | 31.5 | 37.8 |
| 1000 | 6.9 | 9.5 | 12.5 | 16.3 | 20.2 | 23.4 | 28.0 |
| 1250 | 5.5 | 7.6 | 10.0 | 13.0 | 16.1 | 18.6 | 22.3 |
| 1500 | 4.6 | 6.3 | 8.3 | 10.8 | 13.4 | 15.5 | 18.5 |
| 2000 | 3.5 | 4.7 | 6.2 | 8.1 | 10.0 | 11.6 | 13.9 |
| 2500 | 2.8 | 3.8 | 5.0 | 6.5 | 8.0 | 9.3 | 11.1 |
| 3000 | | 3.2 | 4.1 | 5.4 | 6.7 | 7.7 | 9.2 |
| 4000 | | | 3.1 | 4.1 | 5.0 | 5.8 | 6.9 |
| 5000 | | | | 3.2 | 4.0 | 4.6 | 5.5 |

In the design of two-way, undivided, single carriageway roads, the safe passing sight distance (SPSD) should be considered as mentioned in chapter four. Where SPSD is to be maintained around any curve, a large lateral clearance is to be maintained as given in Table 5.8, which may be uneconomical. Radii given in there may be used on two-way, undivided, single carriageway roads, where full SPSD is possible to be maintained within the right of way.

To maintain the decision sight distance (DSD) the minimum setback distances are given in Table 5.9, they have been calculated using the worst case DSDs as described in paragraph 4.6. For special situations, the setback distances may be determined from the following relationship:

$$x = R - \sqrt{\left[R^2 - D^2/4\right]}$$

where
  x  = offset (m)
  R  = radius of curve (m)
  D  = sight distance (m)

**TABLE 5.10**

**Minimum Transition Lengths for Different Speeds and Curve Radii**

**Plain and Rolling Terrain**

| Curve Radius (meters) | Design Speeds (kph) | | | | | |
|---|---|---|---|---|---|---|
| | 100 | 80 | 65 | 50 | 40 | 35 |
| | Transition Lengths (meters) | | | | | |
| 45 | | | | | NA | 70 |
| 60 | | | | NA | 75 | 55 |
| 90 | | | | 75 | 50 | 40 |
| 100 | | | NA | 70 | 45 | 35 |
| 150 | | | 80 | 45 | 30 | 25 |
| 170 | | | 70 | 40 | 25 | 20 |
| 200 | | NA | 60 | 35 | 25 | 20 |
| 240 | | 90 | 50 | 30 | 20 | NR |
| 300 | NA | 75 | 40 | 25 | NR | |
| 360 | 130 | 60 | 35 | 20 | | |
| 400 | 115 | 55 | 30 | 20 | | |
| 500 | 95 | 45 | 25 | NR | | |
| 600 | 80 | 35 | 20 | | | |
| 700 | 70 | 35 | 20 | | | |
| 800 | 60 | 30 | NR | | | |
| 900 | 55 | 30 | | | | |
| 1000 | 50 | 30 | | | | |
| 1200 | 40 | NR | | | | |
| 1500 | 35 | | | | | |
| 1800 | 30 | | | | | |
| 2000 | NR | | | | | |

**Mountainous Terrain**

| Curve Radius (meters) | Design Speeds (kph) | | | | |
|---|---|---|---|---|---|
| | 50 | 40 | 30 | 25 | 20 |
| | Transition Lengths (meters) | | | | |
| 14 | | | | NA | 30 |
| 20 | | | | 35 | 20 |
| 25 | | | NA | 25 | 20 |
| 30 | | | 30 | 25 | 15 |
| 40 | | NA | 25 | 20 | 15 |
| 50 | | 40 | 20 | 15 | 15 |
| 55 | | 40 | 20 | 15 | 15 |
| 70 | NA | 30 | 15 | 15 | 15 |
| 80 | 55 | 25 | 15 | 15 | NR |
| 90 | 45 | 25 | 15 | 15 | |
| 100 | 45 | 20 | 15 | 15 | |
| 125 | 35 | 15 | 15 | NR | |
| 150 | 30 | 15 | 15 | | |
| 170 | 25 | 15 | NR | | |
| 200 | 20 | 15 | | | |
| 250 | 15 | 15 | | | |
| 300 | 15 | NR | | | |
| 400 | 15 | | | | |
| 500 | NR | | | | |

"NA" = Radius or less radius is not applicable at that speed

"NR" = Transition curves are not required at the radius or higher

## 5.9   COMPUTER-AIDED DESIGN OF HORIZONTAL ALIGNMENT

This session is in continuation of Section 4.7.1 of Chapter 4, for computer-aided processing of topographic survey data, which along with the computer applications for the "Design of Horizontal Alignment" are explained in Chapter 5 in the "The Guide for Computer Application Tutorials" part of the book, available for download from the website www.roadbridgedesign.com.

Refer to menu "Book Tutorials" of the websites at: "www.roadbridgedesign.com," and at "www.techsoftglobal.com/download.php."

## BIBLIOGRAPHY

*A Policy on Geometric Design of Highways and Streets 2001*, AASHTO.

*Overseas Road Note 6, A Guide to Geometric Design*, Overseas Unit, Transport and Road Research Laboratory, Department of Transport, Crowthorne, Berkshire, United Kingdom.

*A Guide on Geometric Design of Roads*, Arahan Teknik (Jalan) 8/86, JKR, Malaysia.

*A Guide to the Design of At Grade Intersections*, JKR 201101-0001-87, JKR, Malaysia.

*A Guide to the Design of Interchanges*, Arahan Teknik (Jalan) 12/87, JKR, Malaysia.

*Manual on Traffic Control Devices Road Marking & Delineation*, Arahan Teknik (Jalan) 2D/85, JKR, Malaysia.

*Guidelines for Presentation of Engineering Drawings*, Arahan Teknik (Jalan) 6/85, Pindaan 1/88, JKR, Malaysia.

*Design Standards, Interurban Toll Expressway System of Malaysia*, Government of Malaysia, Malaysia Highway Authority, Ministry of Works and Utilities.

*Design Manual for Roads & Bridges*, Ministry of Public Works and Kuwait Ministry, State of Kuwait.

*Geometric Design Manual for Dubai Roads*, Dubai Municipality, Roads Department.

*Geometric Design Standards for Rural (Non-Urban) Highways*, IRC 73-1980, IRC.

*Vertical Curves for Highways*, IRC SP 23-1993, IRC.

*Guidelines for the Design of Interchanges in Urban Areas*, IRC 92-1985, IRC.

*Lateral and Vertical Clearances at Underpasses for Vehicular Traffic*, IRC 54-1974, IRC.

*Guidelines for the Design of High Embankments*, IRC 75-1979, IRC.

*Type Design for Intersection on National Highways*, Ministry of Surface Transport, Govt. of India.

*Rural Roads Manual, Indian Roads Congress*, Special Publication 20.

*Manual of Standards and Specifications, for Two-Laning of State Highways on BOT Basis*, IRC SP 73-2007, IRC.

*Manual of Specifications & Standards for Four-Laning of Highways Through Public Private Partnership*, IRC SP 84-2009, IRC.

*Geometric Design Standards for Rural (Non-Urban) Highways*, IRC 73-1980, IRC.

*Highway Engineering* by Paul H. Wright and Karen K. Dixon, by John Wiley & Sons. Inc.

*Principles and Practice of Highway Engineering* by Dr. L. R. Kadiyali, by Khanna Publishers.

*A Course in Highway Engineering* by S. P. Bindra, by Dhanpat Rai & Sons.

*Principles of Highway Engineering and Traffic Analysis* by Fred L. Mannering, Walter and Scott, by Wiley India.

# 6 Design of Vertical Alignment

## 6.1 GENERAL

In most situations the topography primarily governs the vertical alignment of the highway including the tie-ins with the profiles of all intersecting roads. The consideration for crossing utilities, structures, adjacent property line, and earthworks, are also important.

## 6.2 DESIGN CONSIDERATIONS

The vertical alignment consists of longitudinal upward or downward gradients joined by vertical curves, the design of which needs proper care to provide safe travel at a uniform design speed. The factors which influence the design of the vertical alignment are:

- Safety.
- Design speed.
- Topography and adjacent land use.
- Horizontal alignment.
- Earthwork balance in cut and fill.
- Road class.
- Drainage.
- Levels of access to local properties.
- Vehicle operating characteristics.
- Cost.

All the above factors must be considered along with the design of horizontal and vertical alignment. A design will otherwise result in lower speeds and capacity of the road, and subsequently a reduction in safety.

Some specific guidance is given in this chapter. In addition to this, there are some general considerations for the design of a safe and economical vertical alignment for high-speed roads which are mentioned below:

- A smooth profile with gradual changes in elevations consistent with the terrain topography and class of road is always preferable.
- A roller coaster or hidden dip type road profile is always avoidable.
- A smoothly rolling profile is more economical in construction than a straight profile, without compromising for operating needs and aesthetics.

- In flat terrain, the longitudinal and transverse falls are to be considered which should be adequate for drainage of water from road surfaces in areas where rainwater is known to stand above the adjoining ground level.
- Where the natural ground water table is immediately below the ground surface, the vertical profile should be so designed that the bottom of the lowest layer of the pavement structure is at least 0.5 m above that water level. In case the water table is of permanent nature, then the elevation should be increased to 1.0 m, to eliminate the possibility of capillary rise of the water into the pavement layers. In rocky areas the profile should be 0.3 m above the rock level in order to avoid rock cutting.

If possible, the length of the vertical curves may be longer than the required length for the stopping sight distance, this enhances aesthetics. Superelevation runoff, when occurring on vertical curves which are designed to minimum standards, needs proper care to ensure the minimum vertical curvature in all lanes and edges in the carriageway. At intersections any steep gradient must be avoided for substantial lengths on either side. The convention for denoting an uphill gradient as positive (+ve) and downhill gradient as negative (–ve) to be followed in all design work.

## 6.3   VERTICAL CURVES

A vertical curve provides smooth changes in the gradient on the longitudinal profile of the road. Vertical curves should be provided at all vertical intersection points (VIPs) where two different grades meet, except at intersections. On local streets where the change in grade has less than a 1.0% provision, vertical curves may be avoided.

A crest curve is a convex shape which reduces upgrade and/or increases downgrade, whereas a sag curve is a concave shape which increases upgrade and/or decreases downgrade. Applicability of crest and sag curves are illustrated in Figure 6.1.

A parabolic curve with an equivalent axis centered on the VIP is normally used in highway design. The rate of change of gradient at successive points on the curve is constant for equal increments of horizontal distance along the curve. So, the rate of change of gradient is equal to the grade difference (G) divided by the vertical curve length (VCL). The reciprocal of this is the "K" value equal to VCL/G. VCL therefore can be obtained from equation:

$$VCL = K \times G$$

Elevations along the curve for distance "X" may be obtained from the formula:

$$X_e = PC_e + (G_1 \times X) + (G \times X^2)/(200 \times V \times C)$$

where
    $X_e$ = Elevation at a distance "X" along the curve
    $PC_e$ = Elevation at point BVC.

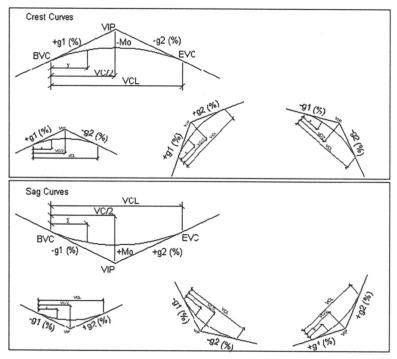

gl & g2 = Tangent Gradients (%)
G = gl–g2 = Algebraic Grade Difference
VIP = Vertical Intersection Points
VCL = Vertical Curve Length
BVC = Beginning of Vertical Curve
EVC = End of Vertical Curve
K = VCL/G
Mo = Maximum Normal Distance (Offset) from VIP to Curve

**FIGURE 6.1** Situations for Applications of Crest and Sag Vertical Curves.
*Source:* Exhibit 3-41, Types of Vertical Curves, A Policy on Geometric Design of Highways and Streets 2011, AASHTO

Height "$M_o$" may be obtained from the relationship:

$$Mo = G \times VCL/800$$

The position of high or low points may be calculated from the relationship:

$$X = (gl \times VCL)/G$$

In case X is negative or X > VCL, then the curve does not have a high or low point.

In the case of crest curves, the visibility requirements determine the minimum value of "K" for its application. In the case of sag curves, Stopping Sight Distance (SSD) is achieved within the length illuminated by headlights at night and is the determining factor. Recommended values of "K" are given in Table 6.1, which will normally meet the requirements of visibility.

SSD should always be checked because the horizontal alignment of the road, crossfall, superelevation, signs, and structures adjacent to or above the carriageway, are all related to the curvature to determine the visibility.

In situations where the decision sight distance (DSD) is to be maintained over a crest curve, significantly flatter curves are to be used, the related "K" values are given in Table 6.2.

**TABLE 6.1**
**Minimum Vertical Curvature for Divided Roads**

| Design Speed (km/hr) | Minimum 'K' Values | |
| --- | --- | --- |
| | **Crest Curves** | **Sag Curves** |
| 30 | 2 | 6 |
| 40 | 4 | 9 |
| 50 | 7 | 13 |
| 60 | 11 | 18 |
| 70 | 17 | 23 |
| 80 | 26 | 30 |
| 90 | 40 | 38 |
| 100 | 53 | 45 |
| 110 | 75 | 55 |
| 120 | 97 | 63 |

*Source:* Table 6.1, Design Manual for Roads & Bridges, Ministry of Public Works and Kuwait Ministry, State of Kuwait

**TABLE 6.2**
**Minimum "K" Values for Crest Curves for DSD (Worst Case)**

| Design Speed (km/hr) | Minimum 'K' Values for Crest Curves |
| --- | --- |
| 50 | 59 |
| 60 | 85 |
| 70 | 117 |
| 80 | 153 |
| 90 | 200 |
| 100 | 247 |
| 110 | 286 |
| 120 | 341 |

*Source:* Table 6.2, Design Manual for Roads & Bridges, Ministry of Public Works and Kuwait Ministry, State of Kuwait

**TABLE 6.3**

**Minimum "K" Values for Crest Curves for Safe Passing Sight Distance (SPSD) on Two-way, Undivided, Single Carriageway Roads**

| Design Speed (km/hr) | Minimum 'K' Values for Crest Curves |
|---|---|
| 30 | 48 |
| 40 | 87 |
| 50 | 142 |
| 60 | 200 |
| 70 | 280 |
| 80 | 347 |
| 90 | 450 |
| 100 | 534 |

*Source:* Table 6.3, Design Manual for Roads & Bridges, Ministry of Public Works and Kuwait Ministry, State of Kuwait

For undivided, single carriageway roads, where the horizontal alignment is designed to allow passing, any crest curve in the alignment should also provide safe passing sight distance. If the horizontal alignment does not allow passing, there is no justification in providing passing on crest curves. "K" values to allow passing on crest vertical curves on two lane undivided, single carriageway roads are given in Table 6.3. Where passing is not allowed, the minimum values of "K" for divided roads, as given in Table 6.1, may be used for undivided, single carriageway roads.

## 6.4 GRADIENT

### 6.4.1 MAXIMUM GRADIENT

Maximum gradient mostly depends on the characteristics of vehicles, particularly of trucks. There are two main considerations, which are the maximum gradient and the length over which it occurs. Commonly used gradients are described in Table 6.4.

In residential areas, the desirable maximum gradient should be 3%, whereas in industrial areas the gradients can be up to 6%. At-grade separated interchanges, the

**TABLE 6.4**

**Maximum Gradient**

| Road Class | Maximum Gradient (%) |
|---|---|
| Special Roads | 4 |
| Primary Roads | 6 |
| Secondary Roads | 6 |
| Local Roads | 8 |

*Source:* Table 6.4, Design Manual for Roads & Bridges, Ministry of Public Works and Kuwait Ministry, State of Kuwait

**TABLE 6.5**
**Critical Grade Lengths**

| % Upgrade | Maximum Gradient (%) |
|-----------|----------------------|
| 2% | 900 |
| 3% | 500 |
| 4% | 350 |
| 5% | 260 |
| 6% | 215 |
| 7% | 180 |
| 8% | 160 |

*Source:* Table 6.5, Design Manual for Roads & Bridges, Ministry of Public Works and Kuwait Ministry, State of Kuwait

maximum grade for on and off ramps can be up to 2% greater than the maximum gradient permitted on the main lane.

At approaches of at-grade intersections, signalized intersections, and roundabouts, the gradients may be either up or down and should not exceed 2% for a minimum length of 15 m before the STOP or Give Way line.

Maximum lengths of sustained gradients are based on a speed reduction for trucks of 15 km/hr for which the guidelines are given in Table 6.5.

### 6.4.2 MINIMUM GRADIENT

Drainage considerations demand that the road cannot be level. Surface water runs off from crown to edge, but if the road edge is flat then there will be ponding over a large area at the curb. To resolve this issue, it is essential to arrange for the main line profile to have a longitudinal gradient. The minimum desirable longitudinal gradient for effective drainage is 0.5%. For local roads an absolute minimum of 0.3% may be used.

## 6.5 VISIBILITY

At locations in the alignment with both horizontal and vertical curvature it is essential to check that there is no obstruction to visibility by safety barriers, median curbs, bridge piers, etc.

At sag curves, the visibility is usually obstructed by over bridges and signs, if these features exist. This should be checked for the upper bound of the visibility envelope at relevant sight distance.

At crest curves, if the line of sight is across the landscaped verge, then a maximum overall height of the landscaped verge should be kept at a height below 0.5 m.

## 6.6    CHOICE OF LONGITUDINAL PROFILE

Although the vertical alignment is mainly controlled by the geometric standards, it is still influenced by earthwork considerations and the materials in the cuttings. A balance should be maintained between cut & fill, shrink & swell, and suitable & unsuitable material.

## 6.7    VISUAL APPEARANCE OF VERTICAL GEOMETRY

This section is related to the design of horizontal alignment, which is described in the last chapter. Blending with the horizontal alignment, the best design of vertical alignment consists of a series of well-modulated vertical curves, this is so proportioned so that it can avoid problems as mentioned below:

- The sag curve, unlike the crest curve, is visible for its whole length at a time and plays an important role in making the alignment well harmonized. The sag curve, when joining two straights, must be sufficiently large so that the appearance of a kink is not there.
- Tangents of short lengths, if applied between two sag curves, result in a bad profile. Figure 6.2 describes types of vertical profiles to be avoided in the design.

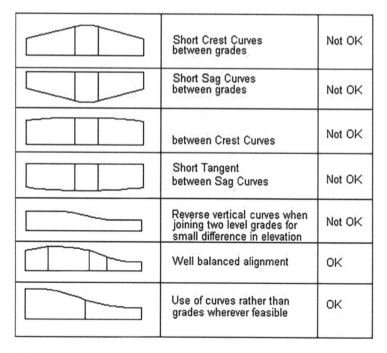

| | | |
|---|---|---|
| | Short Crest Curves between grades | Not OK |
| | Short Sag Curves between grades | Not OK |
| | between Crest Curves | Not OK |
| | Short Tangent between Sag Curves | Not OK |
| | Reverse vertical curves when joining two level grades for small difference in elevation | Not OK |
| | Well balanced alignment | OK |
| | Use of curves rather than grades wherever feasible | OK |

**FIGURE 6.2**    Acceptable and Unacceptable Vertical Alignment.
*Source:* Figure 6.2, Design Manual for Roads & Bridges, Ministry of Public Works and Kuwait Ministry, State of Kuwait

## 6.8   COMBINING HORIZONTAL AND VERTICAL ALIGNMENT

By integrating the horizontal and vertical geometry we get the satisfactory alignment and a road with three dimensional features. The horizontal and vertical alignment should occur in phases, with corresponding elements in the horizontal and vertical planes beginning and ending approximately together.

But it is not always possible to keep the horizontal and vertical elements entirely in phase with each other, but the amount by which the elements are out of phase should be kept small, this will also help for a pleasing visual appearance. A substantial length of overlap, compared to the length of the element, may make for better integration of the alignment geometry.

Longer and coincident curves are preferred but in case the existing conditions do not allow their use, it may not be possible to achieve significant improvement by the use of longer overlapping curves. In Figure 6.2 a number of desirable and undesirable combinations are described.

## 6.9   VERTICAL CLEARANCES

Minimum vertical clearance is defining the maximum height available under a flyover or an underpass for any vehicle with or without its load for not coming into contact with the overhead structure.

The preferred clearance for a new construction is 6.0 m and the minimum clearance is 5.5 m. This is applicable for all the lanes on the carriageway, including the shoulder and edge strips. The allowance for 200 mm of pavement overlay must be considered in addition to the minimum vertical clearance.

The vertical clearance is also applicable to sign gantries, overhead structures, overhead cables, suspended lighting, etc., which are hanging over the road. In case the road is underneath any bridge on sag curve, vertical clearance needs to be increased to allow for the limiting effect of the sag. The details are given in Table 6.6.

### TABLE 6.6
### Additional Clearance to Be Provided on Sag Curves

| Sag K Value | Additional Headroom (m) |
|---|---|
| 4 and 5 | 0.12 |
| 6 and 7 | 0.08 |
| 8 and 9 | 0.06 |
| 10–12 | 0.05 |
| 13–17 | 0.04 |
| 18–25 | 0.03 |
| 226–50 | 0.02 |
| 51–100 | 0.01 |
| Over 100 | Nil |

*Source:*   Table 6.6, Design Manual for Roads & Bridges, Ministry of
            Public Works and Kuwait Ministry, State of Kuwait

Where a specific minimum clearance is mentioned, it may be required by some public utility than the greater value specified and the standard vertical clearance is to be maintained.

Where the road passes below overhead cables or high-tension electrical power lines the provision for vertical clearance should be maximized by positioning the road as close as possible to the pylon or tower, where the cables are at a higher elevation than at the center point between two pylons or towers.

"A" in the above table is the algebraic difference in grades expressed as percentage.

### TABLE 6.7
### Sight Distances for Various Speeds

| Design Speed (kph) | Sight Distance (metres) | | |
|---|---|---|---|
| | Stopping | Intermediate | Overtaking |
| 20 | 20 | 40 | |
| 25 | 25 | 50 | |
| 30 | 30 | 60 | |
| 35 | 40 | 80 | |
| 40 | 45 | 90 | 165 |
| 50 | 60 | 120 | 235 |
| 60 | 80 | 160 | 300 |
| 65 | 90 | 180 | 340 |
| 80 | 120 | 240 | 470 |
| 100 | 180 | 360 | 640 |

*Source:* Table 4, Vertical Curves for Highways, IRC SP 23-1993, IRC

### TABLE 6.8
### Lengths of Vertical Curves for Different Speeds When Length of Curve Is Greater Than the Sight Distance

| Design Speed (kph) | Length of Summit/Crest Curves (metres) | | | Length of Valley/Sag Curves (metre) for Headlight Distance |
|---|---|---|---|---|
| | Stopping Sight Distance | Intermediate Sight Distance | Overtaking Sight Distance | |
| 20 | 0.9A | 1.7A | | 1.8A |
| 25 | 1.4A | 2.6A | | 2.6A |
| 30 | 2.0A | 3.8A | | 3.5A |
| 35 | 3.6A | 6.7A | | 5.5A |
| 40 | 4.6A | 8.4A | 28.4A | 6.6A |
| 50 | 8.2A | 15.0A | 57.5A | 10.0A |
| 60 | 14.5A | 26.7A | 93.7A | 15.0A |
| 65 | 18.4A | 33.8A | 120.4A | 17.4A |
| 80 | 32.6A | 60.0A | 230.1A | 25.3A |
| 100 | 73.6A | 135.0A | 426.7A | 41.5A |

*Source:* Table 6, Vertical Curves for Highways, IRC SP 23-1993, IRC

**TABLE 6.9**
**Minimum Lengths of Vertical Curves**

| Design Speed (kph) | Maximum Grade Change (%) Not Requiring a Vertical Curve | Minimum Length of Vertical Curve (metres) |
|---|---|---|
| Up to 35 | 1.5 | 15 |
| 40 | 1.2 | 20 |
| 50 | 1.0 | 30 |
| 65 | 0.8 | 40 |
| 80 | 0.6 | 50 |
| 100 | 0.5 | 60 |

*Source:*   Table 7, Vertical Curves for Highways, IRC SP 23-1993, IRC

## 6.10   COMPUTER-AIDED DESIGN OF VERTICAL ALIGNMENT

(This session is in continuation of Chapter 5)

The computer applications for the "Design of Vertical Alignment" are explained in Chapter 6 in the "The Guide for Computer Application Tutorials" part of the book available for download from the website www.roadbridgedesign.com.

For downloading the "Book Tutorials" and other items visit the websites at: www. roadbridgedesign.com and www.techsoftglobal.com/download.php.

## BIBLIOGRAPHY

*A Policy on Geometric Design of Highways and Streets 2001*, AASHTO.
*Overseas Road Note 6: A Guide to Geometric Design*, Overseas Unit, Transport and Road Research Laboratory, Department of Transport, Crowthorne, Berkshire, United Kingdom.
*A Guide on Geometric Design of Roads*, Arahan Teknik (Jalan) 8/86, JKR, Malaysia.
*A Guide to the Design of At Grade Intersections*, JKR 201101-0001-87, JKR, Malaysia.
*A Guide to the Design of Interchanges*, Arahan Teknik (Jalan) 12/87, JKR, Malaysia.
*Manual on Traffic Control Devices Road Marking & Delineation*, Arahan Teknik (Jalan) 2D/85, JKR, Malaysia.
*Guidelines for Presentation of Engineering Drawings*, Arahan Teknik (Jalan) 6/85, Pindaan 1/88, JKR, Malaysia.
*Design Standards, Interurban Toll Expressway System of Malaysia*, Government of Malaysia, Malaysia Highway Authority, Ministry of Works and Utilities.
*Design Manual for Roads & Bridges*, Ministry of Public Works and Kuwait Ministry, State of Kuwait.
*Geometric Design Manual for Dubai Roads*, Dubai Municipality, Roads Department.
*Geometric Design Standards for Rural (Non-Urban) Highways*, IRC 73-1980, IRC.
*Vertical Curves for Highways*, IRC SP 23-1993, IRC.
*Guidelines for the Design of Interchanges in Urban Areas*, IRC 92-1985, IRC.
*Lateral and Vertical Clearances at Underpasses for Vehicular Traffic*, IRC 54-1974, IRC.
*Guidelines for the Design of High Embankments*, IRC 75-1979, IRC.

*Type Design for Intersection on National Highways*, Ministry of Surface Transport, Govt. of India, 1995.

*Rural Roads Manual, Indian Roads Congress*, Special Publication 20.

*Manual of Standards and Specifications, for Two-Laning of State Highways on BOT Basis*, IRC SP 73-2007, IRC.

*Geometric Design Standards for Rural (Non-Urban) Highways*, IRC 73-1980, IRC.

*Highway Engineering* by Paul H. Wright and Karen K. Dixon, by John Wiley & Sons. Inc.

*Principles of Highway Engineering and Traffic Analysis* by Fred L. Mannering, Walter and Scott, by Wiley India.

# 7 Design of Cross Section Elements

## 7.1 GENERAL

This chapter considers the geometric elements of a typical road cross section. The limits of the road cross section are governed by the width of the available right of way. This is normally determined at the planning stage.

## 7.2 BASIC CROSS SECTION ELEMENTS

The basic elements of a road cross section are as follows:

- Limits of right of way (Section 7.3).
- Side slopes (Section 7.4).
- Verges (Section 7.5).
- Service reservations (Section 7.6).
- Shoulders and curb clearances (Section 7.7).
- Clearances to structures (Section 7.8).
- Clearances to safety barriers (Section 7.9).
- Lane widths (Section 7.10).
- Median widths (Section 7.11).
- Cross slopes (Section 7.12).
- Gutters and drainage ditches (Section 7.13).
- Other elements within the cross section (Section 7.14).

Typical cross sections (TCS) for urban and rural stretches are shown in the Figures 7.1–7.5.

## 7.3 LIMITS OF RIGHT OF WAY (ROW)

The right of way limits are the outer boundaries of the cross section. The width of the right of way is proposed at the planning stage, the chosen width should permit the design of a well-balanced cross section, taking into account the road class, the projected traffic flows, the topography, the surrounding land uses, and any other relevant parameters, such as a provision for grade separation.

Table 7.1 summarizes the typical provision of the overall right of way widths for various road classes. The values are for guidance only and may be modified to allow for the space taken up by earthworks, utilities, or structures such as bridges or tunnels.

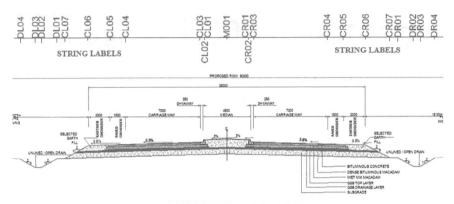

FIGURE 7.1    TCS 01 Cross Section Elements.

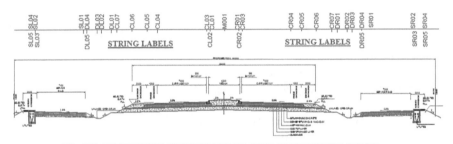

FIGURE 7.2    TCS 02 Cross Section Elements.

All road furniture such as signing, lighting, barriers, services, utilities, and structures, should be positioned within the right of way.

## 7.4    SIDE SLOPES

Side slopes fall into two categories, embankment slopes and cutting slopes. Cuttings have a back slope leading from the surrounding terrain to a drainage ditch, and a fore slope leading up from the ditch to the verge and pavement. The design considerations for a fore slope are the same as for an embankment, whereas a back slope is designed as a cutting.

Side slopes serve two primary functions, enabling the vertical alignment of the road to be achieved and providing structural stability to the road itself. Where side slopes exist, they also serve a secondary function, they provide a surface over which out of control vehicles may travel and recover. Their design, therefore, also seeks to minimize the risks of overturning such vehicles.

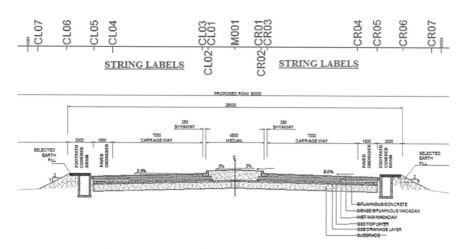

TCS - DUAL CARRIAGEWAY WITH FOOTPATH / COVERED DRAIN AND WITHOUT SERVICE ROAD IN NON URBAN SECTIONS

TYPICAL CROSS SECTION TYPE : TCS 03

**FIGURE 7.3** TCS 03 Cross Section Elements.

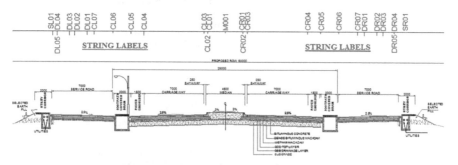

TCS - DUAL CARRIAGEWAY WITH FOOTPATH / COVERED DRAIN AND SERVICE ROAD IN BUILT UP URBAN SECTIONS

TYPICAL CROSS SECTION TYPE : TCS 04

**FIGURE 7.4** TCS 04 Cross Section Elements.

The angle of the side slopes depends on the slope material. Rock cuttings in hilly areas can be stable at relatively steep angles. Embankments of granular material require shallow angles. In areas prone to wind-blown sand, slope angles should be avoided as they create eddies that lead to sand drifts on the pavement.

An adequate geotechnical investigation should be carried out. The investigation will provide scientific guidance regarding the maximum slope for cut and fill, and the criteria for benching or erosion protection, if required.

In general, embankment side slopes should fall away from the verge at a slope of 2 in 1 (H:V) or flatter. It is usual to provide a safety barrier where embankment slopes are steeper than 2 in 1, or where the overall height of the slope is greater than 6 m. Flatter slopes are preferable, provided that there is adequate fall for drainage.

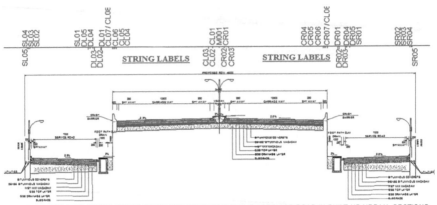

TYPICAL CROSS SECTION : TCS 05

**FIGURE 7.5** TCS 05 Cross Section Elements.

**TABLE 7.1**
**Preferred Width and Range of the Right of Way (m)**

| | Urban | | | |
| --- | --- | --- | --- | --- |
| **Road Class** | **Residential** | **Commercial** | **Industrial** | **Rural** |
| Local roads | 20 (16–25) | 20 (16–25) | 20 (20–25) | 20 (16–20) |
| Secondary roads | 36 (25–40) | 36 (25–40) | 36 (30–40) | 36 (25–40) |
| National highways | | 60 (40–70) | | |
| Special roads | | 70 (70–100) | | |

Cutting slopes are determined by the nature of the material in which they are excavated. For a cut depth up to 1.5 m the slope may be 2:1 (H:V), for a depth from 1.5 m to 4.5 m the slope may be 1.5:1 (H:V), for a depth from 4.5 m to 7.5 m the slope may be 1:1 (H:V). In rock, slopes may be steeper than 1 in 1 (H:V).

If there is insufficient width to provide side slopes in accordance with these guidelines, the use of retaining walls with fill slopes, and gabion walls with cut slopes, or some method of slope stabilization should be considered.

The edges of cuttings and embankment slopes should be rounded and smoothed to meet the existing topography.

The intersection of slope planes in the cross section should be rounded to simulate natural contours. The rounding and smoothing of slopes help to minimize sand drifting, and the wash out of sand or other loose material from embankment edges. In rock cuttings, ditches and debris verges should be provided. These will facilitate surface water runoff and create a safe landing and catchment area for any possible rock fall. The additional width also serves as a useful area for rock face maintenance.

Side slopes under the back spans of open-span, over bridges should be paved, and the aesthetics and economics of the over bridge, rather than other considerations, will normally dictate the slope, 1 in 1½ (67%) is generally regarded as a maximum.

## 7.5  VERGES

The verge acts as a buffer zone between the edge of the pavement (curb or back of the shoulder) and either the side slope or the surrounding physical features. The verge is normally unpaved in rural areas. In urban areas, the verge may include landscaping and a paved footway.

The verge provides stability to the edge of the pavement construction, reducing the chances of damage due to erosion. It also accommodates road furniture, such as signs, signals, lighting, and structures. Utilities, such as electricity and water, are laid underground alongside roads and should be allowed for within the highway right of way. Such services are usually laid in the verge, which may need to be significantly wider than would normally be required for traffic safety reasons.

Verge widths vary from a desirable minimum of 2.25 m (for traffic safety reasons) up to the limits of the right of way. Generally, a paved verge is designed with a 2.5% crossfall toward the road pavement for drainage purposes. However, with wider paved verges, crossfall should be designed toward specific drainage collection points located within the verges themselves.

It is important to ensure that road furniture or landscaping within the verge does not impinge on the sight distances required for the design speed of the road. Isolated, slender obstructions can be ignored, but massive or continuous obstructions need to be identified and appropriate measures taken to achieve the sight distance standards. Particular care should be taken at intersections, where the number of signs and other items of street furniture is greater than on the open road.

If soakaways are to be installed within the verge, this may also have an influence on its width, particularly if services are present.

## 7.6  SERVICE RESERVATIONS

Due investigation of utilities should be made at the outset of the design process by direct liaison with the relevant authorities. The necessary width of the service reservation should then be agreed with the Ministry of Public Works, Ministry of Communications, Ministry of Electricity and Water, and the municipality, prior to the commencement of the design.

Although the recommended cross sections are designed to allow for the inclusion of services, it may be that the width required by the utility authorities is greater than the width that the designer can provide within the right of way. Under such circumstances, it is important to reach a proper agreement with all the relevant parties before the design is finalized.

## 7.7  SHOULDERS AND CURB CLEARANCES

### 7.7.1  Shoulders

The addition of a paved outer shoulder to the outer edge of a road has many advantages and is usually warranted on the basis of the following factors:

- Provides a place for safe stopping in the event of mechanical difficulty, flat tire, or other emergency, with minimal disruption to traffic flow.
- Offers a clear route for emergency vehicles to reach the scene of an accident.
- Provides space that may enable the escape of a potential accident situation.
- Improves storm water drainage by allowing water to be discharged further from the running lanes, thereby preventing ponding on the traveled way during heavy rain.
- Increases sight distance on horizontal curves and lateral clearance to signs and other obstructions.
- Creates a feeling of openness that helps to reduce driver stress.
- Provides structural support to the pavement edges.
- Provides additional running lanes for diversions and space for road maintenance operations.

Outer shoulders are not usually required on urban local or secondary roads, because the curbs provide structural support, and disabled vehicles can generally find safe places to stop in driveways and side streets. Nevertheless, the adoption of outer shoulders on secondary roads in industrial areas can be beneficial.

On special and primary roads, inner shoulders are commonly provided on the median side of the pavement, but these are normally narrower than the outer shoulders. Shoulders should generally be constructed to the same design as the traveled way, so that the shoulders can carry traffic during maintenance operations. Outer shoulders should normally be designed to the same crossfall as the adjacent running lane. When providing shoulders, consideration should be given to some means of discouraging vehicles from casually using them as an extra lane during times of high traffic flows. This practice compromises the safety of the road and fuels driver frustration. "Rumble strips" (a series of raised or lowered strips) perpendicular to the flow of traffic can be provided. The requirements for shoulders are summarized in Table 7.2.

### 7.7.2  Curb Clearances

Where a curb is provided, there is a tendency for drivers to steer a distance away from it. This phenomenon is termed "shying". The greater the speed, the greater the distance drivers will shy away from the curb. Shying is accommodated by the provision of a suitable curb clearance additional to the width of the adjacent lane. Where there is a shoulder, there is no need to provide curb clearance.

A curb clearance of 0.6 m should be generally be added to the width of the lane adjacent to curbed edges on roads with a design speed exceeding 80 km/h. In some instances, it may be beneficial to delineate the clearance to the curb by means of a

**TABLE 7.2**
**Shoulders and Curb Clearances**

| Road Type | Median Edge | | Outer Edge | |
| --- | --- | --- | --- | --- |
| | Inner Shoulder (m) | Curb Clearance (m) | Outer Shoulder (m) | Curb Clearance (m) |
| Local roads | | | | |
| Rural | n/a | n/a | 0.3 | n/a |
| Urban | n/a | n/a | none | none |
| Secondary roads | | | | |
| Rural divided | none | none | 0.3 | n/a |
| Rural undivided | n/a | n/a | 0.3 | n/a |
| Urban | none | none or 0.6 | none | none or 0.6 |
| Primary roads | | | | |
| Rural | 1.2 | none | 3.0 | n/a |
| Urban | 0.6 | none | 1.2 | none |
| Special roads 1.2 | none | 3.0 | none | |

painted edge line. Under such circumstances, the curb should be set 0.6 m back from the painted line, which is coincident with the edge of the adjacent lane. Table 7.2 provides a summary of the normal requirements for curb clearances.

## 7.8   CLEARANCES TO STRUCTURES

It is important that structures and other obstructions are set back adequately from the edge of the traveled way. The width of the necessary "clear zone" is dependent principally on the design speed of the road, but also varies according to the side slope of the earthworks, if any. Table 7.3 sets out the relevant values for clearance to structures.

**TABLE 7.3**
**Clear Zone Width (m)**

| Design Speed (km/h) | Embankments | | | Cuttings | | |
| --- | --- | --- | --- | --- | --- | --- |
| | Side Slope 1:5* | Side Slope 1:6 or Flatter | At Grade | Side Slope 1:6 or Flatter | Side Slope 1:5 and 1.4 | Side Slope Steeper Than 1:4 |
| 60 | 5.5 | 5 | 5 | 5 | 5 | 5 |
| 70 | 8.5 | 6.5 | 6.5 | 6.5 | 6 | 5 |
| 80 | 8.5 | 6.5 | 6.5 | 6.5 | 6 | 5 |
| 90 | 10 | 7.5 | 7.5 | 7.5 | 6.5 | 5.5 |
| 100 | 13.5 | 10 | 10 | 8.5 | 8 | 6.5 |
| 110 | 14 | 10.5 | 0.5 | 9 | 9 | 7.5 |
| 120 | 14.5 | 11 | 11 | 9.5 | 10 | 8.5 |

* safety barrier is provided where side slope exceeds 1:5

**TABLE 7.4**
**Desirable Minimum Lateral Clearance to Safety Barriers**

| Design Speed (km/h) | Setback of Safety Barrier from Edge of Travelled Way (m) |
|---|---|
| 50 | 1.1 |
| 60 | 1.4 |
| 70 | 1.7 |
| 80 | 2.0 |
| 90 | 2.2 |
| 100 | 2.8 |
| 110 | 2.8 |
| 120 | 3.2 |

These distances are measured from the nearest edge of the traveled way and therefore include the width occupied by shoulders, service reservations, and verges.

Where the "clear zone" cannot be kept completely free from obstructions, safety barriers should be provided to protect the driver of an errant vehicle from colliding with the structure or other obstruction.

Additionally, adequate sight distances should be maintained throughout the length of a route. This may necessitate the further setting back of structural walls, piers, abutments, etc., and may require safety barriers to be set back further than normal. Refer to chapters one, five and six.

## 7.9   CLEARANCES TO SAFETY BARRIERS

As a general rule, safety barriers should be placed as far from the traveled way as possible, however, it is also desirable to maintain a uniform clearance in order to provide the driver with a certain level of expectation. Table 7.4 sets out the relevant minimum clearances.

Ends of barriers should be flared away from the road.

## 7.10   LANE WIDTHS

Lane widths have a great influence on the safety and comfort of driving. In particular, the lane width on an undivided road must be sufficient to provide adequate clearance between passing vehicles.

The standard lane width for roads is 3.5 m. However, this may be reduced to an absolute minimum of 3.0 m where standard lane widths cannot be provided and with the approval of the project authority. In urban or industrial areas, a lane width of 3.75 m is more appropriate, as set out in Table 7.5. In some instances, the lane width may be widened to accommodate the maneuvering requirements for parking in an adjacent parking lane.

**TABLE 7.5**
**Normal Lane Widths (m)**

|  | Residential | Commercial | Industrial |
|---|---|---|---|
| Urban local street | 4.0 | 4.0 | 4.0 |
|  | (5.0 if one-way) | (5.0 if one-way) |  |
| Urban secondary | 3.7 | 3.7 | 3.75 |
| Urban primary | 3.7 | (service roads 5.0) |  |
| Rural local road |  | 3.7 |  |
| Rural secondary |  | 3.7 |  |
| Rural primary |  | 3.7 |  |
| Special roads |  | 3.7 |  |

At signalized intersections, lane widths may be reduced to the absolute minimum of 3.0 m. Edge lines are provided within the curb clearance or shoulder width, and lane lines are included within the lane width.

## 7.11   MEDIAN WIDTHS

### 7.11.1   PROVISIONS

Medians are used to separate opposing traffic lanes on multi-lane roads. (Separate advice regarding the outer separation between a service road and the main line is given in Section 7.14.2).

Medians provide protection from interference by opposing traffic, minimize headlight glare, provide additional space for crossing or turning vehicles within at-grade intersections, allow pedestrian refuge in urban areas, and may provide space for utilities or the creation of future additional lanes.

Medians may range in width from as little as 1.2 m in an urban area, to 20 m or more in a rural area with street lighting, drainage, and landscaping. Median widths depend on the extent of the right of way available and the functional requirements of the median.

On special roads, medians are normally provided with safety barriers, to eliminate head-on collisions. Rural divided roads may be similarly treated.

It is recommended that urban medians should be curbed. Rural medians should generally be provided with a 0.6 m or 1.2 m shoulder and not curbed, a depressed median is preferred as this improves drainage of the road. A curbed median is desirable where there is a need to control left turn movements, and is also used where the median is to be landscaped.

Special attention should be given to the drainage of medians. If the median is curbed and paved, the median surface should be designed to have slopes of 2% and should fall away from the center of the median. Non-paved medians should fall toward the center at a rate of 1 in 6 (17%) when self-draining, but consideration should be given to the provision of additional storage capacity or outlets to allow for storm conditions. Paved medians may require positive drainage systems by incorporating manholes or culverts. All drainage inlets in the median should be designed

## TABLE 7.6
### Minimum Median Widths for Certain Functions (m)

| Conditions | At Signalized | Elsewhere Intersections |
|---|---|---|
| Minimum to accommodate signal heads | 1.5–2.0 | n/a |
| Minimum (kerbed) to separate traffic safely | 1.2 | 1.2 |
| Minimum to accommodate pedestrians safely | 2.0–3.5 | 3.5 |
| Minimum to provide left turn lane | 4.5 | 4.8 |
| Minimum to provide U-turn | n/a | 5.0 |
| Desirable for U-turn | n/a | 10.7 or more |
| Minimum for provision of effective landscaping | n/a | 8.0 |

"n/a" means not applicable – other considerations govern

with the top flush with the ground, and the culvert ends provided with safety gratings, so that they will not be hazardous to out of control vehicles that stray into the median.

It is common practice to landscape medians, this provides a better environment and reduces driver stress. Careful consideration should be given to the choice of planting to ensure that Stopping Sight Distance (SSD) is not impaired. Furthermore, the upkeep of the landscape and growth of the plants should be designed for minimal maintenance.

Where two abutting sections of road have different median widths, a smooth transition should accommodate this difference. The transition should be as long as possible for aesthetic reasons and should preferably occur within a horizontal curve. Table 7.6 sets out the minimum widths for certain functional requirements of medians.

### 7.11.2 NARROW MEDIANS

Narrow medians are those less than 4.0 m wide and are used where there is a need to provide a divided road, but where the available right of way does not permit greater median width. They are not wide enough to accommodate effective left turn lanes.

The minimum median width to provide a safe pedestrian refuge (away from signalized intersections) is 3.5 m. The pedestrian's freedom to cross at locations with a narrower median should be actively discouraged by the use of physical obstacles such as guardrails.

It is recommended that narrow medians are not used on rural roads, and in urban areas, a narrow median should not even be considered if it is possible to provide an intermediate or wide median at that particular location, while maintaining acceptable standards for the remaining cross section elements.

### 7.11.3 INTERMEDIATE MEDIANS

Intermediate width medians are those in the range of 4.0–8.0 m. They are generally wide enough to provide for a left turn lane. A 6.0 m wide median permits the introduction of some landscaping.

## TABLE 7.7
## Median Widths (m)

|  | Urban | Rural |
|---|---|---|
| Secondary roads | Minimum 2.0 | Minimum 4.0 |
|  | Normal provision 6.0 | Normal provision 6.0 |
| Primary roads | Minimum 2.0 | Minimum 6.0 |
|  | Normal provision 8.0–10.0 | Normal provision 8.0–10.0 |
| Special roads | Minimum 2.0 | Minimum 6.0 |
|  | Normal provision 8.0–10.0 | Normal provision 8.0–10.0 |

*Note:* Narrower widths may be appropriate at signalized intersections. Median widths may increase to suit visibility requirements.

### 7.11.4 WIDE MEDIANS

Medians 8.0 m wide or more provide space for effective landscaping and may be used for signing, services, and drainage. Wide medians may also be used to absorb level differences across the road reserve. Wide medians should not be implemented at the expense of reduced verge widths.

### 7.11.5 NORMAL WIDTHS FOR MEDIANS

The normal median provision is shown on Table 7.7, together with the minimum requirements.

Every effort should be made to provide the normal provision values in Table 7.7.

## 7.12 CROSS SLOPES

Embankments aligned at right angles to the road can create significant safety hazards for out of control vehicles that have strayed off the pavement. The recommended maximum slopes are set out in Table 7.8 below.

Where available land is at a premium, side slopes may need to be steepened beyond the values given in Table 7.8. At these locations, safety barriers parallel to the main line can be provided to contain errant vehicles. For further details refer to "The Roadside Design Guide".

## TABLE 7.8
## Maximum Cross Slopes

| Condition | Maximum Slope |
|---|---|
| Desirable maximum – all locations | 1 in 10 (10%) |
| Absolute maximum – special roads | 1 in 6 (17%) |
| Absolute maximum – rural primary roads | 1 in 6 (17%) |
| Absolute maximum – urban primary and secondary roads | 1 in 5 (20%) |
| Absolute maximum – all other locations | 1 in 4 (25%) |

## 7.13 GUTTERS AND DRAINAGE DITCHES

Where roads are curbed, the part of the pavement adjacent to the curb acts as a gutter, collecting rainwater, and conveying it to gullies spaced at appropriate intervals. Where a curb clearance is provided, the gullies can be located within that zone, but where the curb abuts the edge of the traveled way, consideration should be given to the use of side-entry gullies, or the adoption of combined curb/drainage units.

Drainage ditches are generally provided between the back slope and fore slope of a cutting, and often at the toe of an embankment. The design of surface water drainage systems is not within the scope of this manual and advice should be sought from the Ministry of Public Works.

## 7.14 OTHER ELEMENTS WITHIN THE CROSS SECTION

### 7.14.1 AUXILIARY LANES

Auxiliary lanes are additional to the normal through lanes and are introduced in specific locations to serve a particular purpose. This purpose may be one or more of the following:

- As a speed change lane.
- As a climbing lane.
- As a turning lane.
- As additional storage space.
- As a method of maintaining lane balance.

Speed change lanes are used either for acceleration or deceleration, and their design is dealt with grade separations and interchanges or with at-grade intersections.

Climbing lanes may be introduced on steep up-gradients, or on sustained up-gradient of lesser severity. Critical gradient lengths, above which the provision should be considered are given in "Chapter 6: Design of Vertical Alignment."

Turning lanes permit turning vehicles to undertake the necessary maneuver, clear of the through traffic.

Additional storage space is required at some at-grade intersections, and design issues associated with widening for this purpose are dealt with signalized intersections.

Lane balance issues are dealt with grade separations and interchanges.

### 7.14.2 SERVICE ROADS

Service roads are roads that run roughly parallel with, and are connected to, the main through route of primary roads and special roads. They are generally of lower design speed and preferably restricted to one-way traffic flow.

Service roads segregate the higher speed through traffic from the lower speed local traffic and reduce the number of access points onto the main line. The provision of service roads reduces the interference to traffic flow on the main line, makes the best use of road capacity and results in a safer road. Service roads may also provide an alternative route if maintenance is required on the main line, or in case of an emergency.

The width of the service road should be at least 5.0 m and is dependent on the type and turning requirement of the traffic, i.e., whether light vehicles, buses, delivery lorries, or heavy goods vehicles are expected to use it. Further considerations include the type and number of access points, and the type and nature of street parking, if required.

Service road connections to primary roads should be designed as at-grade intersections, while those for special roads should be designed as off-ramps and on-ramps.

Where service roads are provided, there is a need for a separation between them and the main line. This is known as the outer separation, and its absolute minimum width is 1.2 m. This distance allows for the provision of a central pedestrian guardrail only and is not sufficient to accommodate any traffic signs. If traffic signs or other street furniture are to be placed in the outer separation, the desirable minimum width is 2.0 m. A wider outer separation, giving greater scope for landscaping, thus enhancing the appearance of the road and its adjacent development, is preferred.

The designer is referred to AASHTO (American Association of State Highway and Transportation Officials) for further detailed explanations and guidelines for the layout of service roads and areas.

## 7.14.3 BRIDGES

Bridges within grade-separated intersections should be designed using the normal parameters contained within this manual, unless it is uneconomical, in which case each situation should be considered on its own merits:

- The designer should establish the clearance requirements and the applicable design speeds, controlling gradients, and vertical curvature limits, before beginning the preliminary design.
- Bridges with long spans, large angles of skew, tapers or splays, small radius curvature, or large superelevation, should be avoided as they are likely to be costly and difficult to construct.
- It should be possible to continue the full standards of the adjacent sections of the route across the bridge.
- Aim to provide a straight structure but if horizontal curvature is unavoidable, then the bridge should be on a circular curve rather than a transition, and the radius should be as large as possible.
- Avoid tapers and flared ends but if this is not possible, aim to start such changes in cross section at a pier position.
- Aim to provide bridges on straight grades (maximum 6% and minimum 0.5% to permit longitudinal drainage) rather than on vertical curves, if this is not possible, do not adopt a crest curve of less than K = 30.

- Avoid sag curves on bridges, they are unattractive visually and cause difficulties with drainage.
- Aim for bridges to have symmetrical spans, this is often achieved by ensuring that both abutments are at the same elevation.
- Variation in the profile of one curb line relative to the other is to be avoided, it leads to a deck that appears warped and is more difficult to construct. If it must occur, the variation should be applied uniformly over the length of the deck.
- The combination of horizontal and vertical geometry must be carefully considered in order to visualize the aesthetics of the final design.
- The presence of bridge parapets may obstruct visibility splays. If this is the case, the road or bridge geometry should be altered accordingly.
- The forward visibility requirements on a sag curve underneath a bridge should always be checked.

For further details on bridges, refer to the Kuwait Bridge Design Manual.

### 7.14.4 TUNNELS

The design of major tunnels is a specialist subject and lies outside the scope of this manual.

Elsewhere on the road network, shorter lengths of tunnel or underpasses may be required, and these should be designed using the normal parameters contained within this manual.

If it proves uneconomical, each situation should be considered on its own merits. The following general guidance is given:

- The tunnel should be as short as practicable.
- The tunnel should be straight, if possible.
- For maximum driver comfort, the aim should be for the tunnel layout to be to the same design speed as the remainder of the route.
- The mainline cross section (lane, shoulder, edge strip, and median widths) should desirably be continued through the tunnel.
- Horizontal curvature in tunnels restricts forward visibility and widening on the inside of the curve is generally required if proper SSD is to be maintained.
- Full vertical clearance should be maintained.
- Vertical curvature can also restrict visibility, and the relevant sight envelopes given in chapter four should be provided.
- When selecting grades for tunnels, consideration should be given to driver comfort, and also to ventilation requirements.
- The design should avoid the need for traffic signs to be provided within the tunnel.
- Merging, weaving, or diverging movements within a tunnel are highly undesirable. On-slips and off-slips should not be provided within tunnels, nor for 300 m beyond the ends of the tunnel.

- Closed circuit television coverage connected to a constantly manned control room should be provided. Emergency telephones and firefighting points should be provided.
- A raised emergency pavement or similar curbed area (minimum width 0.8 m) needs to be provided for drivers of stalled vehicles, and for maintenance operatives.

## 7.15   COMPUTER-AIDED DESIGN OF ROAD CROSS SECTIONS

(This session is in continuation of Chapter 6)

In this session we shall see the selection of a typical road cross section and its application as a uniform cross section from start to finish of the highway project. The design of the cross section is based on the alignment of the centerline of the road passing through the center of the road cross section.

- Selecting a typical road cross sections for applying along the alignment of the highway.
- Modify the dimensions of the selected typical cross section as applicable.
- Create ground cross sections by XSEC by following the reference line (string).
- Create side slopes in cut (with side drain) or fill situations by INTERFACE analysis.
- Draw cross sections of proposed highway along with existing ground.
- Define various pavement layers separately on either side carriageway with paved shoulder.

The computer applications for the "Design of Cross Section Elements" are explained in Chapter 7 in the "The Guide for Computer Application Tutorials" part of the book available for download from the website www.roadbridgedesign.com.

For downloading the "Book Tutorials" and other items visit the websites at: www.roadbridgedesign.com and www.techsoftglobal.com/download.php

## BIBLIOGRAPHY

*A Policy on Geometric Design of Highways and Streets 2001*, AASHTO.

*Overseas Road Note 6, A Guide to Geometric Design*, Overseas Unit, Transport and Road Research Laboratory, Department of Transport, Crowthorne, Berkshire, United Kingdom.

*A Guide on Geometric Design of Roads*, Arahan Teknik (Jalan) 8/86, JKR, Malaysia.

*A Guide to the Design of At-Grade Intersections*, JKR 201101-0001-87, JKR, Malaysia.

*Guidelines for Presentation of Engineering Drawings*, Arahan Teknik (Jalan) 6/85, Pindaan 1/88, JKR, Malaysia.

*Design Standards, Interurban Toll Expressway System of Malaysia*, Government of Malaysia, Malaysia Highway Authority, Ministry of Works and Utilities.

*Design Manual for Roads & Bridges*, Ministry of Public Works and Kuwait Ministry, State of Kuwait.

*Geometric Design Manual for Dubai Roads*, Dubai Municipality, Roads Department.

*Geometric Design Standards for Rural (Non-Urban) Highways*, IRC 73-1980, IRC.

*Rural Roads Manual, Indian Roads Congress*, Special Publication 20.

*Manual of Standards and Specifications, for Two-Laning of State Highways on BOT Basis*, IRC SP 73-2007, IRC.

*Manual of Specifications & Standards for Four-Laning of Highways Through Public Private Partnership*, IRC SP 84-2009, IRC.

*Geometric Design Standards for Rural (Non-Urban) Highways*, IRC 73-1980, IRC.

# 8 Estimation of Earthwork and Pavement Quantities

## 8.1 GENERAL

### 8.1.1 ESTIMATION OF AREAS AND VOLUMES

The calculations for the measurement of areas of land, volumes, and other quantities are connected with engineering and building works, are discussed in this chapter. Areas are considered first, since the computation of areas is used in the calculation of volumes and are dealt with a mechanical device – like "planimeter," for areas enclosed by straight lines and irregular figures. In this chapter we shall consider the calculations of areas and volumes for irregular shapes.

### 8.1.2 IRREGULAR SHAPES

The following methods are used to determine the area of irregular curvilinear figures such as ponds, lakes, stockpiles, and the areas enclosed between formation lines and natural boundaries.

## 8.2 "GIVE AND TAKE" LINES

The whole area is divided into triangles, or trapezoids, by replacing the irregular boundaries by straight lines. The straight lines are so arranged that any small areas excluded from the survey by the straight line are balanced by including other small areas outside the survey (see Figure 8.1).

The positions of the lines are visually estimated by making and applying a stencil of thin straight-edges, and faint lines are drawn on the plan. The areas created by the triangles or trapezoids are calculated by the methods as described previously.

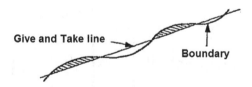

**FIGURE 8.1**   "Give and Take" Lines.
*Source:* Figure 9.7, Textbook "Surveying" by A. Bannister, S. Raymond, ELBS with Longman

## 8.3 COUNTING SQUARES

A piece of tracing paper with squares is overlaid on the drawing. The number of full squares and parts of squares which are enclosed by the shape under consideration are now counted. Based on the scale of the drawing and the size of the squares on the overlay, the total area of the figure is calculated.

## 8.4 TRAPEZOIDAL RULE

Figure 8.2 describes an area enclosed by a survey line and a boundary. The survey line is divided into a number of small intercepts of equal length "d" and the offsets O1, O2, etc., which are measured either directly on the ground or to scale from the plan. Assuming "d" is short enough for the length of the boundary between the off-sets (which are assumed to be straight), thus the area is divided into a series of trapezoids:

$$\text{Area of trapezoid}(1) = \left[(O_1 + O_2)/2\right] \times d$$

$$\text{Area of trapezoid}(2) = \left[(O_2 + O_3)/2\right] \times d$$

$$\text{Area of trapezoid}(6) = \left[(O_6 + O_7)/2\right] \times d$$

Summing up, we get

$$\text{Area} = (d/2) \times \{O_1 + 2O_2 + 2O_3 + \cdots + O_7\}$$

To obtain the generalized formulation with "n" number of offsets, we get

$$\text{Area} = d \times \left\{\left[(O_1 + O_n)/2\right] + O_2 + O_3 + \cdots + O_{n-1}\right\}$$

To determine the area of a narrow strip of ground, this method may be used by drawing a straight line along the strip as shown in Figure 8.3, and then measuring

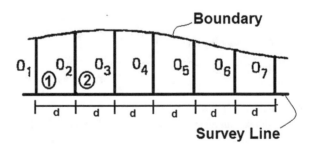

**FIGURE 8.2**  Counting Squares.
*Source:* Figure 9.8, Textbook "Surveying" by A. Bannister, S. Raymond, ELBS with Longman

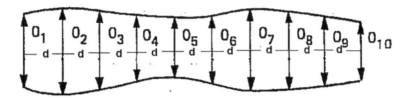

**FIGURE 8.3** Trapezoidal Rule.
*Source:* Figure 9.9, Textbook "Surveying" by A. Bannister, S. Raymond, ELBS with Longman

offsets at equal intercepts along this. By the same consideration it will be seen that the area is given by:

$$\text{Area} = d \times \left\{ \left[ (O_1 + O_n)/2 \right] + O_2 + O_3 + \cdots + O_{n-1} \right\}$$

**Example 8.1:** Calculate the area of the plot shown in Figure 8.3, where the offsets from the scaled plan are at intervals of d = 20 m, and are:

| Offset | $O_1$ | $O_2$ | $O_3$ | $O_4$ | $O_5$ |
|---|---|---|---|---|---|
| Length (m) | 15.26 | 18.22 | 22.64 | 19.28 | 17.26 |
| Offset | $O_6$ | $O_7$ | $O_8$ | $O_9$ | $O_{10}$ |
| Length (m) | 18.56 | 17.14 | 18.37 | 16.26 | 17.58 |

$$\text{Area} = 20 \times \left[ (15.26 + 17.58)/2 + 18.22 + 22.64 + 19.28 \right.$$
$$\left. + 17.26 + 18.56 + 17.14 + 18.37 + 16.26 \right]$$
$$= 20 \times 164.15$$
$$= 3283 \text{m}^2$$
$$= 0.3283 \text{hectare}$$

## 8.5 SIMPSON'S RULE

This method assumes that the irregular boundary is composed of a series of parabolic arcs and gives greater accuracy than other methods. In this calculation, the figure under consideration be divided into an even number of equal strips.

Referring to Figure 8.2, consider the first three offsets, which are shown enlarged in Figure 8.4.

The portion of the area contained between offsets $O_1$ and $O_3 = \text{ABGCDEA}$

$$\text{Trapezoidal area[ABFCD]} + \text{Rectangular area[BGCFB]}$$
$$= (O_1 + O_2)/2 \times 2d + 2/3 \times (\text{area if circumscribing parallelogram})$$
$$= (O_1 + O_3)/2 \times 2d + 2/3 \times 2d \times (O_2 - O_1 + O_2)/2$$
$$= d \times \left\{ 3O_1 + 3O_3 + 4O_2 - 2O_1 - 2O_3 \right\}$$
$$= d/3 \times \left[ O_1 + 4O_2 + O_3 \right]$$

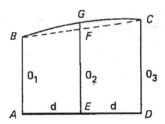

**FIGURE 8.4** Simpson's Rule. *Source:* Figure 9.10, Textbook "Surveying" by A. Bannister, S. Raymond, ELBS with Longman

For the next pair of intercepts, the area contained between offsets $O_3$ and $O_5$.

$$= \frac{d}{3} \times \left[ O_3 + 4O_4 + O_5 \right]$$

For the final pair of intercepts, the area contained between offsets, O5 and O7

$$= \frac{d}{3} \times \frac{X}{3} \left[ O_5 + 4O_6 + O_7 \right]$$

Summing up, we get

$$\text{Area} = \frac{d}{3} \times \left\{ \left( O_1 + O_7 \right) + 2 \left( O_3 + O_5 \right) + 4 \left( O_2 + O_4 + O_6 \right) \right\}$$

We therefore get the generalized formulation as:

$$\text{Area} = \frac{d}{3} \times \left( D + 2O + 4E \right)$$

where
  D = sum of first and last offsets
  O = sum of the remaining odd offsets
  E = sum of the even offsets
  Therefore, the Simpson's Rule gives the area enclosed by a curvilinear figure divided into an even number of strips of equal width. Hence, the area is equal to one-third the width of a strip, multiplied by the sum of the two extreme offsets, twice the sum of the remaining odd offsets, and four times the sum of the even offsets.

## 8.6 CALCULATIONS FOR EARTHWORK VOLUMES

The excavation, removal, and dumping of earth, is a frequent operation with civil engineering projects. In the construction of a canal, for example, a trench of suffi-cient width is excavated to given depths and gradients, the earth being stored in some convenient place, which is usually by the side of the trench. In basement excavation,

**Example 8.2:** Using tape measurements, the following offsets were drawn from a fence to a survey line:

| Chainage (m) | 0 | 25 | 50 | 75 | 100 |
|---|---|---|---|---|---|
| Offset (m) | 0 | 6.51 | 9.85 | 8.28 | 11.22 |
| Chainage (m) | 125 | 150 | 175 | 200 | 225 |
| Offset (m) | 12.23 | 10.25 | 5.17 | 1.78 | 0 |

Calculate the area between the fence and the survey line.

It is to be noted that the term "chainage" refers to a cumulative increase in distance measured from a starting point on the line, i.e., zero chainage at the start point. When there are ten offsets, which is an even number and since Simpson's Rule can be applied to an odd number of offsets for creating even number of strips only, it will be used here to calculate the area contained between the first and ninth offsets. The residual triangular area between the ninth and tenth offsets is calculated separately. It is convenient to create a table for the working, as shown below.

| Offset No. | Offset | Simpson Multiplier | Product |
|---|---|---|---|
| $O_1$ | 0 | 1 | 0 |
| $O_2$ | 6.51 | 4 | 26.04 |
| $O_3$ | 9.85 | 2 | 19.70 |
| $O_4$ | 8.28 | 4 | 33.12 |
| $O_5$ | 11.22 | 2 | 22.44 |
| $O_6$ | 12.23 | 4 | 48.92 |
| $O_7$ | 10.25 | 2 | 20.50 |
| $O_8$ | 5.17 | 4 | 20.68 |
| $O_9$ | 1.78 | 1 | 1.78 |
| $O_{10}$ | 0 | | 0 |
| | | | $\Sigma = 193.18$ |

$$\text{Area}\left(O_1 - O_9\right) = \frac{25}{3} \times 193.18 = 1609.83 \ m^2$$

$$\text{Area}\left(O_9 - O_{10}\right) = \frac{25}{3} \times 1.78 = 3.56 m^2 = 1613.39 m^2 = 1613.39 \times 0.0001 \text{ hectare}$$
$$= 0.161339 \text{ hectare}$$

for being a cut job all the material dug out will require to be carted away. For embankments formed in the construction of highways, railways, and earth dikes, the earth is required as a fill job and will have to be brought from some other place, called a borrow area.

The payment for the work is paid to the contractors on the basis of the calculated volume of material handled. It is therefore essential that the engineer or surveyor should be able to make good estimates of the volumes from the earthworks.

The three general methods for calculating earthworks are:

- By cross sections
- By contours
- By spot heights

For road work the calculation of earthwork is generally done by using the first method of cross sections, whereas the calculations of earthwork for sight leveling and grading or stockpiles are done either by using the contours or spot heights. As the first method is relevant to the present context of this book, it is therefore discussed in detail in the following sections.

## 8.6.1 Volume from Cross Sections

In this method, cross sections are taken at right angles to some convenient reference line which runs longitudinally through the earth works and, although it is capable of general application, it is most commonly used on long narrow works such as roads, railways, canals, embankments, and pipe excavations, etc., the volumes from earthworks between successive cross sections are calculated from a consideration of the cross sectional areas, which in turn are measured or calculated by the general methods already given, i.e. by planimeter, division into triangles, counting squares, etc.

In long constructions which have constant formation width and side slopes it is possible to simplify the computation of cross sectional areas by the use of formulas and is useful for road, railway, and long embankments, etc., and formulas will be given for the following types of cross section:

- Sections level across
- Sections with a crossfall
- Sections part in cut and part in fill
- Sections of variable levels

### 8.6.1.1 Sections Level Across (Figure 8.5)

For vertical depth or height at centerline of "h" units from the original level to the final formation level, where the side slopes are "m" units as the horizontal projection for every single unit of vertical rise, then in "h" units the side gives a horizontal projection of "m × h" units.

$$AB = BC = w = (b/2) + (m \times h)$$

$$\text{Area of cross section} = \frac{1}{2}\left[2\left(\frac{b}{2} + mh\right) + b\right]h$$

$$= h \times (b + m \times h)$$

(8.1)

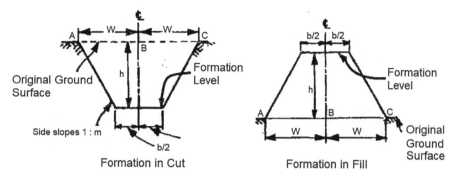

**FIGURE 8.5** Volume from Cross Sections.
*Source:* Figure 9.11, Textbook "Surveying" by A. Bannister, S. Raymond, ELBS with Longman

**Example 8.3:** At a station, an embankment is formed on level ground and has a height at its centerline of 4.10 m, the width of the formation is 14.0 m.
Find:

(i)     the area of cross section, given that the side slope is vertical to horizontal = 1:2.0.

(ii)    the side widths.

(i)     H = 4.10, b = 14.0 m, m = 2.0

$$\text{Therefore Area is} = h \times (b + m \times h)$$
$$= 4.10 \times (14.0 + 2.0 \times 4.10)$$
$$= 4.10 \times 22.2$$
$$= \textbf{91.02m}^2$$

(ii)

$$\text{Either side width is} = b/2 + m \times h$$
$$= 14.0/2 + 2.0 \times 4.10$$
$$= 7.0 + 8.2$$
$$= \textbf{15.2m}$$

## 8.6.1.2   Sections with Crossfall

Referring to Figure 8.6, the existing ground has a transverse slope relative to the centerline, and since the section is not symmetrical about the centerline, the side widths are not equal.

Now

$$C_1 B = \frac{w_1}{k}$$

This is the difference in level between $C$ and $B$ due to a gradient of 1 in $K$ over a distance of $w_1$.

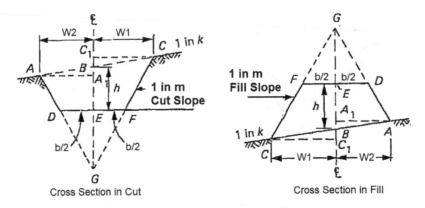

**FIGURE 8.6** Sections with Cross Fall.
*Source:* Figure 9.12, Textbook "Surveying" by A. Bannister, S. Raymond, ELBS with Longman

Similarly

$$A_1 B = \frac{w_2}{k}$$

Also, if the side slopes intersect at G, then GE will be the vertical difference in levels for a horizontal distance of b/2.

Hence

$$GE = \frac{b}{2m}$$

Since triangles $C_1CG$ and $EFG$ are similar:

$$\frac{CC_1}{EF} = \frac{CC_1}{GE}$$

$$\frac{w_1}{\dfrac{b}{2}} = \frac{\dfrac{b}{2m} + h + \dfrac{w_1}{k}}{\dfrac{b}{2m}}$$

Therefore

$$w_1 = \left(\frac{b}{2} + mh\right)\left(\frac{k}{k-m}\right) \tag{8.2}$$

Also

$$\frac{AA1}{DE} = \frac{GA1}{GE}$$

$$\frac{w_2}{\dfrac{b}{2}} = \frac{\left(\dfrac{b}{2m}\right) + h + \dfrac{w_2}{k}}{\dfrac{b}{2m}}$$

Hence

$$w_2 = \left(\frac{b}{2} + mh\right)\left(\frac{k}{k+m}\right) \tag{8.3}$$

The area of the cut or fill in the embankment is the area ACFDA = area BCG + area ABG – area DFG

$$= \frac{1}{2}W_1\left(\frac{b}{2m} + h\right) + \frac{1}{2}W_2\left(\frac{b}{2m} + h\right) - \frac{1}{2}b\frac{b}{2m}$$

$$= \frac{1}{2m}\left\{\left(\frac{b}{2} + mh\right)(w_1 + w_2) - \frac{b^2}{2}\right\} \tag{8.4}$$

Difference in level between C and F

$$= h + \frac{w_1}{k} \tag{8.5}$$

Difference in level between A and D

$$= h - \frac{w_2}{k} \tag{8.6}$$

This type of section is sometimes known as a "two-level section," since two levels are required to establish the crossfall of 1 and K.

### 8.6.1.3  Sections Part in Cut and Part in Fill (Refer to Figure 8.7)

Figure 8.7 shows that the "cut" position on the right side of the center line is similar to the "cut" position of Figure 8.6 (cross section in cut) and hence:

$$w_1 = \left(\frac{b}{2} + nh\right)\left(\frac{k}{k-m}\right) \tag{8.7}$$

Substituting "n" with "m," the "fill" part, on the left side of the centerline is similar to the "fill" part as shown in Figure 8.6 (cross section in fill), except that "h" is effectively negative here. Therefore:

$$w_2 = \left(\frac{b}{2} - mh\right)\left(\frac{k}{k-m}\right) \tag{8.8}$$

**Example 8.4:** Calculate the side widths and cross sectional area of an embankment of a road with the formation width of 14.0 m, and side slopes of 1 vertical to 2 horizontals, when the center height is 4.10 m and the existing ground has a crossfall of 1 in 9 at right angles to the centerline of the embankment.

Referring to Figure 8.6.

$$w_1 = \left(\frac{b}{2} + mh\right)\left(\frac{k}{k-m}\right)$$

where,

$$b = 14.0, K = 9, m = 2, h = 4.10$$

Hence,

$$w_1 = \left[(14.0/2) + (2 \times 4.10)\right] \times \left[9/(9-2)\right]$$
$$= 19.54 \text{m}$$

$$w_2 = \left(\frac{b}{2} + mh\right)\left(\frac{k}{k+m}\right)$$
$$= \left[\left(\frac{14.0}{2}\right) + (2 \times 4.10)\right] \times \left(\frac{9}{(9+2)}\right)$$
$$= 12.44 \text{m}$$

From Equation (8.4),

$$
\begin{aligned}
\text{Area} &= \frac{1}{2_m}\left\{\left(\frac{b}{2} + mh\right)(w_1 + w_2) - \frac{b^2}{2}\right\} \\
&= \frac{1}{4}\left\{12.45(14.94 + 10.67) - \frac{12.50^2}{2}\right\} \\
&= \left(1/(2 \times 2)\right) \times \left\{\left[(14.0/2) + (2 \times 4.10)\right] \times (19.54 + 12.44) - 14.0^2/2\right\} \\
&= 0.25 \times \left[(15.2 \times 31.98) - 98\right] \\
&= 97.024 \text{ m}^2
\end{aligned}
$$

Note that when $h = 0$, $w_2 = \frac{b}{2}\left(\frac{k}{k-m}\right)$

$$
\begin{aligned}
\text{Area of fill} &= \frac{1}{2}h_2 \times \text{DB} \\
&= \frac{1}{2}h_2 \times \left(\frac{b}{2} - kh\right) \\
&= \frac{1}{2}\left(\frac{2w_{2-b}}{2_m}\right)\left(\frac{b}{2} - kh\right) \\
&= \frac{1}{2}\frac{(b/2 - kh)^2}{(k-m)}
\end{aligned}
$$

(8.9)

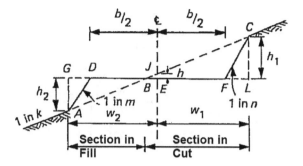

**FIGURE 8.7** Volume in Hill Road Formation.
*Source:* Figure 9.13, Text Book "Surveying" by A. Bannister, S. Raymond, ELBS with Longman

$$\text{Area of cut} = \frac{1}{2} h_1 \, \text{BF}$$

$$= \frac{1}{2} h_1 \times \left( \frac{b}{2} + kh \right)$$

$$= \frac{1}{2} \left( \frac{2w_{1-b}}{2_n} \right) \left( \frac{b}{2} + kh \right) \tag{8.10}$$

$$= \frac{1}{2} \frac{\left( b/2 - kh \right)^2}{\left( k - n \right)}$$

When the cross section is in fill at the centerline, instead of being in cut as in Figure 8.7 then, replacing +h by −h, the following modified formulas is obtained:

$$\text{Area if cut} = \frac{1}{2} \frac{\left( b/2 - kh \right)^2}{\left( k - n \right)}$$

$$\text{Area if fill} = \frac{1}{2} \frac{\left( b/2 + kh \right)^2}{\left( k - m \right)}$$

### 8.6.1.4 Sections of Variable Level

Refer to Figure 8.8 as a "three-level section," in which at least three levels are required on each cross section to calculate the ground slopes. The side width formulas are again:

$$w_1 = \left( \frac{b}{2} - mh \right) \left( \frac{k}{k - m} \right) \tag{8.11}$$

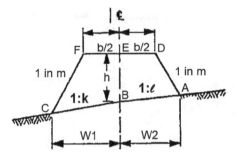

**FIGURE 8.8** Volume on Sections with Varying Levels.
*Source:* Figure 9.14, Textbook "Surveying" by A. Bannister, S. Raymond, ELBS with Longman

**Example 8.5** A road has a formation width of 14.0 m and side slopes of vertical to horizontal as 1: 1.5 in cut, and 1: 2 in fill. The original ground had a cross fall of vertical to horizontal as 1: 4. If the depth of excavation at the center line is 0.6 m, calculate the side widths and the areas of cut and fill.

From Equations (8.7), (8.8), (8.9) and (8.10), we get:

$$K = 4, n = 1.5, m = 2, \quad b = 14.0, h = 0.6$$

$$
\begin{aligned}
w_1 &= \left(\frac{k}{k-n}\right) \times \left(\frac{b}{2} + nh\right) \\
&= \left[\frac{4}{4-1.5}\right] \times \left(\frac{14.0}{2} + 1.5 + 0.6\right) \\
&= 12.64 \text{m}
\end{aligned}
$$

$$
\begin{aligned}
w_2 &= \left(\frac{k}{k-m}\right) \times \left(\frac{b}{2} - mh\right) \\
&= \left[\frac{4}{4-3}\right] \times \left(\frac{14.0}{2} + 2 + 0.6\right) \\
&= 32.8 \text{m}
\end{aligned}
$$

$$
\begin{aligned}
\text{Area if cut} &= \frac{1}{2} \frac{\left(\frac{b}{2} - kh\right)^2}{k-n} \\
&= \left(\frac{1}{2}\right) \times \left[\frac{\left(\frac{14.0}{2} - 4 \times 0.6\right)^2}{4-2}\right] \\
&= 5.29 \text{m}^2
\end{aligned}
$$

$$\text{Area if fill} = \frac{1}{2} \frac{\left(\dfrac{b}{2} + kh\right)^2}{(k-m)}$$

$$= \left(\frac{1}{2}\right) \times \left[\frac{\left(\dfrac{14.0}{2} + 4 \times 0.6\right)^2}{4 - 1.5}\right]$$

$$= 17.67 \, \text{m}^2$$

$$w_2 = \left(\frac{b}{2} + mh\right)\left(\frac{l}{l+m}\right) \tag{8.12}$$

If "BA" falls away from the centerline:

$$w_2 = \left(\frac{b}{2} + mh\right)\left(\frac{l}{l-m}\right)$$

$$\text{Area of cross-section} = \frac{1}{2} w_1\left(h + \frac{b}{2m}\right) + \frac{1}{2} w_2\left(h + \frac{b}{2m}\right) - \frac{1}{2} b \cdot \frac{b}{2m}$$

$$\frac{1}{2}\left\{(w_1 + w_2)\left(h + \frac{b}{2m}\right)\frac{b^2}{2m}\right\} \tag{8.13}$$

$$= \frac{1}{2m}\left\{(w_1 + w_2)\left(mh + \frac{b}{2}\right) - \frac{b^2}{2}\right\}$$

## 8.7  COMPUTATION OF VOLUMES

By calculating the various areas of cross section, the volumes of earth in construction of road, railways, earth dams etc. are computed by one of the following methods:

- Volumes by mean areas
- Volumes by end areas
- Volumes by prismoidal formula

### 8.7.1  VOLUMES BY MEAN AREAS

In this method the volume is calculated by multiplying the mean of the cross sectional areas by the distance between the end sections. If the areas are $A_1, A_2, A_3 \ldots A_{n-1}, A_n$ and the distance between the two extreme sections $A_1$ and $A_n$ is $L$, then:

$$\text{Volume} = V = \frac{A_1 + A_2 + A_3 + \cdots + A_{n-1} + A_n}{n} \cdot L$$

The method is not a very accurate one and gives better accurate results when the cross sectional areas are nearly equal, either in cut or fill.

## 8.7.2 Volumes by End Areas

If "$A_1$" and "$A_2$" are the areas of two cross sections at distance "D" apart, then the volume "V" between the two is given by:

$$V = D \times \frac{A_1 + A_2}{2} \tag{8.14}$$

This expression is correct as long as the area of the section mid-way between $A_1$ and $A_2$ is the mean of the two, either in cut or fill, which can be assumed to be the case when there is no wide variation between successive sections. If there is a variation, then a correction must be applied, and this rectification is dealt with in the next method. In case of irregularities in ground surface, which exists between successive cross sections, and the problems of bulking and depression that are normally associated with earthworks, it is reasonable to use the end areas formula for normal estimating.

For a series of consecutive cross sections, at varying intervals of "D," the total volume will be:

$$\text{Volume} = \Sigma V = \frac{D_1(A_1 + A_2)}{2} + \frac{D_2(A_2 + A_3)}{2} + \frac{D_3(A_3 + A_4)}{2} + \cdots$$

If $D_1 = D_2 = D_3$, etc. $= D$

$$\Sigma V = D \times \left\{ \frac{A_1 + A_n}{2} + A_2 + A_3 + \cdots + A_{n-1} \right\} \tag{8.15}$$

This is sometimes referred to as the trapezoidal rule for volumes.

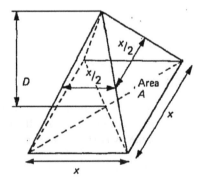

**FIGURE 8.9**   Volume of a Pyramid.
*Source:* Figure 9.15, Textbook "Surveying" by A. Bannister, S. Raymond, ELBS with Longman

**Example 8.6:** An embankment is formed on ground which is level transverse to the embankment but falling at V:H = 1.5:20 longitudinally so that three sections 20 m apart have centerline heights of 6.0 m, 7.5 m, and 9.0 m respectively above original ground level. If the side slopes of V:H = 1:2 are used, determine the volume of fill between the outer sections when the formation width is 8.00 m, by using the trapezoidal rule:

H = 6.0, 7.5, 9.0, m = 2, b = 8.0, L = 20.0

Method 1 – Using Equation (8.1):

$$A = h \times (b + mh)$$
$$A1 = 6.0 \times (8.0 + 2 \times 6.0) = 120.0m^2$$
$$A2 = 7.5 \times (8.0 + 2 \times 7.5) = 172.5m^2$$
$$A3 = 9.0 \times (8.0 + 2 \times 9.0) = 234.0m^2$$

Here, the mid-area $A_2$ is not the mean of $A_1$ and $A_3$:

$$V = (2 \times L/2) \times [A1 + 2 \times A2 + A3]$$
$$= (20/2) \times [120.0 + 2 \times 172.5 + 234.0]$$
$$= 6990.0m^3$$

Method 2 – In the same example, if the method of end areas is applied between the two extreme sections, we get:

$$V = (2 \times L/2) \times [A1 + A3]$$
$$= (2 \times 20/2) \times [120.0 + 234.0^-]$$
$$= 7080m^3$$

Note that, here is an over-estimation of over 1% compared with the first method.

As an extreme case of such a discrepancy, consider the volume of a pyramid (Figure 8.9) of base area "A" and perpendicular height "D":

$$\text{True volume} = \frac{1}{3} \times \text{base area} \times \text{height}$$
$$= \frac{1}{3} \times A \times D$$

$$\text{But by end areas } V = \frac{A + 0}{2} \times D$$
$$= \frac{1}{2} \times A \times D$$

### 8.7.3 Volumes by Prismoidal Formula

Considering the volume from earthworks between two successive cross sections as a "prismoid," the prismoidal formula may be used, which is a more precise formula. It is generally considered that all things being equal, use of this formula gives the most accurate estimate of volume. A prismoid is a solid form of two end faces which must be of parallel planes but not necessarily of the same shape. The other faces connecting these two end faces, i.e., top, bottom, and two sides, must be formed by straight continuous lines running from one end face to the other.

The volume of a prismoid is given by:

$$V = \frac{D}{6}\left(A_1 + 4M + A_2\right) \tag{8.16}$$

Where $A_1$ and $A_2$ are the areas of the two end faces separated by distance "D", "M" is the area of the section mid-way between the end faces.

The theory assumes that in calculating "M," each linear dimension is the average of the corresponding dimensions of the two end areas.

As an example of the application of the prismoidal formula, we may derive the formula for the volume of a pyramid in Figure 8.9:

$$\text{Base area} = A = x^2$$

$$\text{Mid-area} = M = \frac{x^2}{4} = \frac{A}{4}$$

$$\text{Top area } A_2 = 0$$

Therefore,

$$V = \frac{D}{6}\left(A + 4 \cdot \frac{A}{4} + 0\right)$$

$$= \frac{D \cdot A}{3}$$

Similarly, for a wedge as in Figure 8.10:

$$V = \frac{D}{6}\left(ab + \frac{4_{ab}}{2} + 0\right) \frac{D}{6} \times \left[(a \times b) + \frac{4 \times a \times b}{2} + 0\right]$$

$$= \frac{D_{ab}}{2} \quad \frac{D \times a \times b}{2}$$

The volume of a triangular prism, as in Figure 8.10, is also given by the product of the area of the normal section, in this case D × b/2 and one-third of the sum of the three parallel edges:

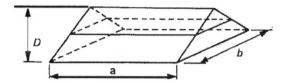

**FIGURE 8.10**   Volumes by Prismoidal Formula. *Source:* Figure 9.16, Textbook "Surveying" by A. Bannister, S. Raymond, ELBS with Longman

$$V = \frac{Db}{2} \cdot \frac{1}{3}(a+a+a) = \frac{Dab}{2}$$

The prismoidal formula is described in most of the standard textbooks on solid geometry.

There are various alternative ways in which the prismoidal formula may be used, and some of these are given below:

a.  Consider, each cross section as the end area of a prismoid of length "D" and estimate the dimensions of the mid-areas at the "D/2" points as the mean of the two corresponding dimensions in the end areas. This is difficult when the sections are irregular.

b.  Where the estimation of the area at the mid-section is difficult, arrange for extra sections to be leveled at the mid-section positions as required. This will result a large increase of field work.

c.  Treat the alternate sections as end areas, i.e., the length of the prismoid becomes "2 × D." Unless the ground profile is regular, both transversely and longitudinally, then the assuming of a volume of earth is prismoidal over such a length and may create errors. This method takes pairs of end areas like, $A_1$ and $A_3$, $A_3$ and $A_5$, $A_5$ and $A_7$, etc., as end areas of successive prismoids, and we get:

$$V_1 = \frac{2D}{6}\left(A_1 + 4A_2 + A_3\right)$$

$$V_2 = \frac{2D}{6}\left(A_3 + 4A_4 + A_5\right), \text{etc}$$

Therefore, where "n" is an odd number:

$$V = \frac{D}{3}\left(A_1 + 4A_2 + 2A_3 + 4A_4 + \cdots + 2A_{n-2} + 4A_{n-1} + A_n\right) \tag{8.17}$$

This is Simpson's Rule for volumes. It can be applied with greater accuracy by considering method (b) in which mid-sections are leveled but note that "D" in Equation (8.17) will then be halved.

d. Calculate the volumes between successive cross sections by the method of end areas and apply corrections to these volumes known as "prismoidal corrections." Such corrections can be derived for regular sections only, e.g., consider sections level across.

If "D" is the spacing of cross sections, "$A_1$" and "$A_2$" are the two end areas, "M" is the mid-area, "$h_1$" and "$h_2$" are the difference in level between ground level and formation at "$A_1$" and "$A_2$" respectively, and "b" is the formation width, then from Equation (8.1):

$$A_1 = h(b + mh_1)$$

$$A_2 = h_2(b + mh_2)$$

Hence,

$$\text{Vend areas} = \left(bh_1 + mh_1^2 + bh_2 + mh_2^2\right)$$

Assuming $h_m = (h_1 + h_2)/2$ (difference in level between ground level and formation at the mid-section)

$$M = \left(\frac{h_1 + h_2}{2}\right)\left\{b + m\left(\frac{h_1 + h_2}{2}\right)\right\}$$

$$= \frac{bh_1}{2} + \frac{bh_2}{2} + \frac{mh_1^2}{4} + \frac{mh_2^2}{4} \frac{mh_1h_2}{2}$$

Hence,

$$\text{Vprismoid} = \frac{D}{6}\left\{bh_1 + mh_1^2 + 4\left(\frac{bh_1}{2} + \frac{bh_2}{2} + \frac{mh_1^2}{4} + \frac{mh_2^2}{4} + \frac{mh_1h_2}{2}\right) + bh_2 + mh_2^2\right\}$$

$$= \frac{D}{6}\left\{3bh_1 + \frac{2mh_1^2}{3} + bh_2 + \frac{2mh_2^2}{3} + \frac{2mh_1h_2}{3}\right\}$$

Hence,

$$V_{EA} - V_p = \frac{D}{2}\left\{\frac{mh_{1^2}}{3} - \frac{2mh_1h_2}{3} + \frac{mh_{2^2}}{3}\right\}$$

$$= \frac{D}{6} \cdot m(h_1 - h_2)$$                                    (8.18)

$$= \text{Prosmoidal correction for a level section}$$

The term $(h_1 - h_2)^2$ must always be positive, therefore, the correction must be deducted from the volume as calculated by the end-areas formula. The correction is simple to apply as it requires no additional information.

Similarly, the "prismoidal corrections" can be derived for other types of section starting from the general expression, as below:

$$V_{EA} - VP = \text{Prismoidal correction(PC)}$$
$$= \frac{D}{2}(A_1 - A_2) - \frac{D}{6}(A_1 + 4M + A_2) \qquad (8.19)$$
$$= \frac{D}{3}(A_1 - 2M + A_2)$$

In the case of a section with a crossfall (Figure 8.6) it has been proved that:

$$A = \frac{1}{2m}\left\{\left(\frac{b}{2} + mh\right)(w_1 + w_2) - \frac{b^2}{2}\right\}$$
$$= \frac{1}{2m}\left(\frac{b}{2} + mh\right)\left(\frac{b}{2} + mh\right)\left(\frac{k}{k+m} + \frac{k}{k-m} - \frac{b^2}{4m}\right)$$

If $h = h_1$, $h_2$ and $(h_1 + h_2)/2$ at the two end sections and mid-section respectively:

$$PC = \frac{D}{3}(A_1 - 2A_m + A_2)$$
$$= \frac{D}{6m}\left(\frac{k}{k+m} + \frac{k}{k-m}\right)\left\{\left(\frac{b}{2} + mh_1\right)^2 + \left(\frac{b}{2} + mh_2^2\right) - 2\left(\frac{b}{2} + m\left(\frac{h_1 + h_2}{2}\right)^2\right)\right\}$$
$$= \frac{Dk}{6m}\left(\frac{k - m + k + m}{(k^2 - m^2)}\right)\frac{m^2 h_1^2}{2} - m^2 h_1 h_2 + \frac{m^2 h_2}{2} \qquad (8.20)$$
$$= \frac{D}{6}\frac{k^2}{(k^2 - m^2)} \cdot m(h_1 - h_2)^2$$

As with the PC for a level section, where $k > m$, the value from this expression will always be positive. As an exercise, it is advised to check that prismoidal corrections for sloping sections which are part in cut and part in fill, are:

$$\text{Fill} \quad PC = \frac{D}{12(k-m)}k^2(h_1 - h_2)^2 \qquad (8.21)$$

$$\text{Cut} \quad PC = \frac{D}{12(k-n)}k^2(h_1 - h_2)^2 \qquad (8.22)$$

$$PC = \frac{D}{12(k-n)}k^2(h_1 - h_2)^2$$

**Example 8.7:** Using the data of Example 8.5 we have:

$$h = 6.0, 7.5, 9.0, m = 2, b = 8.0$$

Using Equation (8.1), A = h × (b + m × h), we calculated:

$$A1 = 6.0 \times (8.0 + 2 \times 6.0) = 120.0 m^2$$

$$A2 = 7.5 \times (8.0 + 2 \times 7.5) = 172.5 m^2$$

$$A3 = 9.0 \times (8.0 + 2 \times 9.0) = 234.0 m^2$$

We solved the problem by the end areas method. Here the volume is to be calculated by using the prismoidal formula:

a. Taking D = 40 m

$$V = (D/6) \times (A1 + 4 \times A2 + A3)$$
$$= (40/6) \times (120.0 + 4 \times 172.5 + 234.0)$$
$$= \mathbf{6960 m^3}$$

b. Taking D = 20 m, and applying the PC to the "end areas" volumes

$$V_1 = (D/2) \times (A1 + A2) = (20/2) \times (120.0 + 172.5) = \mathbf{2925 m^3}$$

$$PC = (D \times m/6) \times (h1 - h2)^2 = (20 \times 2/6) \times (6.0 - 7.5)^2 = \mathbf{15 m^3}$$
$$V_2 = (D/2) \times (A2 + A3) = (20/2) \times (172.5 + 234) = \mathbf{4065 m^3}$$
$$PC = (D/6) \times (h2 - h3)^2 = (20/6) \times (7.5 - 9.0)^2 = \mathbf{7.5 m^3}$$
Total Volume = $\mathbf{7012.5 m^3}$

The answers are not the same because in this example the dimensions of the middle section are not the exact mean of the two outer ones. In practice, the second method is preferable, since it assumes prismoids of shorter length.

**Example 8.8:** A road has a formation width of 14.0 m, and side slopes of 1 in 1.5 cut, and 1 in 2 in fill. The original ground had a crossfall of 1 in 4. If the depths of excavation at the centerline of two sections of 20.0 m apart are 0.5 m and 0.6 m meters respectively, find the volumes of cut and full over this length:

$$b = 14.0, m = 2, n = 1.5, k = 4, h1 = 0.5 m, h2 = 0.6 m, D = 20.0 m.$$

From Equations (8.9) and (8.10):

$$\text{Area of fill} = \frac{1}{2} \cdot \frac{(b/2 - kh)^2}{(k - m)}$$

$$\text{Area of cut} = \frac{1}{2} \cdot \frac{\left(b/2 + kh\right)^2}{k - n}$$

Section 1:

$$A1 = \text{Area of fill} = 0.5 \times \left[(14.0/2) - 4 \times 0.5\right]^2 / (4 - 2) = 6.25$$

$$A2 = \text{Area of cut} = 0.5 \times \left[(14.0/2) + 4 \times 0.5\right]^2 / (4 - 1.5) = 16.2$$

Section 2:

$$A3 = \text{Area of fill} = 0.5 \times \left[(14.0/2) - 4 \times 0.6\right]^2 / (4 - 2) = 5.29$$

$$A4 = \text{Area of cut} = 0.5 \times \left[(14.0/2) + 4 \times 0.6\right]^2 / (4 - 1.5) = 17.67$$

Fill Volume

$$\begin{aligned} V_{EA} &= (D/2) \times (A1 + A3) \\ &= (20/2) \times (6.25 + 5.29) = 115.4 m^3 \end{aligned}$$

From Equation (8.21):

$$\begin{aligned} PC &= \frac{D}{12(k - m)} \cdot k^2 (h_1 - h_2)^2 \\ &= \left[20/(12 \times (4 - 2))\right] \times 42 \times (0.5 - 0.6)^2 = 0.13 m^3 \end{aligned}$$

Therefore $V_P = 115.4 - 0.13 = 115.27 \ m^3$

Cut Volume $\quad V_{EA} = (D/2) \times (A2 + A4) = (20/2) \times (16.2 + 17.67) = 338.7 m^3$

From Equation (8.22):

$$\begin{aligned} PC &= \frac{D}{12(k - n)} \cdot k^2 (h_1 - h_2)^2 \\ &= \left[\frac{20}{12 \times (4 - 1.5)}\right] \times 42 \times (0.5 - 0.6)2 = 0.11 m^3 \end{aligned}$$

Therefore $V_P = 338.7 - 0.11 = 338.59 \ m^3$
i.e., there is an excess of $(338.59 - 115.27) = 223.32 \ m^3$ of cut over fill.

## 8.8 COMPUTER-AIDED ESTIMATION OF EARTHWORKS AND PAVEMENT QUANTITIES

(This session is in continuation of Chapter 7) Computer-Aided Estimation for Earthwork and Pavement Quantities/Volumes.. In this session we shall see the estimation of areas and volumes between every two successive cross sections of the highway. The cross section details are read either from cross section file HDS002. FIL or interface file HDS003.fil. In locations of flyovers or river bridges, there are no earthworks as described in the interface file, so there will be no earthworks in such locations. When a road embankment is formed with earth fill, then its finished surface is above the original ground surface (OGL). In such case, earthworks are for fill only. But if the construction of pavement layers are considered but sufficient fill height is not available, then earthworks for cut are also considered.

The areas are given for cross sections of the formation at each chainage station and also in PLAN for the central part of the formation and either side slopes to estimate the cost for either stone pitching or turfing on the slopes. The computer applications for the "Estimation of Earthwork and Pavement Quantities" are explained in Chapter 8 in the "The Guide for Computer Application Tutorials" part of the book available in "Book Tutorials" of the website www.roadbridgedesign. com. For downloading the items under "Book Tutorials" visit the websites at: www. roadbridgedesign.com. For downloading the other items visit the websites at: www. techsoftglobal.com/download.php.

## BIBLIOGRAPHY

*A Policy on Geometric Design of Highways and Streets 2001*, AASHTO.

*Overseas Road Note 6, A Guide to Geometric Design*, Overseas Unit, Transport, and Road Research Laboratory, Department of Transport, Crowthorne, Berkshire, United Kingdom.

*A Guide on Geometric Design of Roads*, Arahan Teknik (Jalan) 8/86, JKR, Malaysia.

*Design Standards, Interurban Toll Expressway System of Malaysia*, Government of Malaysia, Malaysia Highway Authority, Ministry of Works and Utilities.

*Design Manual for Roads & Bridges*, Ministry of Public Works and Kuwait Ministry, State of Kuwait.

*Geometric Design Manual for Dubai Roads*, Dubai Municipality, Roads Department.

*Geometric Design Standards for Rural (Non-Urban) Highways*, IRC 73-1980, IRC.

*Vertical Curves for Highways*, IRC SP 23-1993, IRC.

*Guidelines for the Design of High Embankments*, IRC 75-1979, IRC.

*Surveying* by A. Bannister, S. Raymond, by ELBS with Longman.

*Solving Problems in Surveying* by A. Bannister, Raymond Baker, by ELBS with Longman.

*Fundamentals of Geographic Information Systems* by Michael N. Demers, by John Wiley & Sons, Inc.

*Textbook on Remote Sensing in Natural Resources Monitoring and Management* by C. S. Agarwal, P. K. Garg, by Wheeler Publishing.

# 9 Design Drawings

## 9.1 GENERAL

The folder "Book Tutorials\chapter 09 design drawings\" and other items are available for download from the web sites at: www.roadbridgedesign.com and www.tech-softglobal.com/download.php.

This folder further contains six folders inside it as: "All Model Drawings, Cross Section Drawings, Plan Drawings by Sheet Layout, Profile Drawings, Schematic Drawings for Alignment, and Schematic Drawings for Super Elevation."

## 9.2 DRAWINGS REQUIREMENT

Refer to: **Model Drawing No. 01 Drawing Sheet.**
All drawings (in CAD format) are to be Al size, 594 mm by 841 mm with a 10 mm, 12 mm and 21 mm border. The standard title block should be used for all drawings. No deviation from this should be allowed.

General notes and legends are to be located on the right-hand side of the drawing sheet and above the title block.

Drawings should contain the northing arrow to be marked at the top right-hand corner of the drawings. The northing arrow to be used as a standard form. No other type of northing arrow is allowed.

All drawings to be submitted (including draft copies) must have the seal and signature of a professional engineer and at least a director of the consulting engineer's firm.

Engineering data should be prepared in a logical way, starting from general information and the scope of the work, to increasingly specific details or items and methodology of work. The recommended sequence for the drawings is set out later in this chapter.

The scales of drawings should be selected to clearly illustrate the information that is to be provided. Generally, to show more details, larger scale of drawing and a greater number of drawings are required, however, the smaller the scale, the more limited the designer is in providing the complete information in one drawing. Accordingly, information may have to be disaggregated so that one drawing may be limited to one subject at a time.

All scales are to be in metric system except for land acquisition plans which have to be in accordance with the respective land office's requirement.

## 9.3 SEQUENCE AND SCALES OF DRAWINGS

The following illustrates the sequence of tender drawings and scales that are to be used in a typical road project:

| Drawing Title | Scale |
|---|---|
| Cover or title page | Not applicable |
| Drawing index | Not applicable |
| Key plan | 1:2,000,000–1:3,000,000 |
| Location plan | 1:25,000–1:250,000 |
| Abbreviation, symbol and legend plan | Not applicable |
| Elements of curve plan | Not applicable |
| Super elevation details plan | Not applicable |
| Typical road cross sections and pavement details plan | Suitable scale |
| Alignment control plan | 1:3,000–1:6,000 |
| Plan and longitudinal profile | Horizontal: 1:1,000 |
|  | Vertical: 1:100 |
| Cross section plans | 1:100 or 1:200 (recommended) |
| Junction details plans | 1:250 or 1:500 |
| Road marking plans | Suitable scale |
| Traffic, guide and temporary sign plans | Suitable scale |

*Source:* Paragraph 5.1, Guidelines for Presentation of Engineering Drawings, JKR, Malaysia

Land acquisition plans will not form any part of the tender drawings but have to be prepared for the purposes of land acquisition. The scale will be as that required by the respective land office, usually 1 inch to 4 chains or 8 chains.

## 9.4   FEATURES OF DRAWINGS

### 9.4.1   COVER OR TITLE PAGE

Refer to: **Model Drawing No. 02 Title Sheet.**
The cover or title page shall follow the above referred model drawing.

### 9.4.2   DRAWING INDEX

Refer to: **Model Drawing No. 03 Drawing Index List of Drawings.**
This drawing lists all the titles and number of each drawing within the contract set by following the above referred model drawing of the department. A standard drawing numbering system should be drawn up by the authority and all drawings must follow this system. For all projects, whether designed Departmentally or by Consultants, the project number must be obtained from the Documentation Department.

## 9.5   KEY PLAN

This drawing should indicate:

- Project location with respect to other states in the country.
- State boundaries.
- Name of states and main towns where the project is located.
- All the major federal routes.
- Northing arrow.
- Suitable scales: 1:2,000,000–1:3,000,000.

The key plan can be combined with the location plan.

## 9.6   LOCATION PLAN

Refer to: **Model Drawing No. 04 Location Plan.**
This drawing shows:

- Specific location of the project with respect to the surrounding areas, limit of project and kilometer post.
- The proposed alignment.
- Any other projects in the vicinity.
- Local towns, villages, rivers, reservoirs, roads, railway tracks and other physical features, etc.
- Boxes encompassing a particular stretch of alignment, details of which are shown in the plan and profile plans, whose reference numbers are indicated next to the boxes.
- Trigonometrical stations with heights.
- Route number.
- Northing arrow.
- Suitable scales: 1:25,000–1:250,000.

## 9.7   ABBREVIATION, SYMBOL, AND LEGEND PLAN

Refer to: **Model Drawing No. 05 Abbreviation Symbol and Legend.**
This drawing shall contain abbreviations, symbols, legends and their corresponding full words, details, and representations. The abbreviations, symbols, and legends must be standardized and should not be changed.

## 9.8   ELEMENTS OF CURVE PLAN

Refer to: **Model Drawing No. 06 Elements of Curve.**
This plan indicates the following:

- Typical horizontal curves with or without transition.
- Typical vertical curve – parabolic curve.
- Clothoid, formulas, or other formulas used, and their notations.

## 9.9   SUPERELEVATION DETAILS PLAN

Refer to: **Model Drawing No. 07 Super Elevation Details.**
This plan shall contain the following:

- Super elevation for curves with or without transition.
- Super elevation for the single and dual carriageway.
- Typical shoulder treatment on super elevation.

## 9.10   TYPICAL ROAD CROSS SECTION AND PAVEMENT DETAILS PLAN

Refer to: **Model Drawing No. 08 Typical Road Cross Section and Pavement Details.**
This drawing shall indicate the following:

- Typical road cross section with embankment/fill and cut (for both flat and rolling terrain where applicable) showing width of carriageway, shoulders, camber, drainage features, etc.
- Super elevated typical cross sections (as applicable).
- Typical cross sections for urban and rural carriageways, with drainage details and access road (as applicable).
- Typical section for ground treatment (as applicable).
- Section details of pavement including details for the construction of new pavement over the existing pavement (as applicable).
- Pavement design details such as design period, cumulative total equivalent standard axles, base year, and design subgrade CBR.

## 9.11   ALIGNMENT CONTROL PLAN

Refer to: **Model Drawing No. 09 Alignment Control Plan.**
This drawing is to indicate the bearings and distances of the various intersection points (I.Ps) together with the details of the horizontal curves and I.Ps. The details shall include the following:

- Points of limit of project and horizontal alignment indicated by all the I.Ps.
- Coordinates of all the I.Ps including the start and end points.
- Distances and bearings between all I.Ps.
- Horizontal curve data (as shown) for I.Ps.
- Reference mark information for all I.Ps.
- Locations and chainages of all ST and TS points.
- Location and chainages of all SC and CS points.
- Temporary benchmarks with their levels.
- All relevant survey departments' benchmarks with their levels.
- All relevant references to boundary stone in setting out the first I.P shall be indicated.
- Northing arrow.
- Suitable scales: 1:3,000–1:6,000.

More than one section of the horizontal alignment can be shown on the same plan using separator lines, but the northing arrow should be indicated for each section.

## 9.12   PLAN DRAWING AND LONGITUDINAL PROFILE DRAWING

Refer to: **Model Drawing No. 10 Plan And Longitudinal Profile.**

## 9.13   PLAN DRAWING

The details shall include the following:

- Horizontal alignment of center line together with the limits of carriageway and shoulder.
- Extend of ROW to be acquired.
- Running chainages along centerline.
- Extend of cut and fill areas and carriageways in distinct shades.
- Location of any proposed culverts and their details (including catchment and discharge figures) and any proposed stream deviation.
- Location of all structures such as bridges and box culverts and their levels.
- Locations of boreholes, temporary or permanent benchmarks and their levels.
- Direction of flow of all proposed side drains and their levels.
- Physical features such as existing roads, paths, tracks, public utilities, etc., and contour lines.
- Vegetation of the area passed through, such as rubber estate, padi fields, swamps, etc.
- Existing buildings, property lines, and types of buildings.
- Match lines.
- Existing streams, rivers (with names), culverts, sumps, side drains, etc., and their flow directions, levels, bridges, and other structures.
- Lot numbers and boundaries the alignment passes through.
- Northing arrow.
- Scale: 1:1000.

## 9.14   LONGITUDINAL PROFILE DRAWING

The presentation shall be on graphical format and shall include the following details:

- Running chainages as per the horizontal alignment.
- Existing ground and proposed finished levels.
- Profile of existing ground and the proposed vertical alignment with gradients.
- Details of vertical I.Ps, including the chainage and data on the vertical curve.
- Locations of the Beginning of Vertical Curve (BVC) and End of Vertical Curve (EVC) are to be indicated.
- Locations of all proposed culverts, side drains (with flow directions), guard-rails, and structures.
- Above the finished levels, details of the super elevation and the horizontal alignment along a linear scale is to be provided.
- Scale: 1:100.

## 9.15   CROSS SECTION PLAN

This plan shall show the cross sections of the proposed road at various chainages along the horizontal alignment. The details shall include:

- The existing ground's sectional profile.
- The proposed finished cross section indicating the extent and shape of cut and fill required.
- The proposed finished road level at the centerline.
- The crossfall or super elevation of the carriageway.
- Any other details deemed useful for clarity.
- Suitable scales: 1:100 or 1:200 (recommended).

Plans from computer printouts can be used but must be presented as Al size drawings.

## 9.16   JUNCTION DETAILS PLANS

Refer to:
Model Drawing No. 11 Junction Details – General Layout.
Model Drawing No. 12 Junction Details – Setting out.
Model Drawing No. 13 Junction Details – Pavement Super elevation.
Model Drawing No. 14 Junction Details – Road Marking Traffic Guide Signs and Traffic Signals.

## 9.17   JUNCTION DETAILS

Junction details should include the following:

- General layout: plan showing details of carriageway widths, islands, limit of ROW, property lots, dwellings and their types, limit of project on minor roads, etc. The shoulder should be shaded for better visual impact.
- Setting out the plan with sufficient details so as to enable the setting out of the junction.
- Drainage and curb details of the junction showing proposed type and position of drain and curbs, culverts, etc. Proposed super elevations of various pavement and longitudinal section can also be indicated.
- Road marking and traffic sign plan showing the type and positions of the proposed road markings, traffic signs and signals (if applicable).
- Suitable scales: 1:250 or 1:500.

## 9.18   TRAFFIC SIGNAL DETAILS

These shall include the location of the traffic signal posts, the type of lantern and the proposed cycle times. Schematic plans showing the various phases should also be included.

## 9.19   ROAD MARKING PLAN

The plan shall show all details of the proposed lanes and pavement markings for the centerline, edge line, continuity line, chevron, give way marking, stop line, pavement direction arrows, etc.

## 9.20   TRAFFIC, GUIDE, AND TEMPORARY SIGN PLANS

Adequate traffic, guide, and temporary signs shall be proposed and shall follow those approved for use by the department.

## 9.21   DRAINAGE PLANS

Refer to: **Model Drawing No. 15 Culvert Schedule.**
The drainage plans shall be sufficient enough to depict all detail pertaining to drainage matters and should include the following:

- Details of interceptor drain, bench drain, roadside drain, shoulder drain, berm drain, toe drain, outfall drain, and subsoil drain, etc.
- Details of precast concrete slab, curb, curb inlet, catch pit, sump, stream deviation, etc.
- Details of culverts (whether reinforced concrete pipe, corrugated metal pipe or box) with details of headwall and wing walls. A culvert schedule with the relevant particulars should also be prepared.

## 9.22   STRUCTURE PLANS

Refer to: **Model Drawing No. 16 Bridge Details.**
These drawings shall normally be applied to bridges with the following requirements.

## 9.23   BRIDGE STRUCTURAL PLANS

The general arrangement and layout should include the following:

- Location plan indicating the specific location of the bridge project with the proposed alignment, local town, villages, rivers, roads, railway tracks, and other physical features. This should include the latitude and longitude bearings and the northing arrow (scale 1:10,000).
- Plan showing horizontal alignment of the center line complete with limits of carriageway, boreholes locations, direction of flow, benchmarks, and physical features such as public utilities, slope protection, drainage behind abutment, etc. Pile layout can also be included here.
- Longitudinal profile showing running chainage as per horizontal alignment, existing ground and the proposed levels, extent of cut and fill, and limits of bridge construction.
- Detailed cross section showing arrangement of beams and girders in typical section and other deck details.
- Notes, legends, etc., which should include the design criteria, relevant codes used, and other useful information. The list of other related drawings should also be inserted.

### 9.23.1  Soil Profile

This should include the borehole details at various depths till the end of boring. Other laboratory test results, such as the plasticity index, plastic limit, and liquid limit, should be incorporated if available.

### 9.23.2  Piles

This should include pile details and joints. The numbers and types of piles for abutment and piers, and their capacities and estimated lengths should be tabulated. The notes should include the materials, type of piles (e.g., frictional or end bearing), working loads, test loads, and other relevant information.

### 9.23.3  Abutment and Wing Wall Details

These should include the typical cross section and elevation, showing details for bearings and pile layout, and showing principal dimensions and description of all structural components. The notes should normally include the concrete strength, cover, etc.

A separate drawing showing reinforcement details shall be prepared with proper legends and notes.

### 9.23.4  Piers

Same as above for abutments.

### 9.23.5  Beams/Girders

These should include sectional, longitudinal, and plan details, indicating all dimensions and cable profiles. The reinforcement and end block details may be on the same or different drawings.

The notes should include among other things the design criteria, ultimate bending moment resistance and shear capacity, brief material specifications, etc.

### 9.23.6  Diaphragms

These may be included in the drawings for beams or together with details of deck slab.

### 9.23.7  Deck Slabs

These should include plan and sectional details and reinforcement detail complete with relevant notes.

### 9.23.8   Handrails/Parapet and Expansion Joints

These should include plan, elevation and sectional details complete with relevant notes on material properties and installation requirements.

### 9.23.9   Water Main Brackets and Other Miscellaneous Details

Same as above.

## 9.24   SEQUENCE AND SCALES FOR BRIDGE STRUCTURES

The following illustrates the sequence of tender drawings and scales that are to be used for a typical bridge structure:

| Drawing Title | Scale | | | |
|---|---|---|---|---|
| General layout | 1:50, | 1:100, | 1:150, | 1:200 |
| Soil profile | Suitable scale | | | |
| Pile details | 1:5, | 1:10, | 1:20, | 1:50 |
| Abutment – concrete details | 1:20, | 1:25, | 1:30, | 1:40 |
| Abutment – reinforcement details | Same as above | | | |
| Pier – concrete details | Same as above | | | |
| Pier – reinforcement details | Same as above | | | |
| Beams and diaphragms | 1:5, | 1:10, | 1:20, | 1:25 |
| Deck slat details | 1:20, | 1:25, | 1:50 | |
| Handrails/parapet and expansion joints | 1:2.5, | 1:5, | 1:20 | |
| Water main brackets etc. | 1:1, | 1:7.5, | 1:25 | |

*Source:*   Paragraph 6.16.2, Guidelines for Presentation of Engineering Drawings, JKR, Malaysia

## 9.25   REQUIREMENT FOR DIMENSIONS

A typical requirement for dimensions is shown below:

- All dimensions shall be expressed in meters or millimeters.
- Unit abbreviations shall be "m" for meter and "mm" for millimeters. Abbreviation should always be given in description notes but not for normal dimensioning.
- Normal dimensions of more than one meter shall be written to three decimal places but with the decimal point omitted. Thus, two meters four hundred and fifty millimeters shall be written as 2450.
- Descriptive dimensions of more than one meter shall be written to three decimal places and unit abbreviation included. Thus, for normal water level is written as: NWL 37.128m.
- Dimensions expressed in millimeter shall be whole numbers, not in decimals.
- Dimension shall be written to be read from the bottom or right-hand side of the drawing.

## 9.26   RELOCATION OF SERVICES PLANS

This plan shall show details for the relocation of:

- Water mains.
- Electrical Posts and Cables.
- Telecom Posts and Cables.

The drawings shall be have sufficient details to properly define the scope of work. They shall show the locations of the existing services and their proposed relocations.

## 9.27   MISCELLANEOUS PLAN

Refer to: **Model Drawing No. 17 Ground Profile.**
Miscellaneous plans are for those details not described above and may consist of the following:

- Street lighting details.
- General subsoil profile.

The streetlight detail should include:

- Locations of all proposed lamp posts, crossings, and feeders pillar boxes.
- Types of lanterns together with details of columns.

## 9.28   DRAWING NUMBERING SYSTEM

The system used is as follows:
**FH/2017/RD/PC/R41/DPR/DD-08**

- FH – federal highway.
- 2017 – year.
- RD – road department.
- R41 – route 41.
- DPR – detail project report.
- DD-08 – design drawing number.

The system shall be used for all projects including all departmental and consultant designs. The recording of serial numbers of projects will be registered by the documentation section.

## 9.29 COMPUTER-AIDED DESIGN DRAWINGS

(This session is in continuation of chapters from 4 thru 8)

The computer applications for the "Design Drawings" are explained in chapter nine in the "The Guide for Computer Application Tutorials" part of the book, available for download from the web site www.roadbridgedesign.com.

For downloading the "Book Tutorials" and other items visit the web sites at: www. roadbridgedesign.com and www.techsoftglobal.com/download.php

In Section 9.29.1, the Creating of drawings is described for the followings:

- Alignment schematic diagram drawings for horizontal and vertical alignments by running 200 ProHALIGN and 300 ProVALIGN for entire length of the project. Alignment schematic diagram drawing for super elevation was already created by 400 ProOFFSET, as described in chapter six for the design of cross section elements.
- Plan drawing, is one single drawing for the entire length of the project.
- Profile drawing, is for either a full length or set of drawings for each kilometer of the project, by including alignment schematic diagram drawings for horizontal alignment, vertical alignment, and super elevation.
- Cross section drawings at a regular interval as desired, but multiple of interval used in the "interface."

### 9.29.1 PROJECT DRAWINGS FOR PLAN, PROFILE, AND CROSS SECTIONS

The procedures described in this section are in continuation from chapter four thru eight. Individually, the processes may be executed by copying data from each of the following folders as available for download from the web site of the book 'User's Guide for Hands On Practice':

- Book Tutorials\Chapter 9 Design Drawings\Schematic Drawings for Alignment.
- Book Tutorials\Chapter 9 Design Drawings\Schematic Drawings for Super Elevation.
- Book Tutorials\Chapter 9 Design Drawings\Plan Drawings by Sheet Layout.
- Book Tutorials\Chapter 9 Design Drawings\Profile Drawings.
- Book Tutorials\Chapter 9 Design Drawings\Cross Section Drawings.

In Section 9.29.2, the procedure will be described to cut the plan drawing of full length into one kilometer pieces.

In Section 9.29.3, the procedure will be described to insert one kilometer pieces of plan and profile drawings within the drawing sheet with specified size and scale.

In Section 9.29.4, various model drawings are described with procedures for creating, maintaining and management, and finally submission, as project drawings are described with various details.

The reader/user may also see the related Tutorial Videos in Channel 'Techsoft Forum' in 'YouTube', where the procedure to create the drawings automatically are also described. The readers/users may write to techsoftinfra@gmail.com.

### 9.29.1.1    Drawings for Alignment Schematics

We select **STEP 10** to create design drawings for project construction. The first page is for **Alignment Schematics**, we click on button **Process Data** and the drawings in DRG format are created for horizontal and vertical alignments in two successive processes, which come one after another. The user has to click the button **Proceed**, as the process is over when the message comes up, click on the button **OK** and finally click on the button **Finish**. This process will happen twice, for processes **ProHalign** and **ProValign**, and data processes created for alignment schematics upon design of horizontal and vertical alignments as mentioned before.

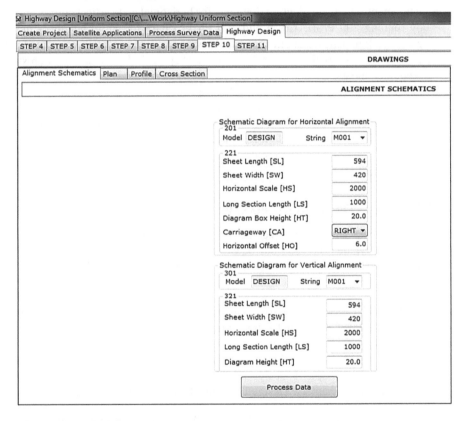

Data below is for creating schematic diagrams for a horizontal alignment diagram to show in profile drawing, and is processed in these steps:

    **PROHEADS**
    **200,PROHALIGN**
    **201,MODEL=DESIGN,STRING=M001**
    **220,DIAGRAM**
    **221,SL=594,SW=420,HS=2000,LS=1000,HT=20.0,CA=RIGHT,HO=6.0,**
    **CHAINAGE,un=KM,f=2**
    **FINISH**

Data below is for creating schematic diagrams for a vertical alignment diagram to show in profile drawing, and is processed in these steps:

**PROHEADS**
**300,PROVALIGN**
**301,MODEL=DESIGN,STRING=M001**
**320,DIAGRAM**
**321,SL=594,SW=420,HS=2000,LS=1000,HT=20.0,CHAINAGE,un=K**
**M,f=2**
**FINISH**

Schematic diagrams for super elevations are created while processing PROOFFSET data for the edges of carriageways and paved shoulders, so as to show in the profile drawing and has already been processed.

### 9.29.1.2   Drawings for Plan

Next, we select tab page **Plan** to create plan drawing for various strings listed in the box. We click on the button **Process Data** and the Plan drawing is created.

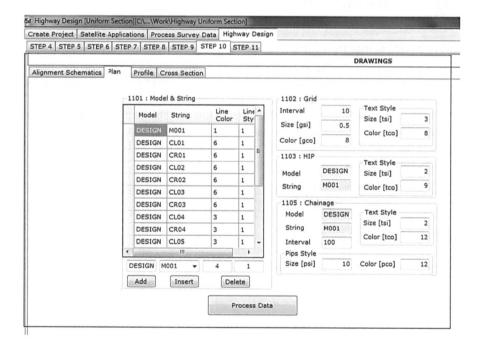

The batch data for "1100 PLAN" as given below is written in the project data file "Project Data File.TXT" and is for creating the plan (top view) of the entire road and is explained in relevant section of "HEADS Major Options" in "HEADS Pro User's Manual."

**HEADS**
**1100,PLAN**
**1101,MODEL=DESIGN,STRING=M001,COLOR=1,STYLE=1**
**1101,MODEL=DESIGN,STRING=CL01,COLOR=6,STYLE=1**
**1101,MODEL=DESIGN,STRING=CR01,COLOR=6,STYLE=1**
**1101,MODEL=DESIGN,STRING=CL02,COLOR=6,STYLE=1**

1101,MODEL=DESIGN,STRING=CR02,COLOR=6,STYLE=1
1101,MODEL=DESIGN,STRING=CL03,COLOR=6,STYLE=1
1101,MODEL=DESIGN,STRING=CR03,COLOR=6,STYLE=1
1101,MODEL=DESIGN,STRING=CL04,COLOR=3,STYLE=1
1101,MODEL=DESIGN,STRING=CR04,COLOR=3,STYLE=1
1101,MODEL=DESIGN,STRING=CL05,COLOR=3,STYLE=1
1101,MODEL=DESIGN,STRING=CR05,COLOR=3,STYLE=1
1101,MODEL=DESIGN,STRING=CL06,COLOR=3,STYLE=1
1101,MODEL=DESIGN,STRING=CR06,COLOR=3,STYLE=1
1101,MODEL=DESIGN,STRING=CL07,COLOR=3,STYLE=1
1101,MODEL=DESIGN,STRING=CR07,COLOR=3,STYLE=1
1101,MODEL=DESIGN,STRING=I001,COLOR=11,STYLE=1
1103,HIP,MODEL=DESIGN,STRING=M001,TSI=2,TCO=9
1105,CHAINAGE,MODEL=DESIGN,STRING=M001,INTERVAL=100,
PSI=10,PCO=12,TSI=2,TCO=12
FINISH

### 9.29.1.3   Drawings for Profile

Next, we select page **Profile** to create the profile drawings for various strings listed in the box. The start and end chainages are already displayed by taking them from the previous design steps. The default length per sheet is for the full length of the road alignment. This may be changed to one thousand in case the length of one kilometer per sheet is desired. The horizontal scale should allow the drawing to fit in to the length of the sheet. The vertical scale should be one tenth of the horizontal scale. We click on button **Process Data** to create the drawings.

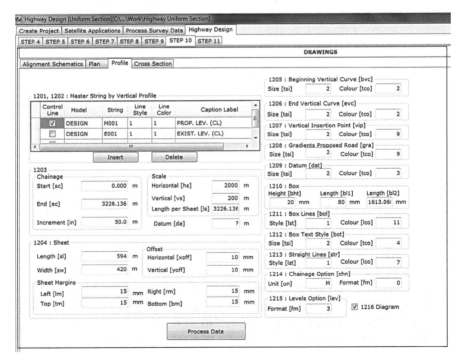

The batch data for "1200 PROFILE" as given below is written in the project data file "Project Data File.TXT" and is for creating the profile drawing (side view) of the entire road and is explained in relevant section of "HEADS Major Options" in "HEADS Pro User's Manual."

**HEADS**
**1200,PROFILE**
**1201,MODEL=DESIGN,STRING=M001,PL=1,PC=1,OL=1,OC=252,CA**
**P=PROP. LEV. (CL)**
**1202,MODEL=DESIGN,STRING=E001,PL=1,PC=1,OL=1,OC=252,CAP**
**=EXIST. LEV. (CL)**
**1203,sc=0.000,ec=3304.888,hs=2000,vs=200,ls=1000.0,in=50.0,da=?**
**1204,sl=594,sw=420,lm=15,rm=15,tm=15,bm=15,xoff=10,yoff=10**
**1205,bvc,tsi=2,tco=2**
**1206,evc,tsi=2,tco=2**
**1207,vip,tsi=2,tco=9**
**1208,gra,tsi=2,tco=9**
**1209,dat,tsi=2,tco=3**
**1210,box,bht=20,bl1=80,bl2=510**
**1211,bol,lst=1,lco=11**
**1212,bot,tsi=2,tco=4**
**1213,str,lst=1,lco=7**
**1214,chn,un=M,fm=0**
**1215,lev,fm=3**
**1216,DIAGRAM**
**FINISH**

### 9.29.1.4   Drawings for Road & Ground Cross Sections

Next, we select page **Cross Section** to create cross section drawings. The start chainage, end chainage, scale, sheet size and other data are given in the page. We can modify the data if desired. We click on button **Process Data** and a set of cross section drawings are created.

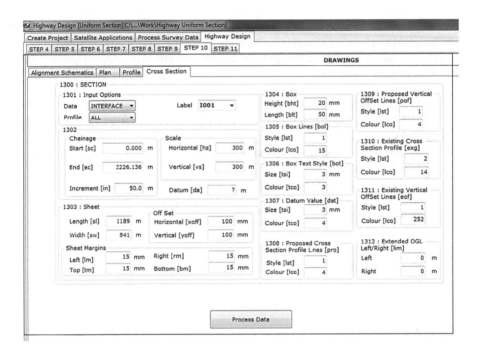

The batch data for "1300 SECTION" as given below is written in the project data file "Project Data File.TXT" is for creating cross section drawing (at regular interval of chainage stations) of the entire road and is explained in relevant section of "HEADS Major Options" in "HEADS Pro User's Manual."

**HEADS**
**1300,SECTION**
**1301,INTERFACE,I001**
**1302,sc=0.000,ec=3304.888,in=50.0,hs=300,vs=300,da=?**
**1303,sl=1189,sw=841,lm=15,rm=15,tm=15,bm=15,xoff=100,yoff=100**
**1304,box,bht=20,blt=50**
**1305,bol,lst=1,lco=15**
**1306,bot,tsi=3,tco=3**
**1307,dat,tsi=3,tco=4**
**1308,pro,lst=1,lco=4**
**1309,pof,lst=1,lco=4**
**1310,exg,lst=2,lco=14**
**1311,eof,lst=1,lco=252**
**FINISH**

### 9.29.1.5   Viewing of Computer-Aided Design Drawings

As the final step, we select tab **STEP 11** to view project data file, survey data file, design report files, and design drawings now available with us.

We can view the design drawings in the CAD Viewer by selecting from the list and save the drawings in desired format. The project data files, survey data files, design report files, horizontal alignment design report, vertical alignment design report, and the volume from earthworks estimation report, may also be opened for viewing and printing.

### 9.29.1.5.1   Design Drawings

To view the design drawings in the CAD Viewer we click on button **View Drawings**, as the message comes click on button **Yes** and the HEADS CAD Viewer opens with all the design drawings and a list on the left side. The drawings may be saved as a selected type compatible with various popular CAD software programs.

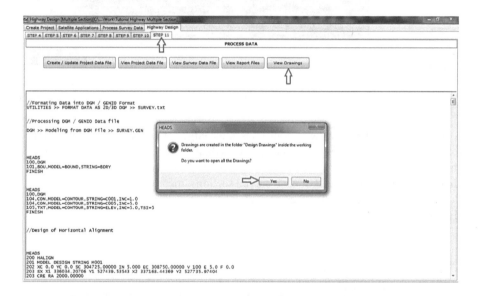

The CAD Viewer opens with a list of all design/construction drawings, from which user can select a drawing and the drawing is displayed. The drawings may be saved as a selected type compatible to various popular CAD software.

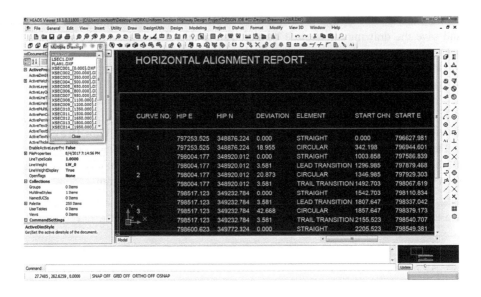

The plan drawing showing various lines at the edges of various components of the road cross section, for example, edges of median, either side of the carriageway, shoulder, tadpoles describing fill and cut sections with side drains for storm water, chainages with start and end of each curved sections, mentioning circular and transition curves, and the details of curves, etc., at every horizontal intersection points (HIPs).

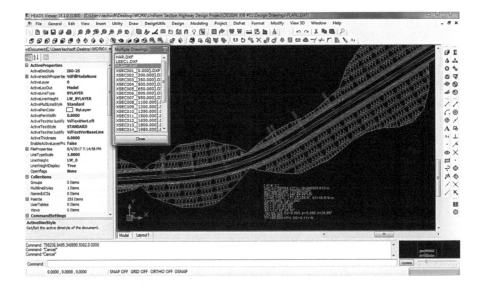

The long section drawing is created as shown below. This drawing describes the vertical profile of ground and design road centerline, or the median bottom edges, as desired by the user. The various design details of the vertical profile of the project highway are displayed. The schematic drawings for design of horizontal alignment, vertical alignment, and super elevations are also included in the drawing. The vertical profile of maximum of twenty strings can be included in the drawing with their chainages and elevations.

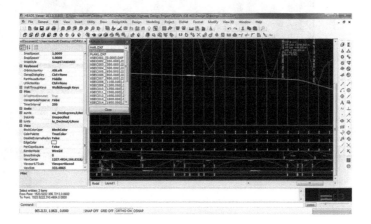

The cross section drawing is created as shown below. This drawing describes the design cross section of the proposed road in fill and cut sections with side drains in case of cut situations.

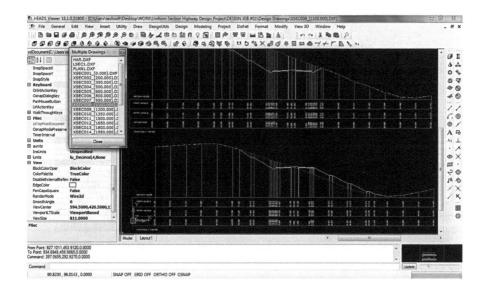

The plan drawing with chainage, HIP and curve details, etc., for multiple cross section highway designs, where the configuration of the road cross section chainages at various length sections of the project highway.

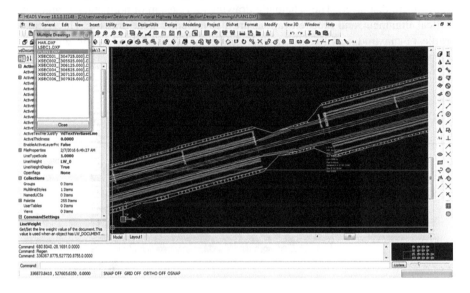

The cross section drawing with the main carriageway, service roads, elevated sections, etc., for multiple cross section highway design, where the configuration of the road cross section chainages at various length sections of the project highway.

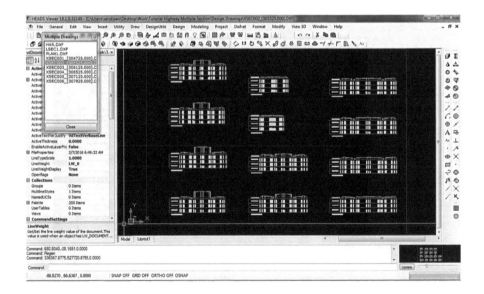

The drawings can be saved in a desired format compatible with various popular CAD software programs and can be printed by using and A4/A3/A2 sized printer or plotter. The saving, printing, and plotting of drawings is possible only with authorized licensed version of HEADS Pro. In the demo or unauthorized version, the saving, printing, and plotting of drawings is not possible.

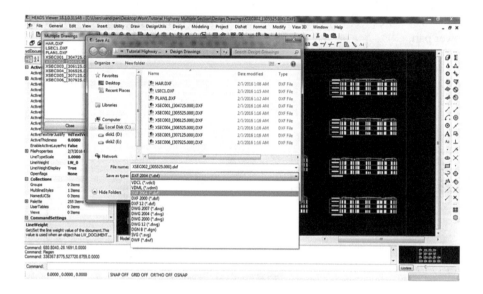

## 9.29.2   CREATE PLAN DRAWINGS BY CUTTING WITH SHEET LAYOUT

In the final Sections 9.3 and 9.4 we learned the processing of project data file and text data files respectively to re-produce the highway design done in workspace mode as described in chapters four thru eight. In chapter nine the model drawing no.10 plan and longitudinal profile is described. This drawing most commonly contains one kilometer length of the plan and profile of the proposed road in a drawing sheet. The commonly used drawing sheet sizes and respective scales for creating plan and profile drawings are as mentioned below:

- For Sheet Size A0 – Scale = 1:1000.
- For Sheet Size A1 – Scale = 1:1500.
- For Sheet Size A2 – Scale = 1:2000.
- For Sheet Size A3 – Scale = 1:3000.

On the drawing sheet, the all-around sheet margin is normally ten millimeters, but on plan and profile drawings the sheet margins on the left, right, and top, of the sheet may be maintained as ten millimeters but the bottom margin should be enough to accommodate the plan in the upper part, with the profile and title block in the lower part of the drawing sheet.

### 9.29.2.1  Draw Match Lines in the Base Drawing of Full Length

At every kilometer a match line is drawn, which commonly appear on every two successive sheets. This will help the user to join the prints of the drawings one after another. If there are large variations in the horizontal alignment, then user has to cut the plan into two or more pieces and arrange them in their placements in the upper part. If there are large variations in the vertical profile, then user has to cut the profile into two or more pieces by changing the "datum" and arrange their placements in the lower part.

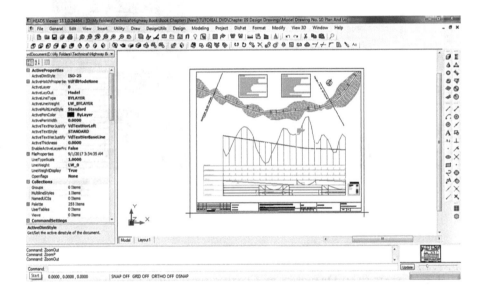

In the previous sections we have re-produced the design by processing the project data file and a set of text data files. Now the working folder containing the total station survey data file, various major option data files numbered from 03 thru 24, various report files as ".REP," various system files as ".FIL" and ".LST," and a folder "design drawings" are shown below:

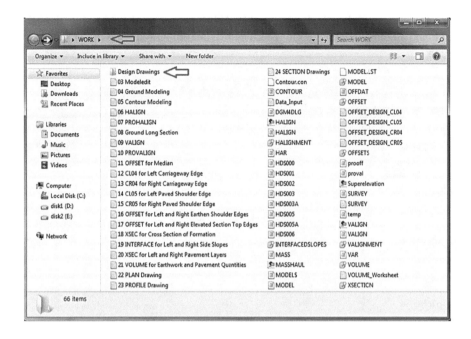

Inside the folder "Design Drawings" there are a set of drawing files, which contains the drawing "PLAN1.DXF." This drawing is for the full length of the project and is to be cut into one kilometer pieces. So PLAN1.DXF is the base drawing.

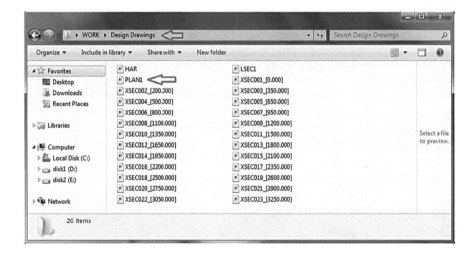

A folder is created on the Desktop as "PLAN Drawings," and the drawing PLAN1. DXF and alignment database file HALIGN.FIL are brought into the folder as shown below:

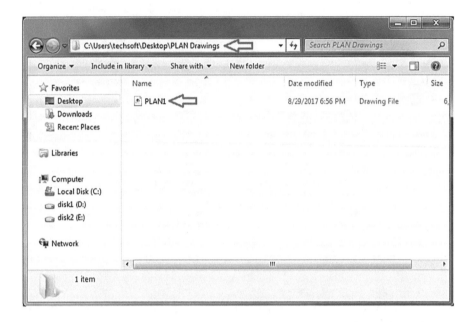

First, we open the HEADS Viewer and open the drawing PLAN1.DXF as shown below:

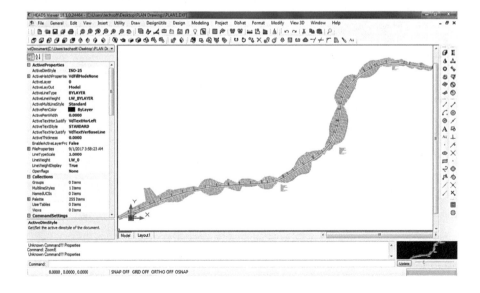

As the length of the road in the drawing is about three kilometers, we draw four straight lines on either side, near chainages 1+ 0.000 and 2+0.000. Click on the icon for "Extend" on the top bar as shown, and click both the straight lines near chainage 1+0.000, we right click the mouse button.

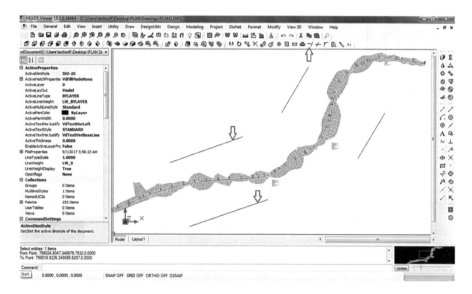

We click on either end of the straight as chainage mark at 1+ 0.000, which will be extended to meet the straight lines drawn by us in the previous step.

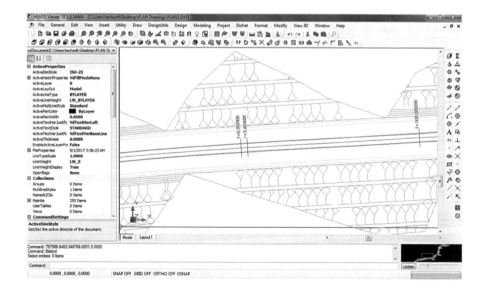

The straight line on chainage 1+0.000 is extended to form the "Match Line" at chainage 1+0.000:

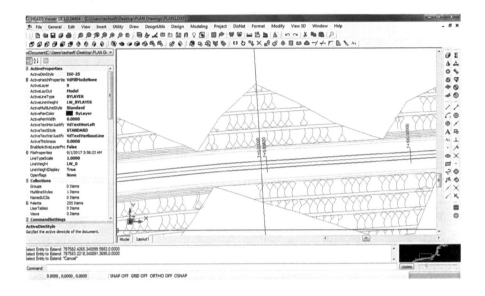

In a similar way, the "Match Line" is also drawn at chainage 2+0.000 and the temporary straight lines are erased:

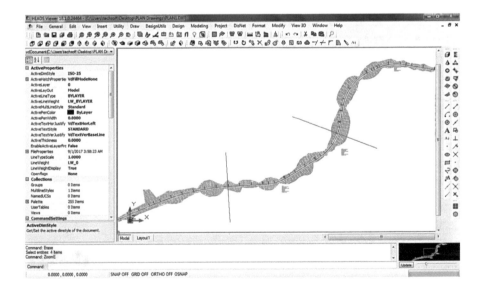

## 9.29.2.2   Making Sheet Layouts on the Base Drawing Within Two Match Lines

Next, we select HEADS Viewer menu item "Design >> Layout >> Apply" as shown below:

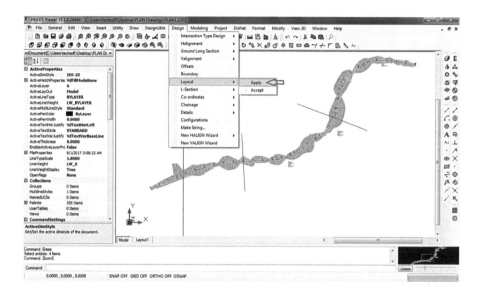

As the dialog box comes up, we select paper type "A3," bottom margin 100 mm, scale 1:3000 and click on button "OK":

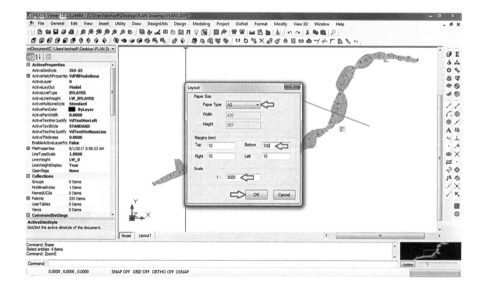

Next, we click at the location as shown below and the drawing sheet is placed:

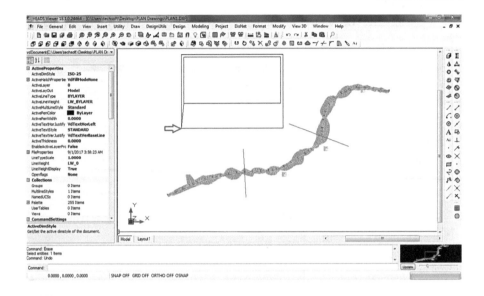

Next, we select the HEADS Viewer menu item "Modify >> Move" and click on any line of the drawing sheet, then we click right and the left mouse button:

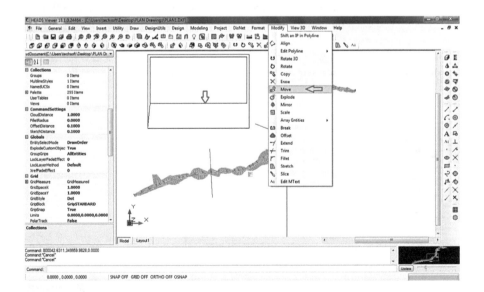

We move the mouse to bring the sheet to the desired location and then click the left mouse button to place it:

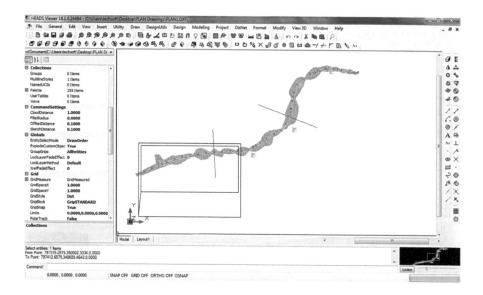

Next, we select the HEADS Viewer menu item "Modify >> Rotate" and click on any line of the drawing sheet, move the mouse slowly to rotate the sheet:

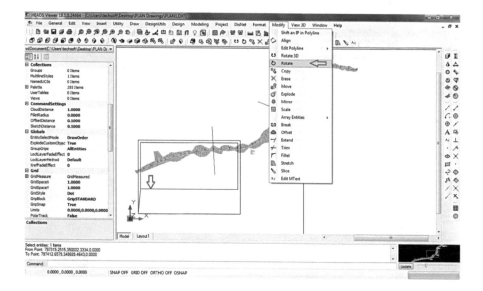

Next, click the left mouse button to place the sheet as desired, as shown below:

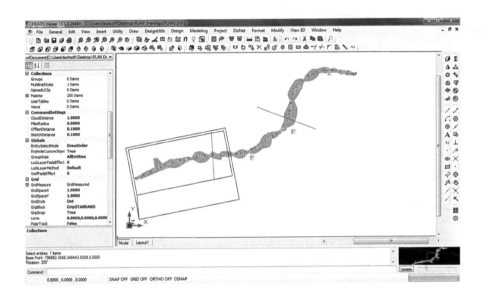

Next, we select the HEADS Viewer menu item "Design >> Layout >> Accept" as shown below, this writes the details of sheet as: X,Y coordinates of left bottom corner, angle of sheet with horizontal, sheet size, scale and all four margins in the file "Layout.FIL."

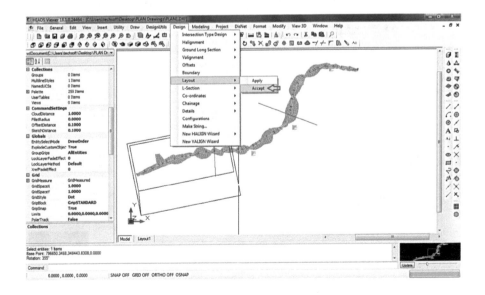

All the steps in Section 9.30.2.2 as described above are repeated twice, and two more drawing sheets are placed as shown below:

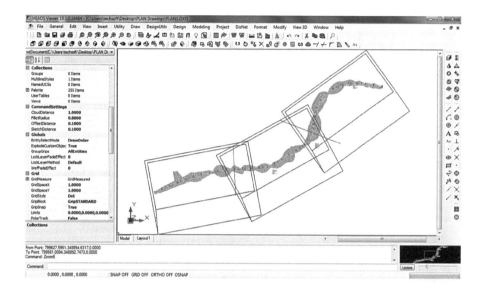

### 9.29.2.3   User's Input Data File with Major Option 1100

Next, in the HEADS Pro main screen menu, we select item "Help >> Contents" as shown below:

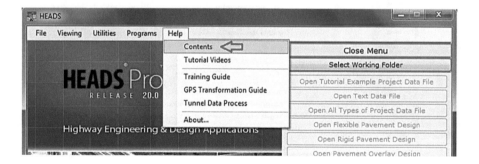

As the "Help" window opens, select the items as shown below, to get the 1100 plan data on right side of the help window:

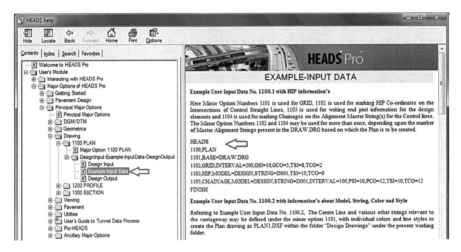

Copy the 1100 plan data as shown below and close the "Help Window":

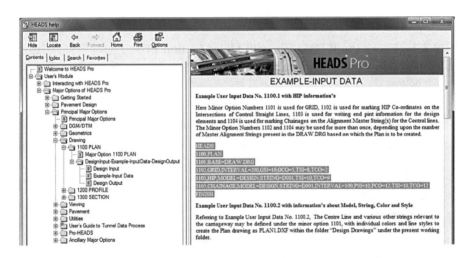

Select HEADS Pro menu item "Utilities >> Notepad."

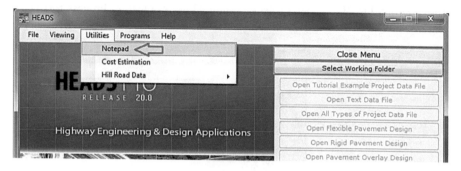

Paste the 1100 plan data in the notepad, as shown below:

```
HEADS
1100,PLAN
1101,BASE=DRAW.DRG
1102,GRID,INTERVAL=200,GSI=10,GCO=5,TSI=8,TCO=2
1103,HIP,MODEL=DESIGN,STRING=D001,TSI=10,TCO=9
1105,CHAINAGE,MODEL=DESIGN,STRING=D001,INTERVAL=100,PSI=10,PCO=12,TSI=10,TCO=12
FINISH
```

The 1100 plan major option data is modified and save in notepad as file "Inp. TXT" in the folder "PLAN Drawings" located on the Desktop, as shown below. Here the base full length drawing is PLAN1.DXF, which is to be cut into required number of pieces for each length of one kilometer as marked by the match lines at the start and end of each kilometer:

```
HEADS
1100,PLAN
1101,BASE=PLAN1.DXF
1102,GRID,INTERVAL=200,GSI=100,GCO=8,TSI=2,TCO=9
1103,HIP,MODEL=DESIGN,STRING=M001,TSI=2,TCO=3
1105,CHAINAGE,MODEL=DESIGN,STRING=M001,INTERVAL=100,PSI=2,PCO=8,TSI=2,TCO=3
FINISH
```

### 9.29.2.4   Open and Process User's Input Data File with Major Option 1100

From HEADS Pro "File" menu select the item "File >> Open & Process text Data File," as shown below:

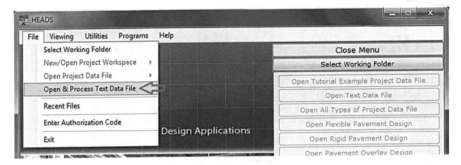

As the selection box comes, select file "Inp.TXT" from within the folder "PLAN Drawings," and click on button "Open."

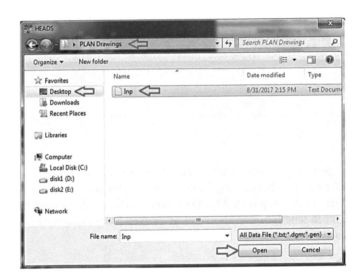

The file is selected and opened in the notepad, close the notepad.

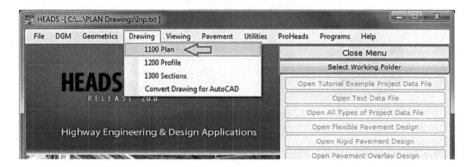

Select HEADS Pro menu item "Drawing >> 100 PLAN."

As the process starts, click on the button "Proceed," as the process finishes the message comes up about the three drawings created specific to the three sheet layouts made earlier, click on the button "OK" and then on the button "Finish," as shown below:

### 9.29.2.5   Viewing All the Plan Drawings each of One Kilometer Length

Select HEADS menu item "Viewing >> Interactive Drawing," as shown below:

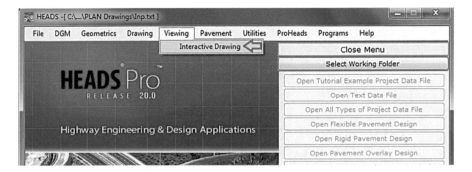

As the HEADS Viewer opens, select viewer menu item "File >> Open Multiple Drawings."

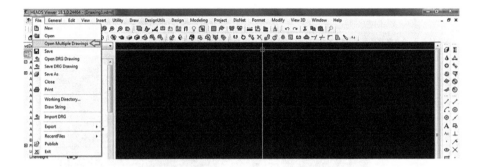

Select the folder "Design Drawings" inside our earlier folder "PLAN Drawings," as shown below:

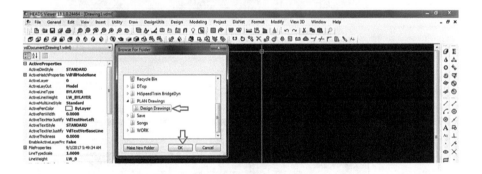

Next, the three plan drawings open with the list as PLAN1.DXF, PLAN2.DXF, and PLAN3.DXF on the left side, click on a drawing to display it, drawing PLAN1. DXF is already opened, as shown below:

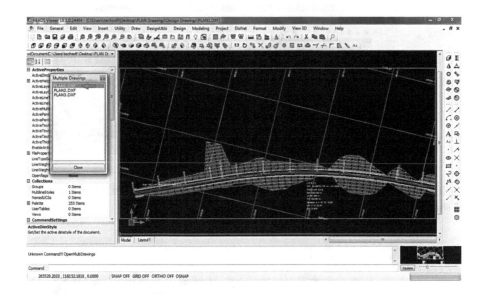

Drawing PLAN2.DXF is selected and opened.

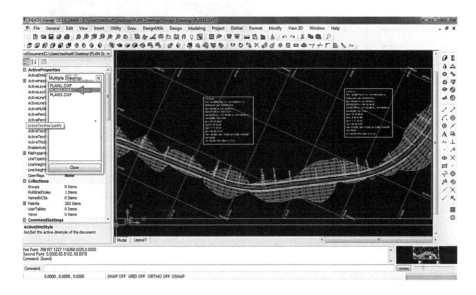

Drawing PLAN3.DXF is selected and opened.

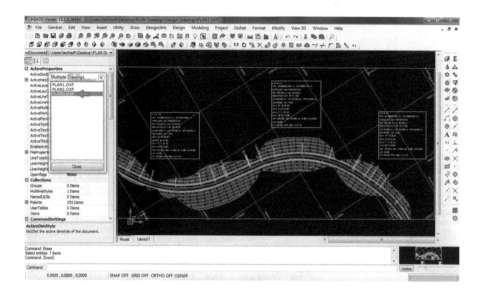

Click on "Close" in the list on the left side, as the message comes up asking whether to save the drawings, click on the button "Yes," as shown, the drawings will be saved in the authorized version of HEADS Pro:

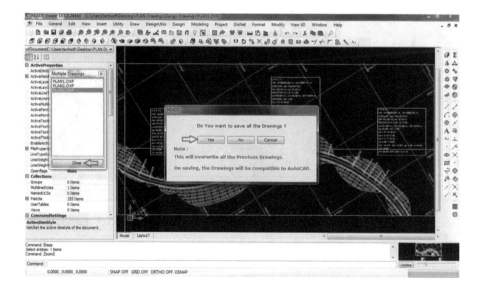

### 9.29.3   CAD BASICS TO CREATE THE PLAN AND PROFILE DRAWINGS

Select the HEADS Viewer menu icon to open a drawing, select drawing "Model Drawing No. 01 Drawing Sheet," click on button "Open" as shown below:

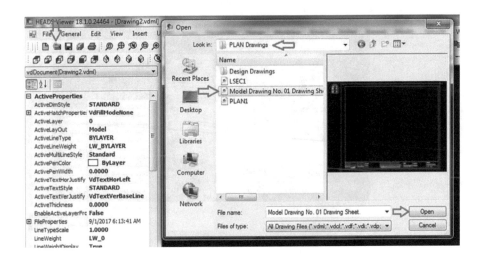

Select Viewer menu item "Insert >> Insert Block," as shown below:

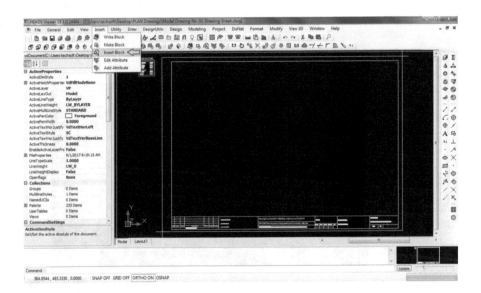

As the "Insert" dialog box comes up, select option "From File," then select "Select all parameters on screen," select button "Select file," from selection box select file "PLAN1.DXF," click on the button "Open," and click on the button "OK."

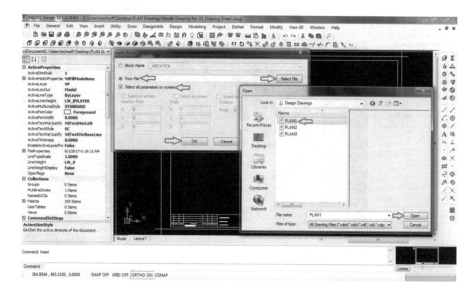

On the command line, at the bottom of the Viewer, the command comes up to select the "Insertion Point," we click on a point on the screen to define the "Insertion Point" as shown below:

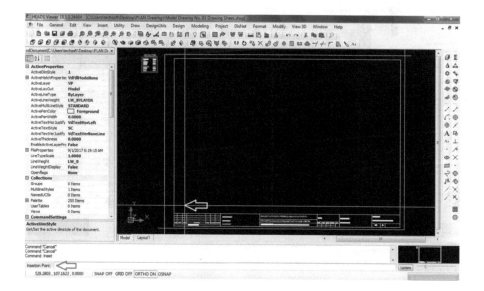

Next, the command comes up to enter the "Scale Factor," we type in the value as "1.0," as shown below and press the "Enter" key:

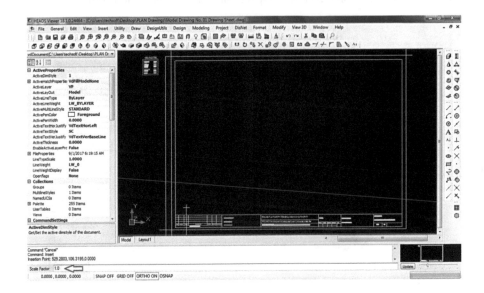

Next, the command comes up to enter the "Rotation Angle," we type in the value as "0.0," as shown below and press the "Enter" key:

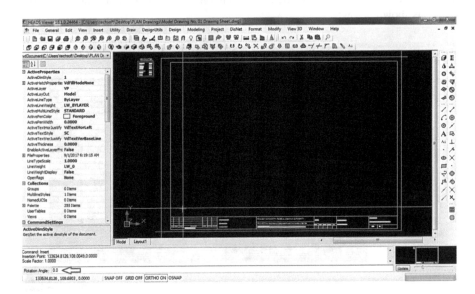

Select the Viewer menu item "View >> Zoom >> Zoom All."

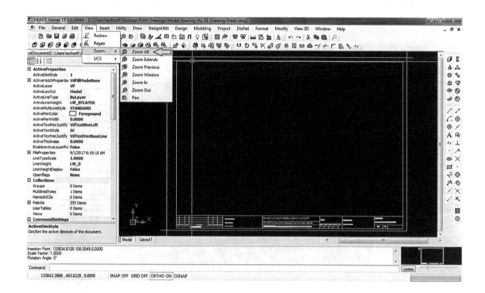

Select the Viewer menu item "Modify >> Move."

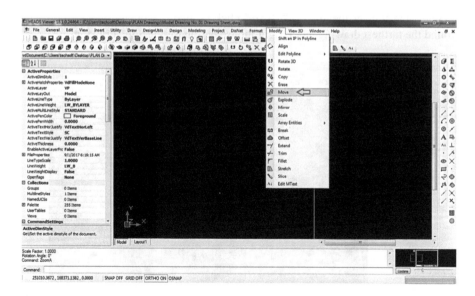

Select the Viewer menu icon "-" to "Zoom-out," the two drawings can be located on the CAD screen, they are marked with two red colored rectangles, as shown below:

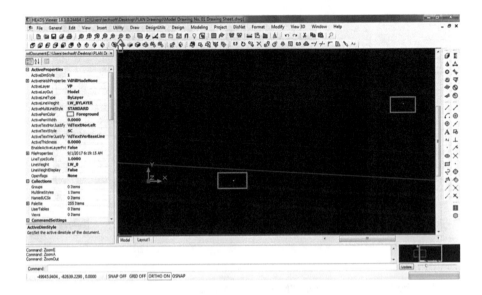

Select the Viewer menu icon "Move," click on two diagonally opposite corners around the farthest drawing, click right mouse button to select the drawing:

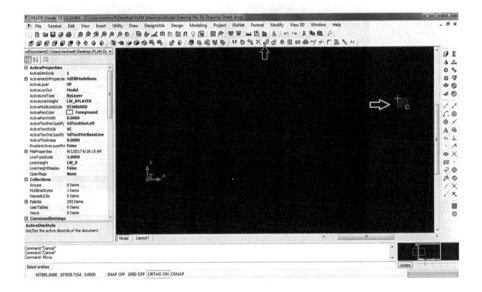

The "ORTHO" is clicked to turn "Off," the farthest drawing is then brought near the other drawing, now the two drawings are close to each other:

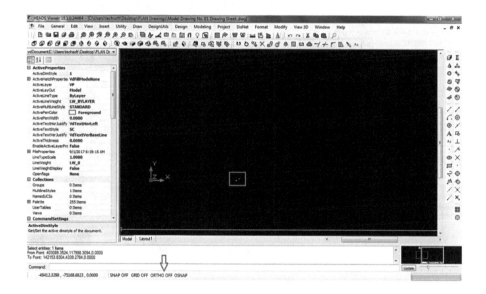

Click on the Viewer icon "Zoom All" and then on the icon "Explode," click on two diagonally opposite corners of the drawing, click right mouse button, the drawing is no longer a block, it is exploded:

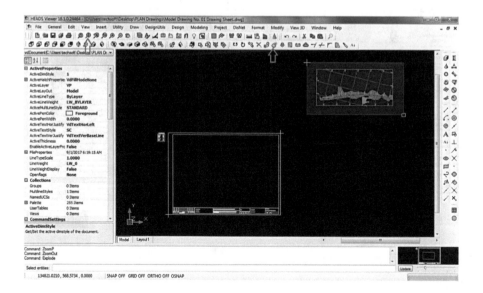

The drawing is moved to inside the drawing sheet, as shown below:

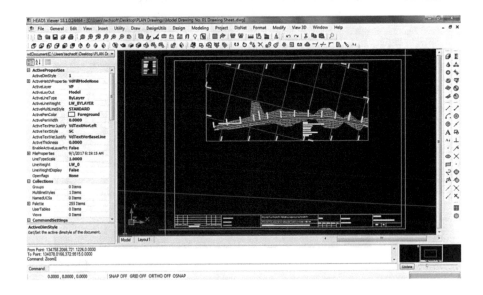

In the similar way, next select the profile drawing for the first one kilometer and insert it in the same drawing, as shown below:

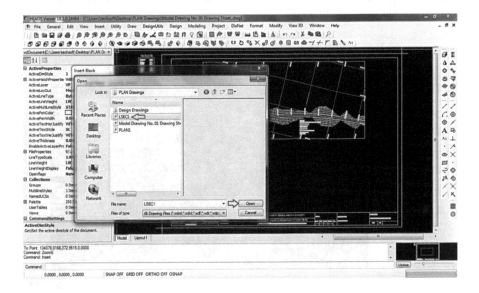

The profile drawing is brought near the drawing sheet, and is to be scaled to match with the plan drawing inside the drawing sheet, as shown below:

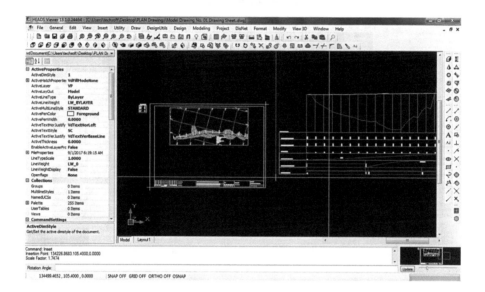

The plan and profile are inserted in the drawing sheet, as shown below:

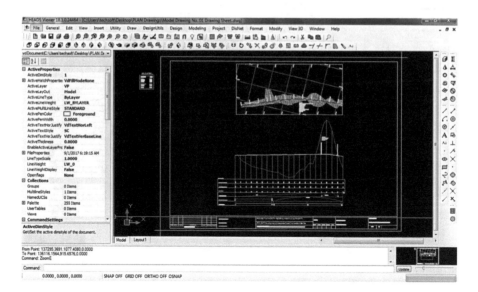

Select the Viewer menu item "File >> Save As."

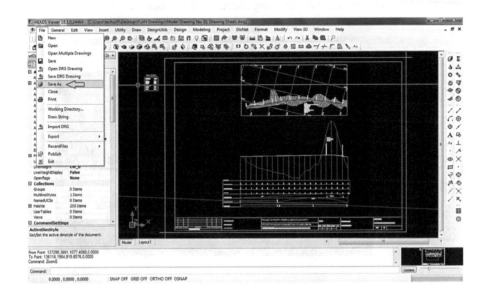

We enter the file name as "Drawing PLAN PROFILE No. 01 OF 03."

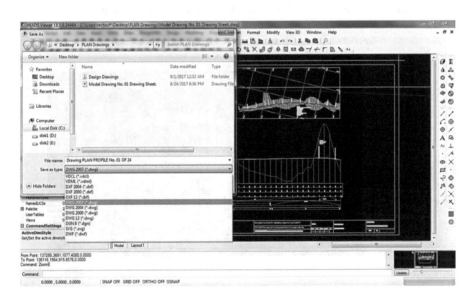

We save as type "DWG" and click on the button "Save."

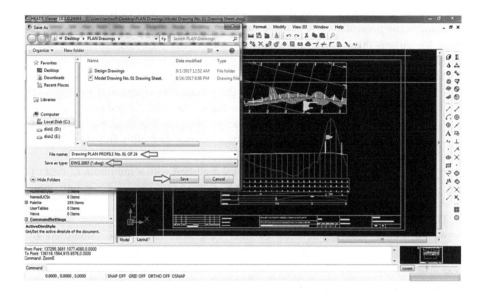

The drawing is saved as "Drawing PLAN PROFILE No. 01 OF 24" in the working folder, as shown below:

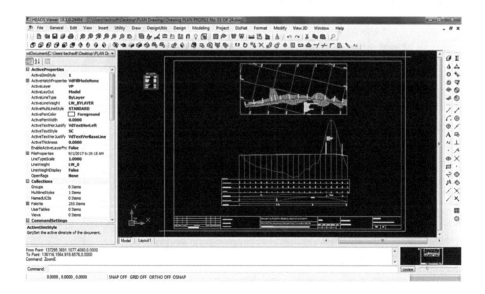

### 9.29.4   MODEL DRAWINGS

In this section some commonly followed guidelines have been provided in order to standardize the presentation of engineering design drawings for road projects undertaken by various departments in various countries. The objectives of setting these guidelines are:

- To maintain a consistent standard in the production of engineering drawings that will clearly and concisely describe the works to be carried out.
- To simplify the draughting process, the checking of drawings, and where possible, reduce the draughting requirements without detracting from the comprehensiveness of the works.
- To produce drawings that can be readily modified with a minimum of redrafting to provide complete set of "As Built" drawings as records for the project.

These guidelines have been compiled by incorporating the good features of design drawings set as design standards in many countries and those submitted by consultants.

### 9.29.5   INDIVIDUAL MODEL DRAWINGS

Create a new folder "WORK" on the Desktop and copy all the seventeen model drawing files from folder "Book Tutorials\Chapter 9 Design Drawings\All Model Drawings" to the folder "WORK" on the Desktop:

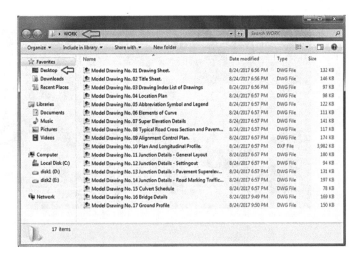

Open HEADS Pro Viewer and select menu item "File>>Open Multiple Drawing."

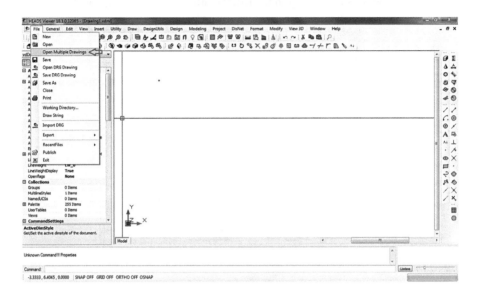

As the selection box comes up, select folder "WORK" on the Desktop.

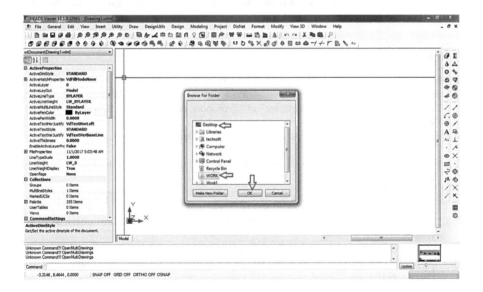

Next, all the model drawings will be opened with a list as shown below, clicking the mouse button on any drawing in the list, the drawing may be displayed.

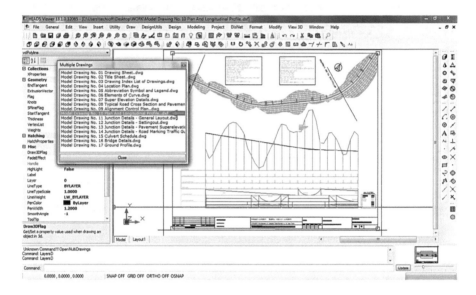

This is the end of the session of the computer-aided design for producing design drawings for the project highway.

## BIBLIOGRAPHY

*A Policy on Geometric Design of Highways and Streets 2001*, AASHTO.

*Overseas Road Note 6, A Guide to Geometric Design*, Overseas Unit, Transport and Road Research Laboratory, Department of Transport, Crowthorne, Berkshire, United Kingdom.

*Manual on Traffic Control Devices Road Marking & Delineation*, Arahan Teknik (Jalan) 2D/85, JKR, Malaysia.

*Guidelines for Presentation of Engineering Drawings*, Arahan Teknik (Jalan) 6/85, Pindaan 1/88, JKR, Malaysia.

*Design Standards, Interurban Toll Expressway System of Malaysia*, Government of Malaysia, Malaysia Highway Authority, Ministry of Works and Utilities.

*Design Manual for Roads & Bridges*, Ministry of Public Works and Kuwait Ministry, State of Kuwait.

*Geometric Design Manual for Dubai Roads*, Dubai Municipality, Roads Department.

*Type Design for Intersection on National Highways*, Ministry of Surface Transport, Govt. of India, 1995.

# 10 Process Project Data File and Text Data Files

## 10.1 GENERAL

During the detail design phase, the stages are schematic design, preliminary design, and final design. The consultants prepare drawings at each of these stages, and then submit them to the project authority, explaining the various options to be considered in the design and their respective merits and demerits. Each design has a construction cost associated with it.

The project authority verifies the design for its merits and demerits with respect to land use along the route, budgetary provisions, financial viability, and public demand.

To finalize a design, modifications are made several times so as to make the design the most suitable and justified for its purpose.

In the process of revisions of the design, modifications are made at some specific locations, while other parts remain unchanged. The computer software must be capable of enabling the design engineer to make the modifications and reproduce the design, along with the revised set of drawings and estimations in the shortest possible time.

## 10.2 PROJECT WORKSPACE

The design processes described in Chapters 4–9 are using the project workspace. The output of the processes in Chapters 5–8 creates the design reports and project data file, along with other outputs. The project data file contains various major options for input data in the proper sequence and may be processed in batch mode at any future time to reproduce the entire design project. So, after any modifications it becomes easy to reproduce.

## 10.3 PROJECT DATA FILE AND BATCH PROCESSING

During the process of design, two data sets are mainly used, the first one is the ground survey data which is supplied by the surveyor who carries out the topo survey and submits the total station survey data and the survey base plan drawings to the design engineer, and the second one is the design data file where the data related to the design of the horizontal alignment, vertical alignment, cross sections, estimation, and drawings, are stored as the backup of the original design, this is called the project data file.

The availability of the project data file makes any modification quite easy, fast, and accurate. Once the modifications are done, the project data file is processed in batch mode and it produces the design reports, drawings, and estimations of the modified design.

During the interaction with the project authority the consultant's engineer can present both the previous and present designs as softcopies of the drawings, and the difference between the two drawings enables the project authority to decide whether to accept the modified design. The batch process of the project data file makes the matter very simple and straight forward for the consultant's engineer by incorporating the modifications at some specific locations while retaining the original work at the rest of the locations.

## 10.4   TEXT DATA FILE AND MANUAL PROCESSING

The text data files are a set of files with data formats and syntax acceptable to the software being used. The data formats and syntax are different for different pieces of software, but the sequence of processing the data for various design tasks is the same in different software programs.

There are two types of software programs available on the market, the first one is a template type, where a typical cross section is defined and it forms the road by moving the cross section template along the horizontal alignment and vertical profile of the road centerline, and the second one is a string modeling type, in which the edges of the central verge or median, carriageways, and shoulders are designed as per the requirements of the configuration of the road cross section, and at every chainage the strings are united in proper sequence from the left most string to right most string in order to form the road cross sections, which depends on the design of various strings in terms of their horizontal spacing and vertical rise or fall.

The template software has limitations with respect to various demands within the design, its primary shortcoming is its inability to change the road cross sections over the length of the road in the following cases:

- The width of the carriageway, median, or verge/shoulder changes.
- The configuration of the road cross section changes from two-lane, four-lane, six-lane and eight-lane.
- The provision of a service road at selected stretches.
- The project is the widening of an existing road.
- The vertical elevations of either carriageway are different.
- There is separation of the carriageways for some lengths.
- There is a vehicular underpass, pedestrian underpass, or new bridge for one of the two carriageways.
- The design of traffic intersections.
- The design of multi-level grade separated interchanges.

In the above situations, template type software is not applicable, whereas the string modeling type is capable of all of the above requirements using its flexible technology. In the string modeling type, the user gets maximum working freedom by using the text data files written in the appropriate format and by using the syntax for various major option data for the software. Creating the desired sets of strings and along with the sequence of operations provides complete power to the user to achieve the desired results.

In the tutorials of this book the technology of the string modeling type is used, and the designs with various requirements are produced successfully, accurately, and in simplest way.

## 10.5 COMPUTER-AIDED DESIGN PROCESS

The computer applications as "Chapter 10 Process Project and Text Data Files" are available under menu "Book Tutorials" of the website at: www.roadbridgedesign. com and www.techsoftglobal.com/download.php

1. At website: www.roadbridgedesign.com under menu "Book Tutorials," the folder "Chapter 10 Process Project and Text Data Files" contains the following folders:

   **Folder**: Chapters 4–8 Uniform Section Highway Design
      Contains the Total Station Survey data file "SURVEY.txt"

   **Folder**: Chapter 9 Design Drawings
      Contains the following folders inside:
      **Folder**: All Model Drawings
      **Folder**: Cross Section Drawings
      **Folder**: Plan Drawings by Sheet Layout
      **Folder**: Profile Drawings
      **Folder**: Schematic Drawings for Alignment
      **Folder**: Schematic Drawings for Super Elevation

   **Folder**: Chapter 10 Process Project & Text Data Files
      **Folder**: Process Project Data File
      **Folder**: Process Text Data Files

   **Folder**: Chapter 11 Intersection Design
      **Folder**: Intersection Design
      **Folder**: Intersection Type Design

   **Folder**: Chapter 12 Interchange Design
      **Folder**: Loop Design
      **Folder**: Full Interchange

   **Folder**: Chapter 13 Flexible Pavement Design
      **Files**: All Pavement Design Data.txt
      **Files**: Flexible Pavement Design Data.txt

   **Folder**: Chapter 14 Rigid Pavement Design
      **Files**: All Pavement Design Data.txt
      **Files**: Rigid Pavement Design Data.txt

   **Folder**: Chapter 15 Drainage Design
      **Folder**: Highway Drainage
      **Folder**: Stream Hydrology

2. At website: www.techsoftglobal.com/download.php under item Download>> HEADS Pro, the folder "HEADS Pro Tutorials" is available for download and contains the following folders:

**Folder**: HEADS Pro Manuals
**Folder**: HEADS Pro Tutorial Videos
**Folder**: HEADS Pro Tutorials

All the processes in the book are based on the data provided in the above folders.

The computer applications are explained in "Chapter 10 Process Project and Text Data Files" in the "The Guide for Computer Application Tutorials" part of the book available for download from the website of the book at: www.roadbridgedesign.com and other tutorials are available under item Download>>HEADS Pro, at the website: www.techsoftglobal.com/download.php.

## BIBLIOGRAPHY

*HEADS Pro User's Manual and Design Manual*, Techsoft Engineering Services, website: www.techsoftglobal.com and www.roadbridgedesign.com

The various tutorial data are available in the two primary folders 'Book Tutorials' and 'Software Tutorials', which exist in the website of this book, as mentioned below:

1. Website: http://www.roadbridgedesign.com, for downloading as follows:
   There is menu item 'Book Tutorials', under this menu item, select 'Computer Aided Highway Engineering', which contains 'User's Guide for Hands On Practice', and all other items of this book.
2. Website: http://www.techsoftglobal.com, for downloading as follows:
   There is menu item 'Software Tutorials', under this menu item, select 'HEADS Pro', which contains various items used in this book, and are to be downloaded. Software 'HEADS Pro R24 Installation Setup' is to be run to install HEADS Pro in the computer.
   Chapter wise various data may be processed by following the step-by-step guide in the 'User's Guide for Hands On Practice', for respective chapters of this book.

# 11 Design of At-Grade Intersections

## 11.1 GENERAL

This section provides an overview of common types of At-grade intersections, describes elements of intersections and requirements for each type.

There are two different traffic systems followed by individual countries in the world: one of these two is "Left-Sided traffic with Right-Hand Drive," which is known as a UK style of traffic system; the other is "Right-Sided traffic with Left-Hand Drive," which is known as a USA style of traffic system. UK, India, Malaysia, Singapore follow the UK style of traffic system, whereas the USA, Europe, South America, Japan, Australia follow the USA style of the traffic system. In this chapter, the UK style of the traffic system is followed, which may be reversed in the case of the USA style of traffic system.

## 11.2 DESIGN CONSIDERATIONS

### 11.2.1 PRINCIPLES OF DESIGN

Intersections are important parts of the road system. Their capacity controls the volume of traffic within the network system. The term intersection in this chapter refers to both intersections where two or more roads cross and junctions where they meet.

Each of these can be further classified as either elemental or multiple. An elemental maneuver occurs when any two one-way, single-lane movements interact. A multiple maneuver occurs when more than two one-way single-lane movements take place and are avoided because these reduce safety as well as capacity and creates confusion for the drivers. The intersection should be designed with an attempt to replace multiple maneuvers with a series of elemental ones.

### 11.2.2 TYPES OF CONFLICTING MANEUVER

There are four basic types of driving maneuver, diverging, merging, crossing and weaving occur at the intersection. The occurrence of potential conflicts at an intersection depends on the:

- Several approaches to the intersection.
- Several lanes on each approach.
- Type of signal control.
- The extent of channelization.
- Movements permitted.

### 11.2.3 Types of At-Grade Intersection Layouts

An at-grade intersection occurs where roads meet or intersect at the same level. The following are the three basic types of layouts of at-grade intersections:

- Unchannelized and unflared.
- Flared.
- Channelized.

#### 11.2.3.1 Unchannelized and Unflared Intersections

These are normally provided where minor roads meet. In urban areas, many local street intersections remain "Unchannelized" for economic reasons. In such cases, traffic can be controlled by signals or regulatory signs, like STOP or GIVE WAY, on the minor roads. It is to be remembered that regulatory signs are not substitutes for channelization.

#### 11.2.3.2 Flared Intersections

A flared intersection is a simple form of "Unchannelized" intersection with additional through lanes or auxiliary lanes, to serve as speed-change or right-turns lanes.

Provisions of speed-change lanes allow vehicles turning left or right, to reduce or increase speed when leaving or entering the through the road without making any conflict with the speed of the through traffic. Right-turn lanes give way the through vehicles to pass on the left side of another vehicle waiting to complete its right-turn at an intersection.

#### 11.2.3.3 Channelized Intersections

In a channelized intersection the paths of travel for various movements are separated and delineated. Raised traffic islands, raised markers, and painted markings are to be provided for channelization. A roundabout is a channelized intersection where traffic moves clockwise around a central island, in the UK type of traffic system and counterclockwise in the USA style of the traffic system.

The zone of the intersection should be adequately illuminated by street lighting or defined by pavement reflectors, signing, etc.

### 11.2.4 Factors Influencing Design

At-grade intersections may cause driving with several points of conflict among the vehicles. The design of the intersection is to improve the traffic flow and reduce all possibilities of accidents. The essential factors influencing the design of an intersection are:

- Traffic volume and characteristics.
- Topography and environment.
- Economics.
- Human factors.

### 11.2.4.1 Traffic

An intersection should provide driving comfort and safety by accommodating design peak traffic volume. Consideration should also be given to operating speeds and turning path requirements at the intersection, the type of traffic control, the needs of pedestrians, buses, and commercial vehicles along with their safety aspects.

### 11.2.4.2 Topography and Environment

The design of an intersection is treated location wise by considering factors including the alignment and grade of the approach roads, adequate visibility, proper access, provision for drainage, the extent of interference with public utilities, and local features, both man-made and natural.

### 11.2.4.3 Economics

Variation of new intersections to existing intersections should be justified by commensurate benefits to traffic.

### 11.2.4.4 Human Factors

In an intersection design, normal tendencies of drivers should be considered for:

- Tend to act according to normal habits.
- Tend to follow natural paths of movement.
- Becoming confused when surprised with unexpected events.

The above factors make it essential that a driver:

- Is made aware well before the presence of an intersection.
- Is made aware of the other vehicles within and approaching the intersection.
- Is made confident to negotiate the intersection correctly and safely.
- Is made to maintain uniformity in the application of traffic engineering devices and procedures.
- Is given adequate reaction and decision time (three seconds between decisions are a desirable minimum).

### 11.2.5 SAFETY

Safety is the primary consideration in the design of an intersection, which can be ensured by the following principles:

- Minimizing the number of points of conflict.
- Minimizing the area of conflict.
- Separation of points of conflict.
- Giving privilege to the major movements.
- Control of speed for turning and thru traffic.
- Provision of refuge areas, traffic control devices, and adequate capacity.
- Definition of paths by adequate signing is to be followed.

### 11.2.6 POINTS OF CONFLICT

The minimization of the number of conflict points can be achieved by prohibiting certain traffic movements and by eliminating some roads from the intersection. In non-urban areas, the conflict points can be separated by channelization or by staggering four-way intersections.

### 11.2.7 AREAS OF CONFLICT

In the intersections where roads are at an acute angle or the opposing legs are with offsets, results excessive area for the intersection. In general, large, paved areas at the intersection invite dangerous vehicle maneuvers and are to be eliminated. The areas of conflict are to be minimized by channelization and realignment.

### 11.2.8 MAJOR MOVEMENTS

Major traffic movements are to be given the privilege to allow direct free flowing alignment. A sudden change in alignment or the entry of a high-speed vehicle to a minor road needs enough time to react by the drivers who have traveled at high-speed for long, uninterrupted distances. Adequate warning on minor approaches should be provided, as the minor movements should be subordinated to major or high-speed movements.

### 11.2.9 CONTROL OF SPEED

The operating speed of traffic through an intersection depends on the following factors:

- Alignment.
- Environment.
- Traffic volume and composition.
- Extent and type of traffic control devices.
- The number of points of conflict, the number of possible maneuvers, and the relative speed of the maneuver.

### 11.2.10 TRAFFIC CONTROL AND GEOMETRIC DESIGN

In intersection design, the use of control devices and other road furniture should be considered as per the appropriate need. For both signalized and unsignalized intersections, most of the criteria for geometric design are the same.

The design of an intersection to be controlled by signals can differ significantly from one requiring only channelization and signs. For example, in the UK type of traffic double right-turn lanes which aim at shortening storage length are effective only at signalized intersections, whereas at unsignalized intersections, the number of vehicles that can depart from the queue is dependent on the frequency of acceptable gaps in the major stream disregarding the number of storage lanes. At a signalized intersection, the left-turn lanes require additional consideration, as queuing vehicles on the left-most lane waiting for the green signal would block the entrance to the left-turning channel, which is much less significant in unsignalized intersections.

## 11.2.11  CAPACITY

Adequate traffic capacity throughout the expected life of the intersection is to be ensured in its design. The design may have to consider separate construction stages before the ultimate development of the intersection is reached.

## 11.2.12  LOCATION OF INTERSECTION

The number, type, and spacing of intersections and median openings have a significant effect on the efficiency of major roads, in terms of capacity, speed and safety.

To ensure adequate visibility and driving safety the locations of intersections should not be at sharp horizontal curves, steep grades or the top of crest vertical curves, or the bottom of sag vertical curves. To determine the intersection spacing the future coordination of traffic signals is also to be considered carefully.

## 11.2.13  SPACING OF INTERSECTIONS

The spacing of intersections is evaluated by considering factors such as weaving length and, storage length required for queuing traffic at signalized intersections, and the lengths of right-turning lanes.

Table 11.1 gives the desirable minimum spacing of each successive intersection for the various categories of the major roads.

## TABLE 11.1
### Desirable Minimum Spacing of Intersections

| Area | Category of Major Road | Spacing (m) |
|------|------------------------|-------------|
| Rural | Expressway | 3.000 |
| | Highway | $V \times 20$ |
| | Primary | $V \times 10$ |
| | Secondary | $V \times 5$ |
| | Minor | $V \times 3$ |
| Urban | Expressway | 1,500 |
| | Arterial | $V \times 3 \times N$ |
| | Collector | $V \times 2 \times N$ |
| | Local street | $V \times 1.5 \times N$ |

*Source:*  Table 1.1, A Guide to the Design of At-Grade Intersections, JKR 201101-0001-87
V = Design speed in km/hr
n = Number of through lane in one direction

## 11.2.14  CHANNELIZATION

Standardizing the design of channelized layouts is never practicable or desirable. Designing the layout of an intersection for a particular site depends on the traffic pattern, traffic volume, the area available, topography, pedestrian movement, parking arrangement, bus stoppages, the ultimate development of the neighborhood, and the layout of the existing roads.

Along with the separating conflicting movements, the channelization has the purpose to:

- Minimize the general area of conflict by causing opposing traffic streams to intersect at (or near) right angles.
- Ensure low relative speed between the conflicting streams, by merging the traffic stream at small angles.
- Control the speed of traffic entering or crossing the intersection.
- Provide a refuge for facilitating the turning or crossing of vehicles.
- Prohibit certain identified turning movements.
- Improve the layout of signalized intersections to make them more efficient.
- Provide safety for pedestrians.
- Define an improved alignment of major movements.
- Properly locate the installation of traffic signals and regulatory signs.

## 11.2.15  EXCESSIVE CHANNELIZATION

Care should be taken for the minimum number of islands to be installed as excessive channelization may lead to:

- Unwarranted obstructions on the road.
- Restrict parking and private access adjacent to the intersection.
- Cause problems of maintenance of pavement and drainage.
- Create confusion for road users.

## 11.3  DESIGN CONTROLS

### 11.3.1  PRIORITY CONTROL

Other than the signalized intersection all intersections shall be designed under the assumption that one of the intersecting roads has priority, and is the mainline, which is of the higher design standard. If the two roads are of the same standard, then the priority road shall normally be that with higher predicted traffic volume.

In case of a T-junction composed of three roads or a staggered intersection, composed of four roads (which may be considered as two T-junctions) the priority road shall be the through road. in a T-junction, if the main traffic flow is on the stem of the T, then a change of layout should be considered.

The two roads crossing each other at the intersection are normally referred to as the major road (priority road) and the minor road.

## 11.3.2 TRAFFIC

The capacities of minor intersections are normally sufficient to meet the expected traffic volumes therefore the detailed traffic forecasts and capacity calculations are normally not required. At an intersection where the major road carries a large volume of thru traffic or where both the roads carry nearly the same volume of traffic may have insufficient capacity for crossing or turning traffic flows, for which types of capacity increasing measure may have to be taken. For those intersections, detailed traffic forecasts must be carried out to provide the necessary data for capacity calculations.

A detailed traffic forecast provides hourly traffic flows in all directions in the design year. The design year shall be either 10 years after the construction of an isolated intersection or like the design year of the highway where the intersections are part of the overall road development project. A staged construction for a 5-year traffic requirement is acceptable for isolated intersections in urban areas. However, for the full design year intersection layout, the available land must be sufficient.

The Peak Hour Factor (PHF) for traffic for intersections in "Urban" areas, should also be determined. In case of non-availability of the data, a PHF value of 0.85 can be used.

## 11.3.3 DESIGN SPEED

The design speed on the major road passing through the intersection should be like that on the open section, outside the intersection. However, all at-grade intersections are not safe at design speeds exceeding 90 km/hr. Therefore, for design speeds exceeding 90 km/hr, either the speed limits at the intersection should be restricted or the at-grade intersection is to be upgraded as a grade-separated interchange.

Vehicles on the minor road can be assumed to enter the intersection at the design speed of the minor road and drivers should be able to perceive the intersection from a distance, not less than the stopping sight distance as given in Table 11.1.

## 11.3.4 DESIGN VEHICLES

The various intersection layouts should be designed for the design vehicles P, SU, or WB-50. Table 11.2 shows a general scheme to select the designed vehicle according to the category of road.

### 11.3.4.1 "P" Design

The "P" design is used at intersections where absolute minimum turns are stipulated such as at local street intersections, the intersection of two minor roads carrying low volumes of traffic, or on major roads where turns are made only occasionally.

**TABLE 11.2**

**Desirable Minimum Spacing of Intersections**

| Area | Category of Road | Design Vehicles |
|---|---|---|
| Rural | Expressway | WB-50 |
| | Highway | |
| | Primary | |
| | Secondary | SU |
| | Minor | SU/P |
| Urban | Expressway | WB-50 |
| | Arterial | |
| | Collector | SU |
| | Local street | SU/P |

*Source:* Table 2.2, A Guide to the Design of At-Grade Intersections, JKR 201101-0001-87

### 11.3.4.2 "SU" Design

The "SU" design is the recommended minimum for all roads. For major highways with important turning movements involving a large percentage of trucks, larger radii and speed-change lanes should be considered.

### 11.3.4.3 "WB-50" Design

The "WB-50" design should be used where truck combinations will make turning movements repeatedly. Where designs for such vehicles are warranted, the simpler symmetrical arrangements of three-centered compound curves are preferred if smaller vehicles make up a sizable percentage of the turning volume. It is also desirable to provide for channelization to reduce the paved area.

### 11.3.5 SELECTION OF INTERSECTION TYPE

The adequate capacity for the traffic flows expected on most intersections is achieved by controlled priority of an at-grade intersection. Various other types of intersection are introduced, where the predicted traffic flows exceed the capacity, these are:

- Roundabouts or rotary.
- Signal controlled intersections.
- Grade-separated intersections or interchanges.
  1. The roads which are connected to an intersection are designed for different design vehicles, the highest road design should primarily be chosen for that intersection. However, if the turning traffic is not high, the lower design vehicle may be used.
  2. Design vehicle is normally applicable only to intersections with two local streets or minor roads where the traffic volumes are low.

The traffic volume is the primary factor that decides the type of intersection. Table 11.3 A mentions the general scheme to select the intersection type according to the traffic volume, in addition to other considerations such as the class of road, lane

## TABLE 11.3A
## Selection of Intersection Type

| | Selection of Intersection Type | | | | | | |
|---|---|---|---|---|---|---|---|
| | Total of Two-Way Traffic on Major Road and Heavier Approach Volume on Minor Road (NTH) | | | | | | |
| Intersection Type | 1000 | 2000 | 3000 | 4000 | 5000 | 6000 | 7000 |
| Stop control | | | | | | | |
| Signalized intersection | | | | | | | |
| Interchange | | | | | | | |
| Roundabout | Mini | Small | Convention | | | | |

*Source:* Table 2.2 A, A Guide to the Design of At-Grade Intersections, JKR 201101-0001-87
*Notes:* Roundabouts are usually ranged in size as follows:
a) Mini – Less than 20 m diameter of inscribed circle, less than 4 m. in diameter QSCaster circle.
b) Small – 20–50 m, 4–25 m
c) Conventional – More than 50 m. More than 25 m

configuration, etc., especially when the traffic volume falls at the upper or lower limit of the applicable range for an intersection type.

Factors like heavy pedestrian volume and frequent accident occurrence may demand signalization. Coordinated traffic control along an arterial street may also govern the selection of the intersection type following the type of other adjacent intersections. Table 11.3 B mentions the general scheme to select the intersection type based on the category of roads meeting at the intersection.

### 11.3.5.1 Roundabouts

Roundabouts may be considered when the total traffic volume (sum of all directions) is up to 6000 vehicles/hour and if the layout is possible to be freely chosen, may be

## TABLE 11.3B
## Selection of Intersection Type

| Expressway | Arterial | Collector | Local Street | | |
|---|---|---|---|---|---|
| IC | IC | | | Expressway | |
| | IC/S.I. | S.I. | S.I./S.C. | Arterial | |
| | | S.I. | S.C. | Collector | |
| | | | S.C. | Local Street | |
| Expressway | Highway | Primary | Secondary | Local | |
| IC. | IC. | IC. | | | Expressway |
| | IC. | IC./S.I. | S.I./S.C. | S.C. | Highway |
| | | S.I. | S.I./S.C. | S.C. | Primary |
| | | | S.C. | S.C. | Secondary |
| | | | | S.C. | Local |

*Source:* Table 2.2 B, A Guide to The Design of At Grade Intersections, JKR 201101-0001-87
IC., interchange; S.I., sinalized intersection; S.C., stop Control

designed to cater for any distribution of turning traffic. The major disadvantage of roundabouts is that the speed around the roundabout is reduced because of the obstruction caused by the central island. Moreover, they require larger land space and capacity according to the demand of each approach which cannot be reliably assigned. When the capacity is exceeded, they also tend to "look up traffic." As such, roundabouts well effective only for a situation where the approaches have a similar level of traffic flow. Roundabouts are not encouraged and should only be provided where there is a problem in signalization for reasons like power supply to traffic signals, or where the number and layout of approach legs are difficult for signal control.

### 11.3.5.2  Signal Controlled Intersections

Signal controlled intersections are suitable for:

- Exceedingly high traffic volume of 8,000 vehicles/hour (vph) or more.
- The required number of approach lanes can be provided.
- There is no interference from other nearby intersections.

Paragraph 11.7 of this chapter gives the general warrants which are to be reached before traffic control signals are installed. The operating of traffic signals requires uninterrupted electricity supply, hence their use is limited only in the developed areas. Therefore, the selection of a priority-controlled intersection is the most economical solution initially, which is prepared for traffic control and the traffic signals may be added at a later stage. Signalized intersections with an adequate number of approach lanes can handle heavy traffic. Their shortcoming is the longer clearance time for vehicles to cross the wide road, leading to less efficiency in the fast handling of traffic.

### 11.3.5.3  Grade-Separated Intersections (Interchanges)

When intersections are grade-separated, they serve exceedingly high traffic volumes with very little interference to the through traffic. For all full access-controlled roads these are mandatory and are essential for roads with design speeds exceeding 90 km/hr. If each of the road crossings has four through lanes or more, grade separation is also recommended. The design of grade separation as the design of interchanges is covered in the next chapter of this book.

According to Category of Roads Crossing in Rural (Non-Urban) Area.

### 11.3.5.4  Combination and Coordination in Successive Intersections

Minor roads at proximity create frequent successive intersections on the major road. They should be treated as follows:

- Local service roads should not be linked directly to the major road but should be connected to collector roads or combined into one by clustering and then linked to the major road at a proper location.
- Local streets should not be linked to the major road near major intersections. If this is unavoidable, then for the UK style of traffic, only the left-turning movements, and the USA style of traffic only right-turning movements should

be allowed. Right-turns, for UK style of traffic and left-turns for USA style of traffic, from the major road and the crossroad should be physically prevented with continuous curbed median and remodeling the entrance to the minor road.

• While planning for a new major road over an existing road network, coordination and adjustment on the layout and spacing of intersections that would be created along the road must be done. Relocation of existing roads with systematic traffic control may be required.

## 11.4   GEOMETRIC STANDARDS

### 11.4.1   General Considerations

The geometric standards which relate to the elements of intersection design are to provide for efficient traffic operations. These standards should be relevant to new junctions and wherever possible, to junctions being improved upon. To improve the existing junctions to the recommended standards is sometimes not possible for various limitations of the site. In such a situation the provision for the best possible sight distances and proper traffic control devices should be made.

### 11.4.2   Horizontal Alignment

The desirable angle between two roads is between 70° and 90° at an intersection. Where roads intersect at angles less than 70° the alignment of the minor road should be modified.

### 11.4.3   Staggered T-Junctions

A four-way intersection has considerably higher traffic conflict points than two three-way junctions and allows higher operating speeds on the minor road. Signalized four-way intersections especially in rural areas should generally be avoided or eliminated and should be replaced by two staggered T-junctions except where large volumes of crossing traffic occur. There a four-way signalized intersection may be better than a pair of staggered T-junctions. Staggered T-junctions may either have a left–right or right–left configuration, based on the layout of the roads joining the major road.

STOP or GIVE WAY signs are necessary on the minor road of unsignalized T-junctions. The minimum desirable distances between staggered T-junctions are mentioned in Figure 11.1. Give way sign is mandatory even to left–right stagger.

### 11.4.4   Vertical Alignment

It is desirable to avoid substantial vertical grade changes at intersections. At all intersections having GIVE WAY and STOP signs or traffic signals, the gradients of the intersecting highways should be as flat as practicable so that these sections can be used as storage space for vehicles stopped at the intersection. Grades above 3% should be avoided on intersecting highways. At sites where such design is unduly

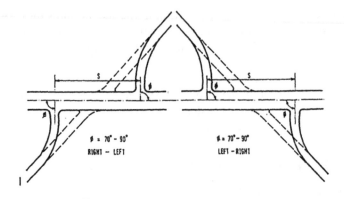

| DESIGN SPEED OF MAJOR RAOD (km/h) | SEPARATION (S) FOR RIGHT / LEFT STAGGER (m) | SEPARATION (S) FOR LEFT / RIGHT STAGGER (m) |
|---|---|---|
| 20 | 60 | 60 |
| 30 | 60 | 60 |
| 40 | 80 | 80 |
| 50 | 100 | 120 |
| 60 | 120 | 160 |
| 80 | 160 | 240 |
| 100 | 200 | 340 |

**FIGURE 11.1** Desirable Separation of Staggered T-Junctions.
*Source:* Figure 3.1, A Guide to the Design of At-Grade Intersections, JKR 201101-0001-87

expensive, there not exceeding 6% with a corresponding adjustment in design factors as detailed in Section 3.5.6, by treating as special cases.

As a general principle, the horizontal and vertical alignments of the major road, as well as its superelevation or crossfall, is maintained unchanged through the intersection, the carriageway of the minor road, and the additional lanes are so designed to fit that of the major road.

The vertical profile of the minor road shall have a gradient limited up to 2% over a section of 25 m from the nearer edge of the major road and the grade shall also in general be connected tangentially (with or without a vertical curve) to the cross-section of the major road. In case of adverse topographic conditions, the grade may be connected to the edge of the carriageway of the major road at an angle, provided that the difference in grade does not exceed 5%.

## 11.4.5  SIGHT DISTANCE

### 11.4.5.1  General

The driver of a vehicle approaching an "at-grade intersection" should have an unobstructed view of the whole intersection and adequate length of the intersecting road to allow control of the vehicle to avoid a collision. At signalized intersections, the unobstructed view may be limited to the area of control. It is advantageous on capacity reasons to increase the sight distances, where practicable along the major road, by up to 50% as this will allow many vehicles to emerge when large gaps in traffic on the main road occur.

For the adequate sight distance of the driver of a vehicle passing through an intersection, three aspects must be considered. First, there must be a sufficient unobstructed view to recognize the traffic signs or traffic signals at the intersection. Second, there must be a sufficient sight distance to make a safe departure after the vehicle has stopped at the stop line. Third, the intersections must be either stop or signal controlled.

### 11.4.5.2   Sight Triangle

To provide drivers to see the traffic appropriately, there should be a zone of sight unobstructed by buildings or other objects across the corners of an intersection. This is known as the sight triangle as shown in Figure 11.2.

Any object within the sight triangle high enough above the adjacent roadways to constitute a sight obstruction should be removed or lowered. Such objects include cut slopes, trees, bushes, and other erected objects and any parking within the sight triangle.

### 11.4.5.3   Sight Distance for Approach

*11.4.5.3.1   No-Stop or No-Signal Control at Intersection*

For this set of conditions, it is assumed that the driver of a vehicle on either road must be able to see the intersection for sufficient time to stop his vehicle if necessary,

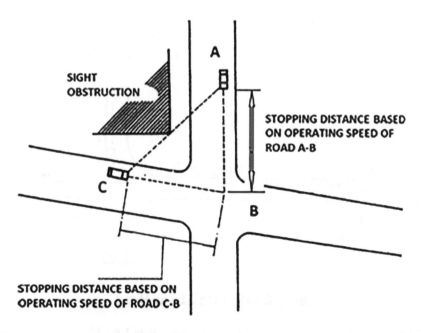

**FIGURE 11.2**   Intersection Sight Triangle.
*Source:* Figure 3.2, A Guide to the Design of At-Grade Intersections, JKR 201101-0001-87

before reaching the intersection. The safe stopping distances for intersection design are the same as those used for the design of the highway.

Where an obstruction is not possible to be removed, except at prohibitive cost, fixes the vertices of the sight triangle at points that are less than the safe stopping distances from the intersection. There the signs showing the safe speed should be so located that the driver can slow down to a speed appropriate to the available sight distance. Referring to Figure 11.3, for a typical case, speed "$V_b$" is known and "$a$" and "$b$" are the known distances to the sight obstruction from the respective paths of vehicles A and B. The critical speed $V_1$ of Vehicle B can then be evaluated in terms of these known factors. Distance "$da$" is the minimum stopping distance for vehicle A. When vehicle A is at a distance "$da$" from the intersection and the drivers of

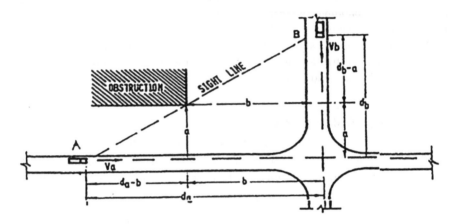

**No Stop Or Signal Control At Intersection**

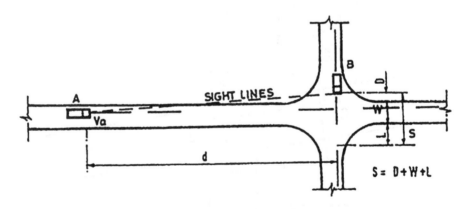

**Stop Control On Minor Road**

**FIGURE 11.3** Sight Distance at Intersection (Minimum Sight Triangle).
*Source:* Figure 3.3, A Guide to the Design of At-Grade Intersections, JKR 201101-0001-87

vehicles A and B first sight each other, vehicle B is at a distance "$d_b$" from the intersection.

By similar triangle,

$$d_b = (a \times d_a)/(d_a - b)$$

and the critical speed "$V_b$" is that for which the stopping distance is "$d_b$." The signs on road B showing the safe speed to approach the intersection should be so located that a driver can reduce his speed to "$V_b$" by the time he reaches the point that is distance "$d_b$" from the intersection. Similar calculations may be used to determine how far back an obstruction needs to be moved to provide sufficient sight distance for safe driving at desired vehicle speed on the respective roads.

For this case, if the major road is with one-way traffic a single sight triangle in the direction of approaching traffic will be adequate. Similarly, if the major road has dual carriageways with no gap in the central reserve then a single sight triangle to the right will be needed. If the minor road serves as a one-way exit from the major road, no sight triangle will be required provided forward visibility for turning vehicles is adequate.

### 11.4.5.3.2  Signalized Intersection

The sight distance is the distance traveled during the total reaction time which is first the interval between the instant that the driver recognizes the traffic signals of the intersection ahead and applies the brakes, and second the distance to stop the vehicle at the stop line with applying the brake. The first reaction time can further be divided into the time required to decide whether the brake should be applied or not, and the time for a reaction after getting the decision. As no data is available on the total reaction time, so, 10 seconds is usually considered. For urban areas, a shorter total reaction time is considered. This is because of frequent occurrences of intersections in urban areas, the drivers are always with anticipation of possible encounters of intersections. The reaction time of 6 seconds for urban areas is usually considered. An allowable maximum acceleration of 0.2 g is taken without excessive discomfort. This is much lower than those used to determine the stopping sight distance, because stops at intersections are quite routine, while stops to avoid possible collision on through road are much less frequent, and more acute deceleration may be acceptable.

Referring to the discussion above, the sight distance for a signalized intersection is given as follows:

$$S = \frac{V \times t}{3.6} + \left\{ \frac{1}{2\infty} \times \left( \frac{V}{3.6} \right)^2 \right\}$$

where
  t = 10 sec (rural)
  t = 6 sec (urban)
  $\infty$ = 0.2 × g = 0.2 × 9.8 = 1.96 m/sec²

### 11.4.5.3.3  Stop-controlled Intersection

Here, time for decision making (as in signalized intersection) is not necessary because every driver must stop, and the reaction time of 2 seconds is taken. Therefore, t = 2 seconds, $\alpha$ = 1.9 m/sec$^2$ are submitted into the above formula.

On the major road, drivers can drive the vehicles without worrying about intersections. Stopping sight distance defined for through road is sufficient. Referring to the discussion above, the criteria shown in Table 11.1 are obtained.

## 11.4.5.4  Sight Distance for Departure

At an intersection where traffic on the minor road is controlled by STOP signs, the driver of a stopped vehicle must see enough distance of the major road to be able to cross before a vehicle on the major road reaches the intersection, Refer to Figure 11.3. The sight distance required along the major highway can be expressed as:

$$d = 0.28v\left(J + t_a\right)$$

where

 d = minimum sight distance along the major road from the intersection, in meters.

 V = design speed of major road, km/hr.

 J = sum of perception time and the time required to shift to first gear or actuate an automatic shift, seconds.

 $t_a$ = time required to accelerate and traverse the distance S to clear the major road, seconds.

The term J represents the time necessary for the driver to look in both directions and to shift gears, if necessary, preparatory to start the vehicle and a value of 2 seconds is assumed. In urban or suburban areas where drivers generally pass through many intersections with STOP sign control, a lower value of 1 ½ or even 1 second may apply. The time "t" which is required to cover a given distance during acceleration depends upon the vehicle acceleration.

The acceleration of buses and trucks is much lower than that of passenger vehicles. On flat grades, the acceleration time for SU (single-unit) and semi-trailer is

---

## TABLE 11.4
### Sight Distance for Intersection Approach

| Design Speed of Major Road (km/hr) | Signal Control | | Stop Control (on Minor Road)* |
|---|---|---|---|
| | Rural | Urban | |
| 100 | 480 | 370 | 260 |
| 80 | 350 | 260 | 170 |
| 60 | 240 | 170 | 105 |
| 50 | 190 | 130 | 80 |
| 40 | 140 | 100 | 55 |
| 30 | 100 | 70 | 35 |
| 20 | 60 | 40 | 20 |

*Source:* Table 3.1, A Guide to The Design of At Grade Intersections, JKR 201101-0001-87)

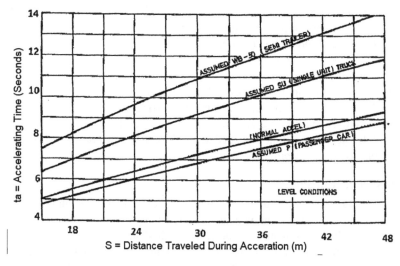

**FIGURE 11.4**   Sight Distance at Intersections (Data on Acceleration from Stop).
*Source:* Figure 3.4, A Guide to the Design of At-Grade Intersections, JKR 201101-0001-87

about 135% and 160% respectively of that for passenger vehicles, the value of "t" can be read directly from Figure 11.4 for nearby level conditions for a given distance "S" in feet.

On the major roads, at the Stop Controlled intersections, the Stopping Sight Distances as specified in the relevant standard for the geometric design of through roads must be satisfied.

Referring to Figure 11.3 the distance S which the crossing vehicle must travel to cross the major road is given by

$$S = D + W + L$$

where

   D = Distance from the near edge of pavement of front of a stopped vehicle
   W = Width of pavement along the path of a crossing vehicle
   L = Overall length of the vehicle
For general design purposes a value of "D = 3 m" is considered. The value of "L," which is the overall length of the design vehicle, can be taken as 5 m, 10 m, and 15 m for passenger cars, single-unit trucks, and semi-trailers, respectively.

To verify whether the sight distance along a major road is adequate at an intersection the distance should be measured from an eye level of 1.15 m to the top of an object of height 1.4 m placed on the pavement.

For divided roads, widths of median equal to or greater than the length of vehicle enable the crossing to be made in two steps. For divided highways with medians less than "L" the median width should be included as part of "W."

Along a major road, the longer distance of (i) the sight distance described here and (ii) the stopping sight distance must be adopted. The former will exceed the latter at higher ranges of design speeds. In case the sight distance along a major road is less than that for departure at an intersection, it is unsafe for vehicles on the major high-way to proceed at the assumed design speed of the highway, and signs indicating the safe approach speed should be provided.

The safe speed may be computed for a known sight distance and the width of pavement on the path of the crossing vehicle. On turning roadways and ramps, at least the minimum stopping sight distance should be provided continuously along such roadways. Where the major road has dual carriageways with a central median width enough to shelter turning vehicles (4.5 m or more) the normal sight triangle to the left of the side road will not be needed but the central median should be clear of obstructions to driver visibility for at least "d" m.

### 11.4.5.5   Effect of Skew

When two roads intersect at an angle much less than a right-angle and increase the angle of intersection by realignment is not justified, some of the factors may need adjustment for the determination of the corner sight distance.

If there is a difficulty for the drivers of approaching traffic to have the view of the intersecting roads makes it undesirable to treat the intersection based on the assump-tions of "No control intersection" even where traffic on both roads is in sight.

At skew intersections treatment by controlled intersection or safe departure whichever is the larger should be used. The distance "S" is larger for oblique than for right-angle intersections, in case of departure. The width of pavement along the path of the crossing vehicle "W," is obtained from Figure 11.5 as the pavement width divided by the sine of the intersection angle.

The distance along the road can be obtained by the formula:

$$d = 0.28 \times V \times (2 + t_a)$$

reading "$t_a$" directly from Figure 11.4.

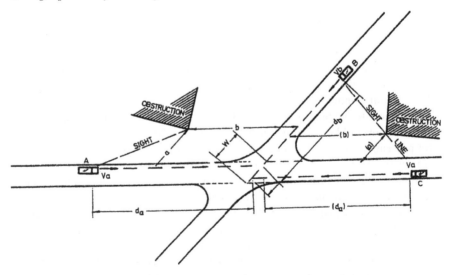

**FIGURE 11.5**   Sight Distance at Intersections (Effect of Skew).
*Source:* Figure 3.5, A Guide to the Design of At-Grade Intersections, JKR 201101-0001-87

## TABLE 11.5
## Effect of Grade on Stopping Sight Distance Wet Conditions

| Design Speed (km/hr) | Correction in Stopping Distance (m) | | | | | |
|---|---|---|---|---|---|---|
| | Decrease for Upgrades | | | Increase for Downgrades | | |
| | 3% | 6% | 9% | 3% | 6% | 9% |
| 30 | – | – | 3 | – | – | 3 |
| 40 | – | 3 | 3 | – | 3 | 6 |
| 50 | – | 3 | 6 | 3 | 6 | 9 |
| 60 | 3 | 6 | 9 | 3 | 9 | 15 |
| 80 | 6 | 9 | – | 6 | 15 | – |
| 100 | 9 | 15 | – | 9 | 24 | – |

*Source:* Figure 3.6, A Guide to The Design of At Grade Intersections, JKR 201101-0001-87

## TABLE 11.6
## Correction Factor for the Effect of Grade on Acceleration Time (Ta)

| Design Vehicle | Minor Road Grade (%) | | | | |
|---|---|---|---|---|---|
| | −4 | −2 | 0 | +2 | +4 |
| Passenger cars (P) | 0.7 | 0.9 | 1.0 | 1.1 | 1.3 |
| Single unit trucks (SU) | 0.8 | 0.9 | 1.0 | 1.1 | 1.3 |
| Semi trailers (WB-50) | 0.8 | 0.9 | 1.0 | 1.2 | 1.7 |

*Source:* Figure 3.7, A Guide to The Design of At Grade Intersections, JKR 201101-0001-87

### 11.4.5.6 Effect of Grades
The differences in stoppinearby level conditions for a giventions are listed in Table 11.5. Grades on an intersection leg should be limited to 3%. In case of departure, the time required to cross on the major road highway is derived by considering the effect of the grade of crossing on the minor road. Normally the grade across an intersection is quite small and need not be considered but when curvature on the major road requires the use of superelevation, the grade across it may be significant. The effect of grade on acceleration can be expressed as a multiplier and to be used with the time "$t_a$" as determined for level conditions for a given distance as described in Table 11.5.

The value of "$t_a$" from Figure 11.4 adjusted by the appropriate factors can be used in the formula

$$d = 0.28 \times V \times (2 + t_a)$$

### 11.4.6 Right-Turn Lanes (Left-Turn Lanes for USA Style of Traffic)

#### 11.4.6.1 General

The provision of right-turn lanes improves the capacity and safety in an intersection, these should be considered in the following cases:

- When the traffic in major road exceeds 600 vehicles/hr.
- At all intersections on divided urban roads with a sufficiently wide median.
- At all intersections on undivided urban roads where right-turning traffic is likely to cause undesirable congestion and hazard.
- At all rural intersections ensuring safety.

#### 11.4.6.2 Design Considerations

The turning path of a semi-trailer should be used for the design of right-turns in the UK system of traffic. The vehicle maneuvering the right-turn must not encroach on the shoulder with its front wheels or opposite side of the road center lines with its rear wheels. The STOP lines, median noses, and "Seagull" island must be located to suit vehicle turning paths. Figure 11.6 illustrates the essential features of design on right-turn lanes.

**A) Layout**

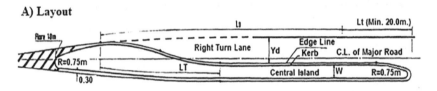

**B) Length Of Taper**

$$LT = \frac{1}{3}\sqrt[V]{Yd}$$

Where
V= Design speed in Km/h

Yd = Width of right turn lane (m)

LT = May be rounded to the nearest multiple of 5m

For design of S-curve
See Figure: 11.3.21 (A) & (B)

**C) Width of Central Island**

| | Yd (m) | Pedestrian Refuge | Signal pedestrian No Pedestrians | No Signal or Pedestrians |
|---|---|---|---|---|
| Desirable | 3-5 | 2.50 | 2.50 | 2.50 |
| Minimum | 3.0 | 2.50 | 1.80 | 1.20 |

**D) Deceleration Length, Lo (M)**

| Gradient in % | | Design Speed in Km/h | | | | | |
|---|---|---|---|---|---|---|---|
| | | 20 | 30 | 40 | 50 | 60 | 100 |
| Uphill | 4 | 20 | 20 | 41 | 54 | 72 | 153 |
| | 2 | 20 | 30 | 45 | 60 | 80 | 170 |
| Level | 0 | 20 | 30 | 45 | 60 | 80 | 170 |
| | 2 | 20 | 30 | 45 | 60 | 80 | 170 |
| Downhill | 4 | 20 | 34 | 53 | 72 | 96 | 204 |

NOTES:
1. The length of the reservoir space shall be rounded upwards to the nearest multiple at 5m.
2. Deceleration lengths for other gradients maybe found by interpolation or up to 5% by extrapolation.
3. All dimensions are in m.

**FIGURE 11.6** Right – Turn Lanes (Mirror Image is Applicable for USA Style of Traffic).
*Source:* Figure 3.8, A Guide to the Design of At-Grade Intersections, JKR 201101-0001-87

## 11.4.6.3  Length of Right-Turn Lanes

The minimum length of a right-turn lane is equal to the length of deceleration for the approach speed. Where storage of waiting vehicles in the queue is required, the length should be increased according to the expected queue length. Storage length can be estimated as follows:

- Signalized intersection.
  Storage length is calculated as

$$L = 1.5 \times N \times S$$

where

N is the average number, if right-turning vehicles are in a cycle of a signal phase (veh.).

S is the average headway in distance (m)

S = 6 m for a passenger car

S = 12 m for other large commercial vehicles.

If the ratio of commercial vehicles to the total vehicles in the queue is not known, S = 7 m may be used.

- Unsignalized intersection
  The effect of traffic fluctuation for the storage length is more significant in unsignalized intersections. The following formula can be applied:

$$L = 2 \times M \times S$$

where

M is the average number of right-turning vehicles in a minute.

At both signalized and unsignalized intersections, the minimum storage length of "20 m" should be provided if the right-turning volume for the above calculation is not possible. A right-turn lane shorter than required would cause the turning vehicles to come upon the parallel through the lane and obstruct the through traffic. In urban areas, various constraints sometimes force the reduction in the length of right-turn lanes. As traffic may not be always at its highest volume, even the shorter lane is effective to some extent. The possible length of the right-turn lane as the constraints allowed should be provided. In this case, the shortage in the length should be adjusted in the taper length to provide the storage length if possible. However, lengths less than half the recommended, should not be used.

The taper is normally formed by an S-curve composed of two circular arcs. Where the right-turn lane is obscured by a crest, then a longer length of the lane is essential to give the driver adequate time to perceive the lane in time to start his deceleration. For new intersections, the estimation of right-turning traffic must be done by gathering information on land development projects and the location of traffic generating facilities along the roads crossing. In most cases, the accuracy of the estimation cannot be satisfactory. New intersections, therefore, should be examined after opening and the design should be improved by observing the actual operating conditions, because the storage length is most difficult to predict, at the time of original construction, it should be prepared for refinement in the future.

### 11.4.6.4  Width of Right-Turn Lanes
Right-turn lanes shall desirably be 3.50 m wide and never be less than 3.0 m wide.

### 11.4.6.5  Seagull Island
A seagull island is triangular in shape and is used to separate right-turning traffic from through traffic in the same carriageway as shown in Figure 11.7.

Adequate length of storage lane is required in approach to the island and a merging taper appropriate to the speed of the through lane must be provided on the departure side.

### 11.4.6.6  Opposed Right-Turns
When two single-lane opposing right-turns are expected to run simultaneously the turning radii and the tangent points should be such that there is a clear width following the table in Figure 11.8.

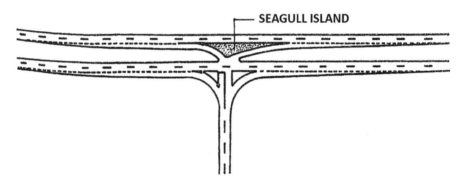

**FIGURE 11.7**  Seagull Island.
*Source:* Figure 3.9, A Guide to the Design of At-Grade Intersections, JKR 201101-0001-87

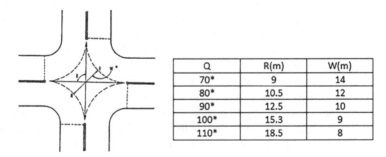

| Q | R(m) | W(m) |
|------|------|------|
| 70* | 9 | 14 |
| 80* | 10.5 | 12 |
| 90* | 12.5 | 10 |
| 100* | 15.3 | 9 |
| 110* | 18.5 | 8 |

**\*Minimum for opposed turns by SU trucks**

**FIGURE 11.8**  Right-Turn Clearance.
*Source:* Figure 3.10, A Guide to the Design of At-Grade Intersections, JKR 201101-0001-87

### 11.4.6.7   Central Island and Median Design

The minimum width of the central island shall follow as listed in Figure 11.6 (C).
Central islands may be made in one of the following ways:

- It is painted as cross hatched areas on the pavement (virtual or ghost island).
- A raised island surrounded by curbs.

Ghost islands should be used where the island is of the width equals to or less than the turn lane. It should also be used in rural intersections in absence of street lighting.
Curbed islands should be provided where the islands are wide. Medians should also be provided as curbed on both sides from the start of the taper of the right-turning lane. In case no turning is present, then medians should be curbed from the start of the larger of the two rounding curves at the central area of the intersection.

### 11.4.7   LEFT-TURN LANES (RIGHT-TURN LANES FOR USA STYLE TRAFFIC)

### 11.4.7.1   General

The type of left-turn lane and its treatment depends on:

- Type and volume of traffic making the turn.
- Restrictions are caused by the surrounding development.
- Operating Speed of the left-turn.

These factors determine the radius of the curb and the width of the left-turn lane for the UK system of traffic. There are two types of treatment for left-turns, these are simple left-turns and separated left-turn lanes.

### 11.4.7.2   Simple Left-Turns

These are usually provided at intersections where:

- The traffic volumes are low.
- Land acquisition costs prevent more extensive treatment.
- The angle of turn prohibits the installation of an island.

At urban intersections, the radius of the curve for the left-turn should be a minimum of 6 m., this allows most commercial vehicles to negotiate the turn at low speeds to avoid encroaching either on the footway with the rear wheels or on the opposite side of the road's centerline with the front wheels.
A radius larger than 10 m increases the speed of turning movements and reduces the safety of pedestrian crossings. It also creates problems in locating signal pedestals and STOP lines. For simple left-turns in urban areas, such radius should only be used after careful consideration of the above. At rural intersections where provision for a pedestrian is not a consideration, larger radius curves may be used. Radii larger than 15 m should not be used without a left-turn island as they form large areas of uncontrolled pavement.

### 11.4.7.3   Separate Left-Turn Lanes

A corner island can be introduced to create a separate left-turn lane, where the volume of left-turning traffic is high, or the skew demands such a layout.

#### 11.4.7.3.1   Design Speed of Left-Turn Lane

For the left-turn lane, the design speed higher than that shown in Table 11.7 should be chosen, by considering

- The turning volume
- Availability of land
- The design speed of the approach road

#### 11.4.7.3.2   The Radius for Separate Left-Turn Lanes

Where environmental and other constraints do not have any direct influence, the radius ($R_1$) of a separate left-turn lane depends on:

- The speed, V, at which vehicles operate
- The superelevation
- The acceptable coefficient of friction "f" between vehicle tires and the pavement

Table 11.8 gives the relationship between these factors.

The values of $R_1$ in the table are calculated from the formula:

$$R_1 = V^2 / \left[ 127 \left( e + f \right) \right]$$

The superelevation at curves in separate turning lanes at intersections, usually has a low value, mainly because of the difficulty of developing the superelevation on a relatively short length of a separate turning lane. A desirable maximum value in rural areas is 0.08. in urban areas, the range should be from 0.04 to 0.06. The values of "f" given in Table 11.8 are greater than those used for open highway design as drivers turning on curves of a small radius at intersections accept a lower level of

---

### TABLE 11.7
### Minimum Design Speeds For Left-Turn Channel Minimum Design Speeds For Left-Turn Channel

| Design Speed of Approach Road (km/hr) | Minimum Design Speed of Left-Turn Lanes (km/hr) |
|---|---|
| 100 | 50 |
| 80 | 40 |
| 60 | 30 |
| 50 | 30 |
| 40 | 20 |
| 30 | 20 |
| 20 | 20 |

*Source:*   Table 3.2, A Guide to The Design of At Grade Intersections, JKR 201101-0001-87

**TABLE 11.8**
**Turning Radii**

| V (km/hr) | f | 0 | 0.02 | 0.04 | 0.05 | 0.08 |
|---|---|---|---|---|---|---|
| | | | | e (m/m) | | |
| | | | | R1 (m) | | |
| 20 | 0.34 | 10 | 9 | 9 | 8 | 8 |
| 30 | 0.28 | 25 | 23 | 22 | 20 | 19 |
| 40 | 3.23 | 55 | 50 | 46 | 43 | 40 |
| 50 | 0.19 | 104 | 93 | 85 | 78 | 72 |
| 60 | 0.17 | 167 | 143 | 135 | 123 | 112 |
| 80 | 0.15 | 315 | 280 | 252 | 229 | 210 |

*Source:* Figure 3.12, A Guide to The Design of At Grade Intersections, JKR 201101-0001-87

comfort. For "$R_1$" within the range of 12–30 m, the turn should be designed to provide for keeping the design vehicle on track. A compound curve with successive radii of $1.5R_1$, $R_1$, and $3R_1$ satisfies such a requirement. For radii, R1 between 30 and 45 m keeping the vehicle on track can be accommodated by using a compound curve with successive radii of $2R_1$, $R_1$, and $2R_1$. Figure 11.10 describes the combination of radii and widths required for keeping the design vehicle on track. For $R_1$ more than 45 m the off-tracking is negligible therefore, a single radius R1 is acceptable.

In the design of the intersection, the method of attainment of superelevation runoff for the open road should be followed.

In the case of a painted island or an island that is either not required or cannot be provided the compound curves are not necessary. For occasional use by semi-trailers, the front wheels can be steered wide enough to prevent the back wheels from running over the curb or running into the shoulder.

A corner island is introduced to create a separate left-turn lane and a three-centered curve is justified, here the combination of radius and angle of turn should provide minimum island area as follows:

- In urban areas, the 8m² area of the island is adequate to provide pedestrians shelter as well as installation of traffic signals.
- In rural areas, a 50 m² area of the island may be considered. Figure 11.9 indicates the combination of radius and angle of turn which provides these minimum island areas.

### 11.4.7.3.3 Width of Left-Turn Lanes

The width of a left-turn lane depends on:

- The radius of the turn.
- Volume and type of turning traffic.
- Whether curbside parking is permitted or prohibited.
- The length of the lane.
- Whether both the edges are curbed.

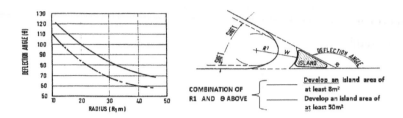

**FIGURE 11.9**   Island Areas.
*Source:* Figure 3.11, A Guide to the Design of At-Grade Intersections, JKR 201101-0001-87

There are three design conditions:

- Single-lane flow for width W1 is the normal application and is used in rural or semi-urban locations where there is a shoulder on the inner edge of the pavement. It may also be applied in urban areas where the inner edge of the lane is curbed but the corner is small.

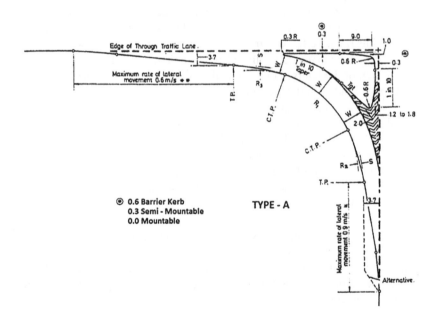

**FIGURE 11.10**   Type – A: Desirable Treatment.
*Source:* Figure 3.13-Type A, A Guide to the Design of At-Grade Intersections, JKR 201101-0001-87. Type – B: Design of Separate Left-Turn Lanes Alternative Treatment (R1 > 45 m), All Dimensions Are in Meters. *Source:* Figure 3.13-Type B, A Guide to the Design of At-Grade Intersections, JKR 201101-0001-87. Type – C: Minimum Treatment. *Source:* Figure 3.13-Type C, A Guide to the Design of At-Grade Intersections, JKR 201101-0001-87

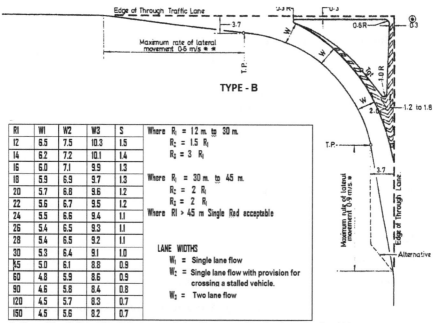

| RI | WI | W2 | W3 | S |
|----|-----|-----|------|-----|
| 12 | 6.5 | 7.5 | 10.3 | 1.5 |
| 14 | 6.2 | 7.2 | 10.1 | 1.4 |
| 16 | 6.0 | 7.1 | 9.9 | 1.3 |
| 18 | 5.9 | 6.9 | 9.7 | 1.3 |
| 20 | 5.7 | 6.8 | 9.6 | 1.2 |
| 22 | 5.6 | 6.7 | 9.5 | 1.2 |
| 24 | 5.5 | 6.6 | 9.4 | 1.1 |
| 26 | 5.4 | 6.5 | 9.3 | 1.1 |
| 28 | 5.4 | 6.5 | 9.2 | 1.1 |
| 30 | 5.3 | 6.4 | 9.1 | 1.0 |
| 45 | 5.0 | 6.1 | 8.8 | 0.9 |
| 60 | 4.8 | 5.9 | 8.6 | 0.9 |
| 90 | 4.6 | 5.8 | 8.4 | 0.8 |
| 120 | 4.5 | 5.7 | 8.3 | 0.7 |
| 150 | 4.5 | 5.6 | 8.2 | 0.7 |

Where $R_1$ = 12 m. to 30 m.
$R_2$ = 1.5 $R_1$
$R_3$ = 3 $R_1$

Where $R_1$ = 30 m. to 45 m.
$R_2$ = 2 $R_1$
$R_3$ = 2 $R_1$

Where RI > 45 m Single Rad acceptable

LANE WIDTHS
$W_1$ = Single lane flow
$W_2$ = Single lane flow with provision for crossing a stalled vehicle.
$W_3$ = Two lane flow

Traffic condition: Sufficient SU vehicles to govern design with some considerations for semi-trailer vehicles

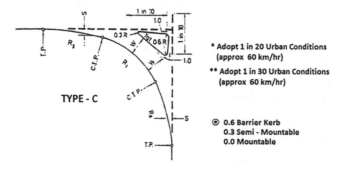

* Adopt 1 in 20 Urban Conditions (approx 60 km/hr)
** Adopt 1 in 30 Urban Conditions (approx 60 km/hr)

⊕ 0.6 Barrier Kerb
0.3 Semi - Mountable
0.0 Mountable

**FIGURE 11.10**   (Continued)

- Single-lane flow with provision for passing a stalled vehicle for width W2 is desirable for urban locations where parking is prohibited, and the corner island has an inner edge longer than 20 m.
- Two-lane flow for Width W3 is to be adopted where traffic volumes require two lanes and parking is prohibited. Width W3 is maintained for the whole length of the left-turn lane.

Design conditions that define the lane width of the left-turn lane are described in Table 11.9 according to the class of road.

**TABLE 11.9**
**Lane Widths For Left – Turn Lane**

| Area | Category of Road | Lane Width |
|------|------------------|------------|
| Rural | Highway | W3/W2 |
| | Primary | W2 |
| | Secondary | W1 |
| | Minor | W1 |
| Urban | Arterial | W3/W2 |
| | Collector | W2/W1 |
| | Local street | W1 |

*Source:*   Table 3.3, A Guide to The Design of At Grade Intersections, JKR 201101-0001-87

### 11.4.8   PAVEMENT TAPERS

#### 11.4.8.1   General

Pavement tapers are provided at the following places:

- At the ends of acceleration and deceleration lanes provided for left and right-turn maneuvers.
- At the ends of the widened carriageway or dual carriageway to assist the merging and diverging of through traffic maneuver.

The widths shown in the table above are determined for the design vehicle "SU" including some consideration for WB-50. A separate study is required if design vehicle "P" is applied for design. If two alternatives are given, the selection is to be done according to the turning volume of traffic.

#### 11.4.8.2   Design Principles

The following are the general design principles on pavement tapers:

- For diverging movements, pavement tapers should provide for a rate of lateral movement of 0.9 m per second.
- For merging movements pavement tapers should provide for a rate of lateral movement of 0.6 m per second. Longer lengths may be provided where traffic volumes are high.
- Care must be exercised for error in designing diverging tapers to ensure that through traffic is not led into an auxiliary lane.
- Care must be exercised for the location of all merging tapers to ensure that there is sufficient sight distance and geometry of the merge.
- Sufficient lengths of straight, horizontal, and vertical alignment should precede diverging tapers to allow three seconds of travel by vehicles at the prevailing speed.
- Diverging and merging tapers should be designed to encourage low relative speed maneuver by the drivers.

### 11.4.8.3 Taper Length

The minimum lengths of pavement taper for diverging and merging movements can be computed by the formula:

$$T_d = \frac{V}{3.6} \times \frac{Y_d}{0.9}$$

$$T_m = \frac{V}{3.6} \times \frac{Y_d}{0.9}$$

where

$T_d$ = Min. length of pavement taper for diverging movements (m)

$T_m$ = Min. length of pavement taper for merging movements (m)

$Y_d$ = Lateral deflection of diverging traffic (m)

$Y_m$ = Lateral deflection of merging traffic (m)

Various types of tapers which may be used are described in Figure 11.14.

### 11.4.8.4 Auxiliary Lanes

*11.4.8.4.1 Deceleration Lanes*

For U.K. type of traffic, the deceleration movements for left-turn should be separated from the through traffic stream. This may be done by providing a diverge taper "Td" followed by a length of a parallel lane in the left-turn approach. The combined length should be equal to the distance "ht," required to decelerate from the approach speed in the through road to the design speed of the left-turn. The length of deceleration lanes are mentioned in Table 11.10.

The ratio from Table 11.11 multiplied by the length from Table 11.10 gives the length of the deceleration lane on a grade.

In urban areas, it is desired that traffic using the left-turn should flow uninterrupted. In case of a queue is formed at the STOP line, a parallel lane long enough for the left-turn vehicles to by-pass the end of the queue should be provided. Figure 11.12 illustrates the provision.

*11.4.8.4.2 Acceleration Lanes*

In urban areas where the through and left-turn movements are simultaneous, there should be a zone that enables the two streams of traffic to merge at a small angle. For a low volume of merging traffic or where traffic signals are installed, this zone may be provided by a merging taper of length "Tm" at the exit of the left-turn.

Where the volume of merging traffic is high and signals are not provided, a driver reaching the exit to the left-turn lane may not find any gap in the through traffic stream. He should therefore be allowed to continue in the direction parallel to the through traffic until a merging opportunity occurs or until he attains the speed of the through traffic to create an opportunity to merge. In such cases, a

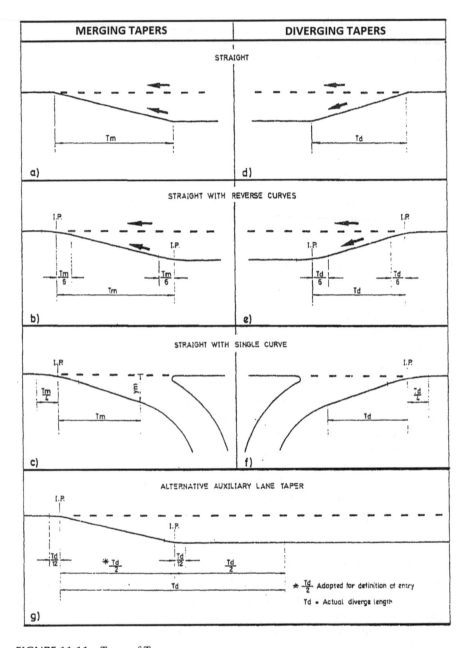

**FIGURE 11.11**   Types of Taper.
*Source:* Figure 3.14, A Guide to the Design of At-Grade Intersections, JKR 201101-0001-87

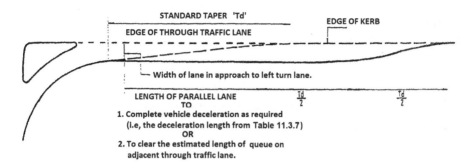

**FIGURE 11.12**   Treatment in Approach to Left-Turns.
*Source:* Figure 3.17, A Guide to the Design of At-Grade Intersections, JKR 201101-0001-87

parallel acceleration lane together with the merging taper of length "Tm" should be considered. The combined length should be equal to the distance required for a vehicle to accelerate from the design speed of the left-turn to the design speed of the through road.

Lengths of acceleration lane are given in Table 11.12. If necessary, a correction for a grade as shown in Table 11.13 should be applied.

### 11.4.8.5   Width of Auxiliary Lanes

Widths of auxiliary lanes shall desirably be 3.5 m but shall not be less than 3.0 m.

## 11.4.9   Island and Openings

### 11.4.9.1   General

There are two types of islands – pedestrian and traffic.

The ratio from Table 11.11, multiplied by length in Table 11.10 gives the length of the deceleration lane on grade. Figure 11.12 illustrates the configuration of a left-turn lane with a deceleration lane with the merging taper and Figure 11.13 illustrates the configuration of an acceleration lane with the merging taper.

The provision of the pedestrian island is for refuge for people waiting for public transport or for crossing wide streets. Traffic islands are divisional or form channelization.

Visibility to approaching traffic, both at day and night, is an essential factor in any island location. Only traffic islands will be the concern here.

### 11.4.9.2   Traffic Islands

The purposes of traffic islands are to:

- Separate opposing streams of traffic.
- Guide traffic away from and past fixed obstructions and other hazardous points.
- Reduce the area of conflicts and control the angles which create the conflicts.
- Provide shelter for turning or crossing vehicles.

## TABLE 11.10
### Length of Deceleration Lanes

| Design speed of Approach Road (km/hr) | Length* of Acceleration Lane (m) | | | | | | |
|---|---|---|---|---|---|---|---|
| | (Including Length of Pavement Taper) | | | | | | |
| | Where Design Speed of Exit Curve (km/hr) is: | | | | | | |
| | 0** | 20 | 30 | 40 | 50 | 60 | 80 |
| 40 | 45 | 40 | 32 | – | – | – | – |
| 50 | 60 | 54 | 45 | 32 | – | – | – |
| 60 | 80 | 74 | 64 | 50 | 28 | – | – |
| 80 | 120 | 112 | 104 | 94 | 82 | 64 | – |
| 100 | 170 | 162 | 154 | 144 | 132 | 118 | 80 |

*Source:* Figure 3.15, A Guide to the Design of At-Grade Intersections, JKR 201101-0001-87

\* Length of level grade (see 11.13 below, for grade correction)

\*\* Length required when a vehicle decelerates to zero speed

*Note:* Where the length of the deceleration lane shown is less than the standard taper "Td," then it (Td) should not be reduced.

## TABLE 11.11
### Correction for Grade

| Grade | Ratio of Length on Grade to | |
|---|---|---|
| | Length on Level | |
| | Upgrade | Downgrade |
| 0–2% | 1.0 | 1.0 |
| 3–4% | 0.9 | 1.2 |
| 5–6% | 0.8 | 1.35 |

*Source:* Figure 3.16, A Guide to the Design of At-Grade Intersections, JKR 201101-0001-87

- Prohibit undesirable and unnecessary traffic movements.
- Control speed of traffic.
- Separate through and turning movements as well as define their respective alignments.
- A place to provide for and protect traffic control devices.

Traffic islands may be created either by pavement markings or by curbs or a combination of both. Large islands in rural areas may be constructed without curbs or with curbs only at the points where separate roadways converge or diverge.

The following aspects should be considered to design the shape, location, and size of islands:

## TABLE 11.12
## Length of Acceleration Lanes

| Design Speed of Road Being Entered (km/hr) | Length* of Acceleration Lane (m) (Including Length of Paveemnt Taper) Where Design Speed of Exit Curve (km/hr) is: | | | | | | |
|---|---|---|---|---|---|---|---|
| | 0** | 20 | 30 | 40 | 50 | 60 | 80 |
| 40 | 65 | 45 | 35 | – | – | – | – |
| 50 | 95 | 75 | 60 | 40 | – | – | – |
| 60 | 135 | 120 | 100 | 75 | 40 | – | – |
| 80 | 230 | 215 | 200 | 180 | 145 | 100 | – |
| 100 | 330 | 315 | 295 | 275 | 250 | 205 | 100 |

*Source:* Figure 3.18, A Guide to the Design of At-Grade Intersections, JKR 201101-0001-87
\* Length of level grade
\*\* Length required when a vehicle decelerates to zero speed
*Note:* Where the length of the deceleration lane shown is less than the standard taper "Td," then it (Td) should not be reduced.

- They should be located and designed to guide traffic for proper line of travel and to make any changes in direction gradual and smooth.
- The approach end of any island should be away by offset from the edge of the adjacent traffic lane and preceded by appropriate pavement markings such as chevron markings. This approach offset should be a minimum of 1.0 m, the

## TABLE 11.13
## Correction For Grade

| L | Ratio of Length of Grade to Length of Level* for: Design Speed of Turning Roadway Curve (km/hr) | | | | | |
|---|---|---|---|---|---|---|
| | Stop | 20 | 40 | 60 | 80 | All Speeds |
| | 3 or 4% Upgrade | | | | | 3–4% Downgrade |
| 40 | 1.3 | 1.3 | – | – | – | 0.7 |
| 50 | 1.3 | 1.3 | 1.3 | – | – | 0.7 |
| 60 | 1.3 | 1.3 | 1.3 | – | – | 0.7 |
| 80 | 1.3 | 1.3 | 1.4 | 1.4 | – | 0.65 |
| 100 | 1.3 | 1.4 | 1.4 | 1.5 | 1.6 | 0.6 |
| | 5 or 6% Upgrade | | | | | 5–6% Downgrade |
| 40 | 1.4 | 1.4 | – | – | – | 0.6 |
| 50 | 1.4 | 1.5 | 1.5 | – | – | 0.6 |
| 60 | 1.5 | 1.5 | 1.5 | | – | 0.6 |
| 80 | 1.5 | 1.5 | 1.6 | 1.9 | – | 0.55 |
| 100 | 1.6 | 1.7 | 1.8 | 2.2 | 2.5 | 0.5 |

*Source:* Table 3.19, A Guide to the Design of At-Grade Intersections, JKR 201101-0001-87

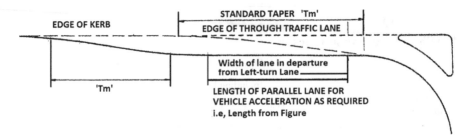

**FIGURE 11.13**  Treatment for Acceleration Lane Taper.
Source: Figure 3.20, A Guide to the Design of At-Grade Intersections, JKR 201101-0001-87

sides of islands should also be away by offset from adjacent traffic lanes by 0.3 m or 0.6 m where semi-mountable or mountable curbs are used. For roads with design speeds exceeding 80 km/hr., the offset should be increased from 0.3 m or 0.6 m to 0.6 m and 1.2 m, respectively.

- Except for large rural islands, other islands should be delineated with semi-mountable type curb. In islands, where the pedestrian refuge is being provided barrier curb should be used.
- In urban areas, raised islands should be of an area not less than 8m². A smaller area may be adopted where traffic signals are to be protected. Islands in rural areas are desired to have a minimum area of 50 m². In rural areas with no street lighting, raised islands should not be used, instead pavement markings should be used to avoid a collision.
- Where an island must provide for stop lines, traffic signals, and pedestrian crossings, the side of the island should have a minimum length of 6 m and a minimum width of 1.2 m at the point where the signal pedestal is installed.
- Figure 11.14 presents the desirable layouts for directional islands.

### 11.4.9.3  Median Islands

Medians are used to divide the carriageways and separate opposing streams of traffic, provide refuge for pedestrians, and reduce the number of crossing conflict along a road.

The following aspects for the design of the median should be considered:

- The approach end of each median island should be set back from the right-hand edge of the adjacent traffic lane by at least 0.3 m and preferably 0.5 m to:
  o  Reduce the possibility of collision with the island.
  o  Relieve the optical illusion of construction in the lane at the start of the island.
- Unless the stopping sight distance is available at the approach end, a median should not start on or beyond a crest. Medians should also not begin on a horizontal curve but may start at or before the first tangent point or 30 m or more after the second tangent point.
- The length of the painted median should precede the approach end of the median so that the approaching driver can notice the median ahead. On

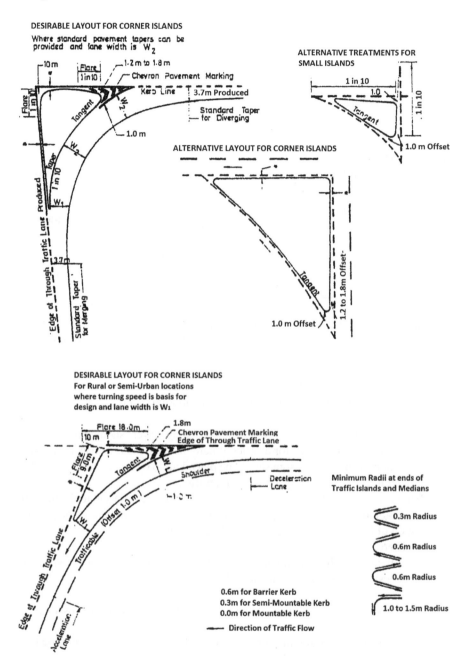

**FIGURE 11.14**   Directional Island.
*Source:* Figure 3.21, A Guide to the Design of At-Grade Intersections, JKR 201101-0001-87

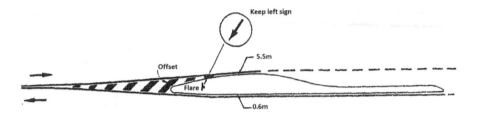

**FIGURE 11.15**   Offset to Median Island.
Source: Figure 3.22, A Guide to the Design of At-Grade Intersections, JKR 201101-0001-87

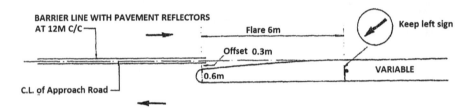

**FIGURE 11.16**   End Treatment for Narrow Median.
*Source:* Figure 3.23, A Guide to the Design of At-Grade Intersections, JKR 201101-0001-87

high-speed roads, any short length of curbed median should be away by offset from the delineated through traffic lane by approximately 0.5 m (see Figure 11.15).

- The first median end encountered by approaching drivers should have a reflectorized KEEP LEFT sign for display. Where the island is less than 1.2 m wide at the approach end, this sign should be placed up to 6 m away from the end to protect it from collision by approaching traffic.
- For a median island when placed in a side road, the end adjacent to the through road should be as narrow as practicable and set back 0.6 m behind the prolongation of the curb line of the through road when:
  o There is no provision for pedestrian crossing.
  o A median with a minimum length of 2 m can be provided between the pedestrian crossing and the through road.

If is not possible, the median should be terminated at the pedestrian crossings.

- Where a median would alter the number of lanes, the treatment as shown in Figure 11.17 should be followed.
- Semi-mountable curbs should be recommended.
- Where curbs cannot be used, painted medians should be recommended as shown in Figure 11.18.

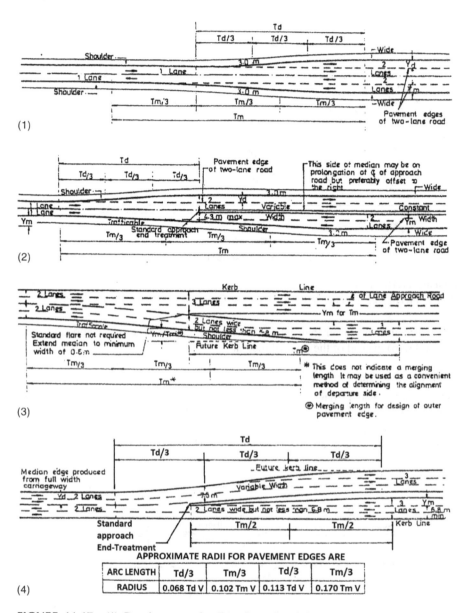

**FIGURE 11.17**  (1) Development of a Four Lane Road from a Two-Lane Road When Widening is on Each Side. (2) Development of a Divided Carriageway from a Two-Lane Road When Widening is on Each Side. (3) Development of a Divided Carriageway from a Four Lane Road When Widening is on Departure Side. (4) Development of a Divided Carriageway from a Four Lane Road When Widening is on Approach Side. Median Terminal Treatments.
*Source:* Figure 3.24, A Guide to the Design of At-Grade Intersections, JKR 201101-0001-87

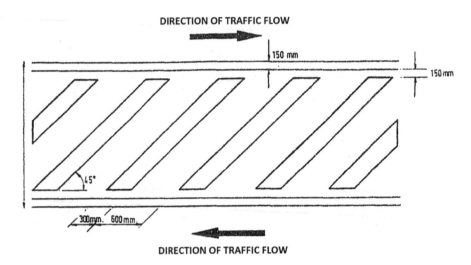

**FIGURE 11.18**   Painted Islands.
*Source:* Figure 3.25, A Guide to the Design of At-Grade Intersections, JKR 201101-0001-87

If the median is narrower than 2 m, a barrier line may be used in the approach (see Figure 11.16) instead of the painted median.

### 11.4.9.4  Median Openings

Where openings are provided in the median, the treatment of the median and its ends should be following the design as shown in Figure 11.19 depending on the width of the median.

The following figures are to describe various Median Terminal Treatments.

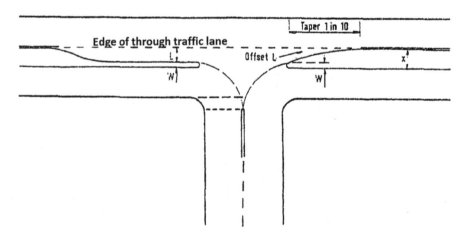

**FIGURE 11.19**   Median Opening.
*Source:* Figure 3.26, A Guide to the Design of At-Grade Intersections, JKR 201101-0001-87

## 11.4.9.5   Outer Separators

Outer Separators are used to separate the main road carriageway with through traffic lanes from service roads. They should be as wide as possible with a desirable with of 5.0 m. Treatment for outer separator openings is described in Figure 11.20.

## 11.4.10   WIDENING OF MAJOR ROAD

Widening of the major road to provide space for the central island should be on a straight portion and to be made symmetrically around the centerline of the road, the widening on a curve portion should be made to the inside of the centerline. The same should be followed where widening of a median is required.

The length of the widening shall be determined by the formula:

$$L_W = V\sqrt{W_{max}}$$

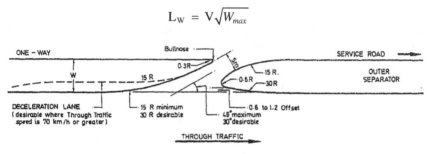

**One-way Entrance to a Service Road**

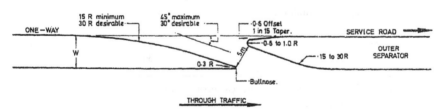

**One-way Exit From a Service Road**

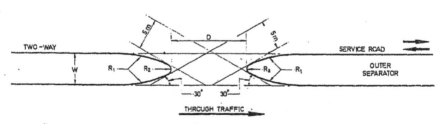

**Two-way Opening Between Through Road and Service Roads.**

**FIGURE 11.20**   Outer Separator Openings.
*Source:* Figure 3.27, A Guide to the Design of At-Grade Intersections, JKR 201101-0001-87

where

$L_W$ = length of the widening in m

V = design speed of major road in km/hr

$W_{max}$ = larger of the two parts of the widening (m) on either side of the centerlinei.e.,

$W_{max}$ = ½ of total widening $W_W$ in the symmetrical case

$W_{max}$ = $W_W$ in the case of one-sided widening

The outer edges of the carriageway are widened over the same length as the central widening even if the required widening is different from the central widening due to changes in lane width.

The widening of both inner and outer edges is adopted to a smooth continuous alignment composed of the usual alignment elements. S or Reverse curves composed of two circular arcs commonly provide a curvature that has acceptable dynamic and visual properties and is recommended. When the road is on a curve, S-curve may produce adverse curvatures, in such a situation the length of widening should be increased, or an alternative curvature should be selected. Figure 11.21 (A) and (B) describes aspects of widening of the major road by S-curves.

A. Widening of Major Road
  1. Without Right-Turn Lane

  L = Length of widening section = V = $\sqrt{W_{max}(m)}$

  V = Design speed (km/hr)

  Wmax. = Larger part of the total widening to either side of the center

  2. Without Right-Turn Lane

| Outer Separator Width W(m) | Radius | |
|---|---|---|
| | R1 (m) | R2 (m) |
| 5–10 | 15 | 0.2 W |
| 11–15 | 23 | 0.2 W |
| 16–25 | 30 | 0.2 W |

  WL1 and WL2 = Lane widths of through lanes.

  Td = Length of taper

  R1 = Turn – in radius (see Figure 11.29 A and B)

B. Radii of S-curve Tangent Center Line

$$R1 = R2 = \frac{L}{4W} \cdot \frac{W}{4}$$

$$Y = \frac{Wx^2}{L^2}\left(3 - \frac{2x}{L}\right)$$

where

R1 and R2 = Radii of the 5 curve in m

L = Length of widening section in m

W = Widening in m

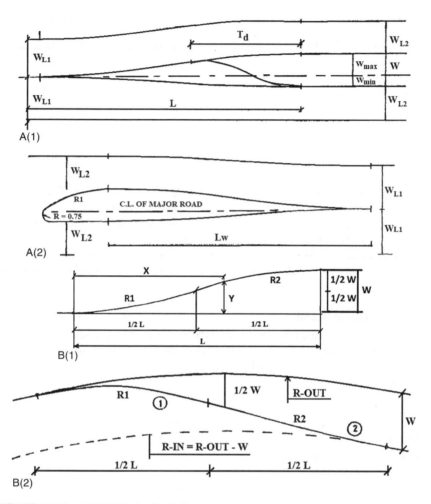

**FIGURE 11.21**   A(1) Widening by S-Curves.
*Source:* Figure 3.28, A Guide to the Design of At-Grade Intersections, JKR 201101-0001-87
A(2) Widening by S-Curves.
*Source:* Figure 3.28, A Guide to the Design of At-Grade Intersections, JKR 201101-0001-87
B(1) Widening by S-Curves.
*Source:* Figure 3.28, A Guide to the Design of At-Grade Intersections, JKR 201101-0001-87
B(2) Widening by S-Curves.
*Source:* Figure 3.28, A Guide to the Design of At-Grade Intersections, JKR 201101-0001-87

X = Distance from straight of curve
Y = Lateral offset from parallel tangent

C.  Radii of S-Curve Center Line on Curve

$$R1 = \frac{1}{\dfrac{1}{R_{OUT}}} + \frac{4W}{L^2} \qquad\qquad R2 = \frac{1}{\dfrac{1}{R_{ON}} - W} - \left(\frac{4W}{L^2}\right)$$

R2 is negative when $L < 2\sqrt{WR_{IN}}$ in which case curve (2) is as shown

R2 is positive when $L > 2\sqrt{WR_{IN}}$ in which case curve (2) turns in the same direction as curve (1)

R2 is infinite when $L = \sqrt{WR_{IN}}$ in which case curve (2) is a straight tangent to the curve (1) and curve with a radius $R_{IN}$

## 11.4.11   Minor Road Treatment

### 11.4.11.1   Types of Treatments

In an intersection, the treatments on the minor road for better traffic control benefits not only the minor road but also the major road. Quick departure of traffic from the major road and smooth merging into it provide maintain a smooth and safe traffic flow on the major road. There are three types of treatments; the description of which is discussed before in paragraph 11.2.3 of this chapter.

The type of the minor road treatments should be selected according to the class of the crossing road and that of the major road to which it is connected, as shown in Table 11.14.

### 11.4.11.2   Guide Islands

Guide islands are provided at the center of the minor road at intersections to define the movements of turning traffic and to control the speed of turning and crossing vehicles. Guide islands also provide space for traffic control devices and refuge for pedestrians.

## TABLE 11.14
## Minor Road Treatment

### a. Rural Area

|  | Minor Road | | | | |
| --- | --- | --- | --- | --- | --- |
| Highway | Primary Road | Secondary Road | Minor Road | | |
| C* | C | F | F/N | Highway | Major road |
|  | C/F | F | N | Primary | |
|  |  | F/N | N | Secondary | |
|  |  |  | N | Minor road | |

### b. Urban Area

|  | Minor road | | | |
| --- | --- | --- | --- | --- |
| Arterial | Collector | Local street | | |
| C | C/F | F/N | Arterial | Major road |
|  | F | N | Collector | |
|  |  | N | Local street | |

*Source:* Table 3.4, A Guide to the Design of At-Grade Intersections, JKR 201101-0001-87
* Normally At-grade intersection should not be adopted C, channelized; F, flared; N, no treatment

A Guide island shall be designed for the following provisions:

- The shape and location of the island shall be such that it can be passed comfortably by the design vehicle both entering and leaving the major road.
- The front end of the island is shaped by following the inner rear wheel paths of the design vehicle while the rear end is shaped to guide the approaching traffic.
- The largest width of the island shall be between 3.0 and 5.0 m while the length shall be from 20 to 35 m.
- The island shall be curved, preferably semi-mountable with an offset of 0.3 m.
- Mandatory KEEP LEFT signs (for UK style of traffic) and KEEP RIGHT signs (for USA style of traffic) shall be placed at both ends of the island. Warning or information signs can be placed if they do not obstruct the visibility of the vehicles.

Figures 11.22 (A) and (B) describes the standard design of guide islands that are to be used.

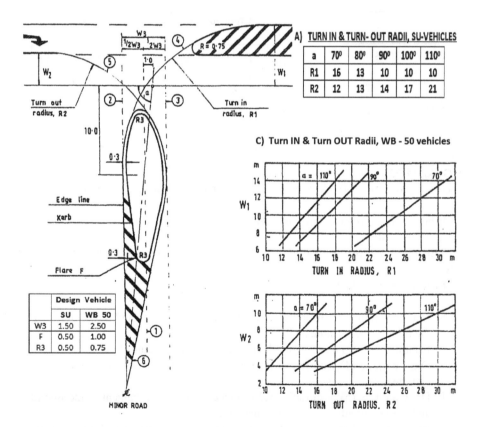

**FIGURE 11.22A**  Standard Design of Guide Island.
*Source:* Figure 3.29A, A Guide to the Design of At-Grade Intersections, JKR 201101-0001-87

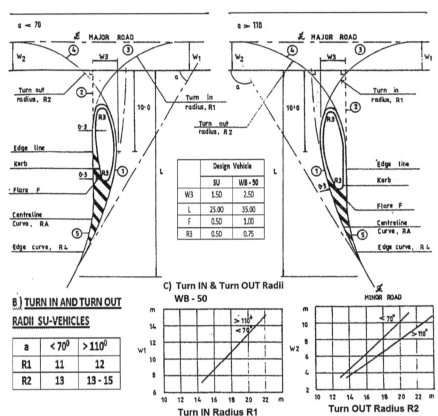

**FIGURE 11.22B** Standard Design of Guide Island.
*Source:* Figure 3.29B, A Guide to the Design of At-Grade Intersections, JKR 201101-0001-87

Where two alternatives are given, Traffic Volume should be considered for the selected alternative.

**The standard design of Guide Island, where Intersecting Angle 70° ≤ a ≤ 110°).**

**Procedure & Notes:**

1. Line (1) cutting the nearer edge of the through the lane and the centerline of the minor road as indicated.
2. Establish line (2) and (3) parallel to line (1) at the distance ½ W3.
3. Draw a circle (4) with radius R1 rounded to whole mj tangent to the line (2) and the nearer edge of the offside through lane.
4. Draw a circle (5) with radius R2 rounded to whole mj tangent to the line (3) and the nearer edge of the right-turn lane.
5. Extend line (2) to intersect the centerline, draw line (6) through the intersecting point tangent to the circle (5).

6. The location of the rear end of the curbed part of the island is determined and the curb line is set 0.3 m behind of circle (4) circle (5) and (6) and tangent between the end circle and the curb line is ehind circle (4). The front and rear ends are rounded by circles with a radius of R3.
7. All dimensions are in meters.

**Procedure & Notes:**

1. Change the centerline using the curve (1) with radius RA > = 50 m. The curve should be perpendicular to the edge of the nearer through lane.
2. Establish line (2) Perpendicular to the edge of the through lane through a point at a distanceW3 from the point on the curve (1) which is 10 m from the edge of the through lane.
3. Draw circle (3) with radius R1 (rounded to the whole m) tangent to the line (2) or circle (1) and the nearer edge of the offside through lane.
4. Draw circle (4) with radius R2 (rounded to the whole m) tangent to circle (1) or line (2) and the nearer edge of the right-turn lane.
5. Draw circle (5) tangent to the centerline within the distance L from the nearer edge of the major road and tangent to the line (2) select the radius R4 as large as possible.
6. The curbed part of the island is now established as described in Figure 11.29 A.
7. All dimensions are in meters.

## 11.4.11.3  Widening of the Minor Road

If no guide islands are present the width of the carriageway shall remain unchanged up to the corners of the intersection. In presence of guide islands, the entry lane shall have a minimum width of 3.5 m and the exit lane a minimum width of 4.5 m past the island.

Provision of right-turn lanes for UK style of traffic (or left-turn lanes for USA style of traffic) and examination of the number of lanes needed are usually emphasized on the major road. However, increasing the number of the right-turning lane for the UK style of traffic (or left-turning lane for USA style of traffic) on minor crossroad especially at signalized intersections also makes the major road benefited. Right-turning vehicles (for UK style of traffic or left-turning for USA style of traffic) departing from two lanes can clear the intersections in a shorter time. The green time allotted to the crossroad can also be reduced. A more favorable split of green time to the major road increases its capacity. This effect is more significant if the widening of the major road is costly and the crossroad is two-lane. Because intersections are usually bottlenecking in any stretch of road, the benefit of any increased capacity extends to the whole stretch. This is more cost effective than a widening of the major road. In this case, however, the green time assigned to the crossroad should not be shorter than 15 seconds and be sufficient for pedestrians who cross the major road in the same phase.

#### 11.4.11.4 Left-Turn Lane on Minor Roads

An auxiliary lane reserved for left-turning traffic (for UK style of traffic or right-turning for USA style of traffic) may be added to the approach of the minor road if the left-turning traffic (for UK style of traffic or right-turning traffic for USA style of traffic) exceeds 50% of the capacity for that movement, or where there are no space constraints. The design of the left-turn lanes (for UK style of traffic or right-turning lanes for USA style of traffic) shall follow the guidelines set out in paragraph 11.4.6.3 of this chapter.

### 11.4.12 SHOULDERS

Shoulder widths, in general, remain unchanged in the intersection area but may be reduced to 2.0 m along the deceleration and turning lanes. Near the intersection where curbed islands are present, the shoulder structure should be of stable gravel, hard or sealed shoulder type. In general, curbs should not be used along the outer carriageway edges.

### 11.4.13 CROSSFALL AND SURFACE DRAINAGE

Crossfalls in the intersection area should be adopted to make provisions for drainage, driving comfort, and visibility. In usual practice, the crossfall of the through lanes in the major road remains unchanged through the intersection.

Crossfalls on auxiliary lanes may follow that of the adjoining through lane or fall to the opposite side as it is required for drainage or side friction criteria. The algebraic difference of crossfalls of two adjoining lanes should not exceed 5%.

The crossfall of the minor road, toward the edge of the through lane, shall be the same as the gradient of the through lane. Where the major road is on a steep grade, this may create an adverse camber for turning vehicles. In such a situation, diverging lanes should be considered. Superelevation of corner lanes in connection with triangular islands on the minor road, in general, shall not exceed 6%.

## 11.5 CAPACITY OF INTERSECTIONS

### 11.5.1 GENERAL

Interrupted traffic flow conditions predominate on most urban roads. The major intersections determine the overall capacity and performance of the road network, irrespective of whether the intersection is signalized or not.

Significant volumes of crossing or turning traffic at minor roads cause interruptions and capacity reductions, which can be reduced by channelization and intersection control.

The capacity of intersection is especially important and to achieve balance, the intersection design should take into account for the capacity of the approach roads.

## 11.5.2 Level of Service

Level of Service is a qualitative measure of the effect of several factors, which include speed and travel time, traffic interruptions, freedom to maneuver, safety, driving comfort, convenience, and operating costs. The concept of level of service is used in the capacity analysis of intersections. The required level of service that is to be used for intersections along the various categories of roads is shown in Table 11.15.

## 11.5.3 Capacity of Unsignalized Intersections

### 11.5.3.1 General

Capacity analysis are seldom required for rural intersections since their traffic volumes are not considerable to make capacity a design consideration. Normally, safety is the major consideration in rural situations. Safety requirements may necessitate the provision of separate lanes for left or right-turning vehicles.

The method of capacity analysis as described below is, therefore, more pertinent for urban intersections. The capacity analysis is based on the Highway Capacity Manual, Special Report 209, Transportation Research Board Washington D.C. The designer is advised to refer to the above publication for a better understanding of the subject.

### 11.5.3.2 Procedure

*11.5.3.2.1 Basic Structures*

The basic structure of the procedure is as follows:

- Define existing geometric and volume conditions for the intersection under study.
- Determine the "conflicting traffic" through which each minor road movement, and the major road right-turn, must cross.

---

**TABLE 11.15**
**Level of Service**

| Areas | Category of Road | Level of Service |
|---|---|---|
| Rural | Expressway | C |
| | Highway | C |
| | Primary | D |
| | Secondary | D |
| | Minor | E |
| Urban | Expressway | C |
| | Arterial | D |
| | Collector | D |
| | Local street | E |

*Source:* Table 4.1, A Guide to the Design of At-Grade Intersections, JKR 201101-0001-87

---

- Determine the size of the gap in the conflicting traffic stream needed by vehicles in each movement crossing a conflicting traffic stream.
- Determine the capacity of the gaps in the major traffic stream to allow to accommodate each of the subject movements that will utilize these gaps.
- Adjust the found capacities to account for impedance and the use of shared lanes.
- Determine the capacity of reserve and from Table 11.18 obtain the level of service for the unsignalized intersection.

Each of these basic analysis steps is discussed in detail in the following sections. The relevant worksheets are to be used for the analysis of unsignalized intersections.

When the level of service as determined is lower than the required level as in Table 11.15, improvements should be considered for channelization, lane use controls, sight distance improvements, etc., and the level of services recalculated. If the level of service is still lower than that as required, then signalization should be considered.

### 11.5.3.2.2  Input Data Requirements

The basic input data requirements are as indicated below:

- Number of lanes and their use.
- Channelization.
- Approach gradient.
- The radius of the curb line and approach angle.
- Sight distances.

Each of these factors have a substantial impact on how gaps in the major road are utilized, and on the size of the gap that is required by the various movements. Volumes must be specified by movement. In general, full hourly volumes are used in the analysis of unsignalized intersections because short-term fluctuations will generally not cause major difficulties at such locations. The engineer may, however, divide all volumes by the peak hour factor (PHF) to choose flow rates for the peak 15-min interval before beginning computations, the volume for movement "I" is designated as "$V_i$".

In conventional practice, subscripts 1–6 are used to define movements on the major road, and subscripts 7–12 to define movements on the minor road. Conversion of vehicles per hour to passenger cars per hour is accomplished by using the passenger car equivalent values as given in Table 11.16.

### 11.5.3.2.3  Conflicting Traffic

At an unsignalized intersection, the nature of conflicting movements is relatively complex. Each subject movement encounters a different set of conflicts that are related to the nature of the movement. These conflicts are mentioned in Figure 11.23 which illustrates the computation of the parameter.

$V_{ci}$ = the "Conflicting Volume" for movement i, is the total volume which conflicts with movement i, expressed as vehicles per hour (vph).

## TABLE 11.16
## Conversion To P.C.U. for Unsignalized Intersection

| Type of Vehicle | Grade (%) | | | | |
|---|---|---|---|---|---|
| | −4% | −2% | 0% | +2% | +4% |
| Motorcycles | 0.3 | 0.4 | 0.5 | 0.6 | 0.7 |
| Passenger cars | 0.8 | 0.9 | 1.0 | 1.2 | 1.4 |
| SU | 1.0 | 1.2 | 1.5 | 2.0 | 3.0 |
| WB-50 | 1.2 | 1.5 | 2.0 | 3.0 | 6.0 |
| All vehicles* | 0.9 | 1.0 | 1.1 | 1.4 | 1.7 |

*Source:* Table 4.2, A Guide to the Design of At-Grade Intersections, JKR 201101-0001-87
* If the composition of vehicles is unknown, these values may be used as an approximation.
*Note:*
Maximum total decrease in critical gap = 1.0 sec
Maximum critical gap = 8.5 sec
For values of average running speed between 50 and 90 km/hr
Interpolate this adjustment is for the specific movement affected by restricted sight distance

| Movement | Conflicting Traffic, Vci | Illustrator |
|---|---|---|
| 1) **LEFT TURN** from minor road. | $\frac{1}{2}$ (Vl)** + Vt* | 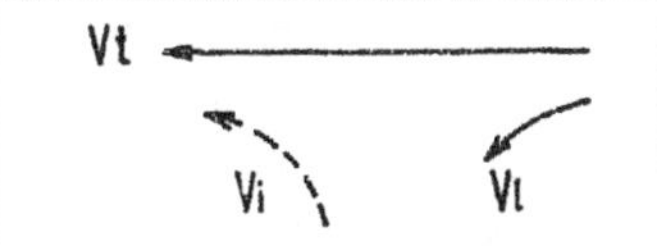 |
| 2) **RIGHT TURN** from major road | V$\ell$*** + Vt | 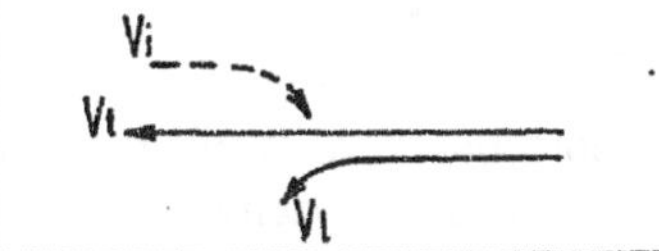 |
| 3) **THROUGH MVT** from minor road | $\frac{1}{2}$(Vla)*** + Vta + Vra + Vlb + Vtb + Vrb | |
| 4) **RIGHT TURN** from minor road. | $\frac{1}{2}$(Vla)** + Vta + Vrd + Vtb*** + Vtb + Vrb + V₀ + V₀l | |

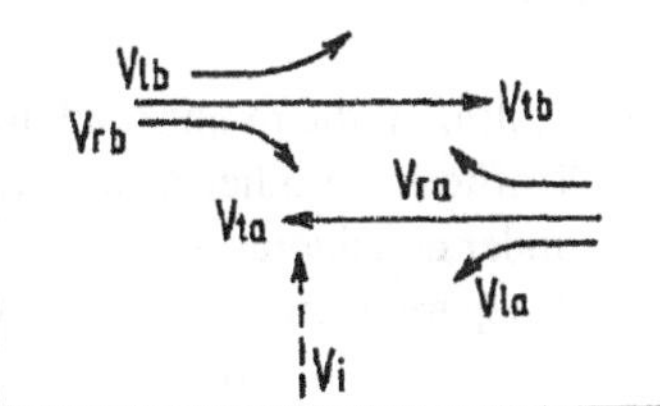

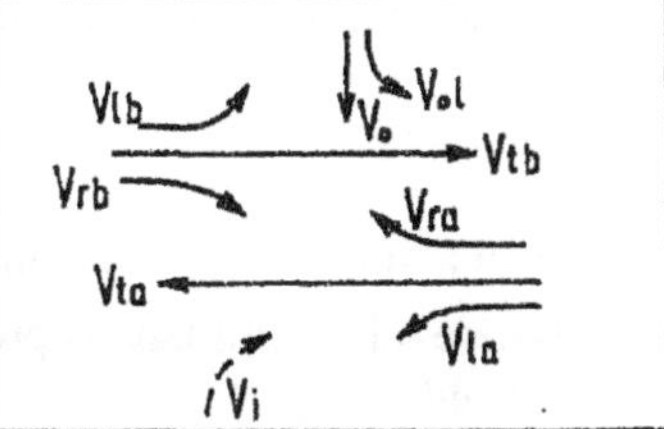

**FIGURE 11.23**   Definition and Computation of Conflicting Traffic Volumes.
*Source:* Figure 4.1, A Guide to the Design of At-Grade Intersections, JKR 201101-0001-87

When using Figure 11.23 to compute the conflicting volumes, the engineer should carefully refer to the footnotes which allow modifications to the equations shown for special cases. In the equations of Figure 11.23, the conflicting traffic volume for movement i, which is denoted $V_{ci}$ is computed in terms of an hourly volume in mixed vph. Subscripts "r" denoted right-turns, "l" for left-turns, "t" for through movements, and "o" for opposing minor road flows.

### 11.5.3.2.4  Critical Gap Size

The "Critical Gap" is defined as the mean time headway between two successive vehicles in the major road traffic stream that is accepted by drivers in a subject movement that must cross or merge with the major road flow. It is denoted as "Tc" and is expressed in seconds. The critical gap depends on several factors viz:

- The type of maneuver being executed.
- The type of control (STOP or GIVE WAY) on the minor road.
- The average running speed on the major road.
- The number of lanes on the major road.
- The geometrics and environmental conditions at the intersection.

The values of the critical gap are selected from Table 11.17 in a two-part process, as below:

- From the first half of the table, the basic critical gap size is selected for the type of movement, control, and major road speed.
- From the second half of the table's adjustments and modifications to the basic critical gap, size is selected for a variety of conditions.

### 11.5.3.2.5  Potential Capacity for a Movement

The potential capacity of a movement denoted as $C_{Pi}$ (for movement i), and is defined as the "ideal" capacity for a specific subject movement assuming the following conditions:

- Traffic on the major road does not block the minor road.
- Traffic from other nearby intersections does not back up into the intersection under consideration.
- A separate lane is provided for the exclusive use of each minor road movement under consideration.
- No other movements impede the subject movement.

Note:

- "Vt" includes only the volume in the left-hand lane.
- Where a left-turn lane is provided on the major road, "Vl" or "Vla" should be deleted.
- Where the left-turn radius into the minor road is large and or where these movements are STOP/Giveway controlled, Vl (Case 2) and Vla and / or Vlb (Case 4) should be deleted. Vlb can also be deleted for multilane major roads.

## TABLE 11.17
## Critical Gap Size Selection Critical Gap Size Selection

### Basic Critical Gap for Passenger Cars (sec)

| Vehicle Maneuver and Type of Control | Average Running Speed, Major Road | | | |
|---|---|---|---|---|
| | 50 km/hr | | 90 km/hr | |
| | Number of Lanes on Major Road | | | |
| | 2 | 4 | 2 | 4 |
| LT from minor road | | | | |
| • Stop | 5.5 | 5.0 | 6.5 | 5.5 |
| • Give way | 5.5 | 5.0 | 6.5 | 5.5 |
| RT from major road | | | | |
| • Stop | 5.0 | | 5.5 | |
| • Give way | 5.5 | | 6.0 | |
| Cross major road | | | | |
| • Stop | 6.0 | 5.5 | 7.5 | 6.5 |
| • Give way | 6.5 | 6.0 | 8.0 | 7.0 |
| RT from minor road | | | | |
| • Stop | 6.5 | 6.0 | 8.0 | 7.0 |
| • Give way | 7.0 | 6.5 | 6.5 | 7.5 |

| Adjustments and Modifications to Critical Gap (sec) | |
|---|---|
| Condition | Adjustment |
| LT from minor road: Kerb radius > 15 m or turn angle <60$^0$ | −0.5 |
| LT from minor road: Acceleration lane provided | −1.0 |
| Restricted sight distance* | Up $t_a$ + 1.0 |
| All movements: Population ≥250.000 | −0.5 |

*Source:* Table 4.3, A Guide to the Design of At-Grade Intersections, JKR 201101-0001-87
* When demand volume exceeds the capacity of the lane, extreme delays will occur with queuing which may cause severe congestion affecting other traffic movements in the intersection. This condition usually warrants improvement to the intersection.

The potential capacity in passenger cars per hour is obtained from Figure 11.24, and is based on the volume of the conflicting traffic, "$V_C$" in vph, and the critical gap, "$T_C$" in seconds. The figure is entered on the horizontal axis with the value of "$V_C$". A vertical line is drawn to the approximate "Critical gap" curve. A horizontal line is drawn from the intersection with the "Critical gap" curve to the vertical axis, the result is obtained, in passenger cars per hour.

### 11.5.3.2.6 Impedance Effects

Vehicles utilize gaps in a prioritized manner at unsignalized intersections. The potential capacity of a lower priority movement can be impeded when traffic becomes congested. The effect of impedance is taken into consideration by multiplying the potential capacity of a movement "$C_{pi}$" by a series of impedance factors "$p_j$" for each impeding movement "j".

Impedance factors "$p_j$" are taken from Figure 11.25 and are based solely on the percent of the potential capacity of the impeding movement used by the existing

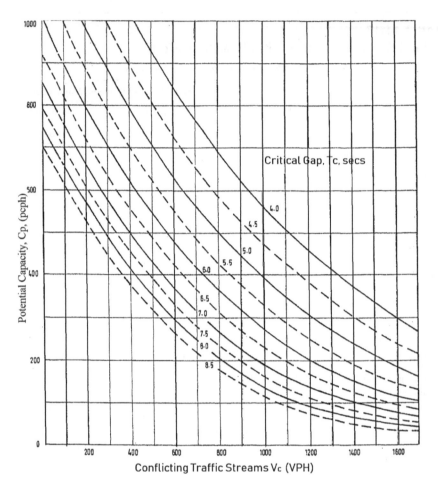

**FIGURE 11.24**   Potential Capacity Based On.
*Source:* Figure 4.2, A Guide to the Design of At-Grade Intersections, JKR 201101-0001-87

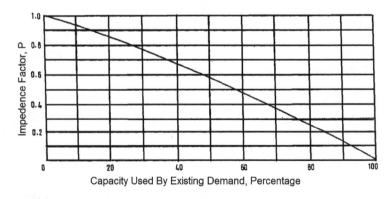

**FIGURE 11.25**   Impedance Factors as a Result of Congested Movements.
*Source:* Figure 4.3, A Guide to the Design of At-Grade Intersections, JKR 201101-0001-87

demand. For example, a right-turn movement from a minor road at a T-intersection is impeded by the right-turn from the major road. The later movement has a potential capacity of 400 pcph and a demand of 200 pcph. Therefore, the major road right-turn uses up 200/400 = 0.50 or percent of its available capacity. From Figure 11.25, an impedance factor of 0.68 is read.

Figure 11.26 explains the computations for the movement capacity, "$c_{mi}$" which is the adjusted capacity of the movement. This however still assumes that the movement has exclusive use of a separate lane.

### 11.5.3.2.7  Shared lane Capacity

Frequently, two or three movements share a single-lane on the minor road approach. At such times vehicles from different movements do not have simultaneous access to gaps. Where several movements share the same lane, and cannot stop side by side at the stop line, the following equation is used to compute the capacity of the shared lane:

$$C_{SH} = V_l + V_t + V_r$$
$$[V_l/c_{ml}] + [v_t/c_{mt}] + [v_r/c_{mr}]$$

Case 1: Right Turn From Minor Road at a T-Intersection (UK-Type of System)

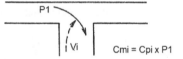

Case 2: Through Traffic From Minor Road at a 4-Leg Intersection (UK-Type)

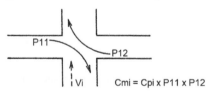

Case 3: Right Turn From Minor Road at a 4-Leg Intersection (UK-Type)

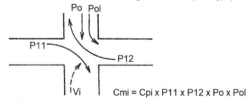

**FIGURE 11.26**  Illustration of Impedance Calculations.
*Source:* Figure 4.4, A Guide to the Design of At-Grade Intersections, JKR 201101-0001-87

where
 $C_{SH}$ = capacity of the shared lane, in pcph
 $V_1$ = volume or flow rate of left-turn movement in shared lane, in pcph
 $V_t$ = volume or flow rate if through movement in shared lane, in pch
 $V_r$ = volume or flow rate of right-turn movement in shared lane, in pch
 $C_{m1}$ = movement capacity of the left-turn movement in shared lane, pch
 $C_{mt}$ = movement capacity of the through movement in share lane, pch
 $C_{mr}$ = movement capacity of the right-turn movement in a shared lane, in pcph

### 11.5.3.2.8  Reserve Capacity
The reserve capacity of the minor road approach lane can be computed by:

$$C_R = C_{SH} - V$$

where
 $C_R$ = reserve capacity of the lane in pcph.
 $C_{SH}$ = shared lane capacity of the lane in pcph.
 $V$ = total volume or flow rate using the lane.

If the shared lane includes only left-turn and through movements, then the terms for right-turning traffic are deleted from both numerator and denominator in the equation in paragraph vii.

Conflicting Traffic Volume and Critical Gap Size.

### 11.5.3.2.9  Level of Service
The level of service (LOS) criteria is related to the reserve capacity and is mentioned in Table 11.18 below.

Direct comparisons of the LOS of an unsignalized intersection with a signalized intersection should not be made because the level of service for unsignalized intersection which is based on reserve capacity is not associated with the delay values considered for signalized intersections.

---

**TABLE 11.18**
**Level of Service Criteria for Unsignalized Intersection**

| Reserve Capacity Delay (PCPH) | Level of Service | Expected to Minor Road Traffic |
|---|---|---|
| 400 | A | Little or no delay |
| 300–399 | B | Short traffic delays |
| 200–299 | C | Average traffic delays |
| 100–199 | D | Long traffic delays |
| 0–99 | E | Very long traffic delays |
| * | F | |

*Source:*  Table 4.4, A Guide to the Design of At-Grade Intersections, JKR 201101-0001-87
* When demand volume exceeds the capacity of the lane, extreme delays will occur with queuing which may cause severe congestion affecting other traffic movements in the intersection. This condition usually warrants improvement to the intersection.

---

### 11.5.3.3   Potential Improvements

It should be noted that the above methodology is not for a formal warrant for the consideration of signalization. Where unacceptable LOS are found, improvements such as channelization, lane use controls, sight distance improvements, multi-way STOP control, etc. should be considered. When such improvements fail to improve the LOS then only signalization is to be considered.

### 11.5.4   CAPACITY OF SIGNALIZED INTERSECTIONS

### 11.5.4.1   General

The capacity of signalized intersections is also based on the concept of LOS. In this case, the stopped delay per vehicle is used as a measure of the LOS. The stopped delay causes driver discomfort, frustration, fuel consumption, and loss in travel time. Table 11.19 gives the LOS criteria for signalized intersections.

When the calculated LOS is unacceptable, the signal timing, phasing, or layout of the intersection should be adjusted and reanalyzed. If the LOS is still unacceptable, then grade separation should be considered.

### 11.5.4.2   Warrants

Warrants for the applications of traffic-controlled signals are laid down in relevant standards of every individual country.

### 11.5.4.3   Intersection Capacity Characteristics

The intersection capacity analysis as detailed below follows that of the Transport and Road Research Laboratory. Capacity analysis can also be done by following the Transportation Research Board's "Highway Capacity Manual – special report 209 – Chapter 9". The analysis is similar in that a basic lane saturation flow is obtained and modified following certain identified operational characteristics. The computer applications are recommended as available in software TransPlan/SICap at the web site of www.techsoftglobal.com,

---

**TABLE 11.19**
**Level of Service Criteria for Signalized Intersection**

| Level of Service | Stopped Delay for Vehicle (sec) |
|---|---|
| A | <5.0 |
| B | 5.1–15.0 |
| C | 15.1–25.0 |
| D | 25.1–40.0 |
| E | 40.1–60.0 |
| F | >60.0 |

*Source:* Table 4.5, A Guide to the Design of At-Grade Intersections, JKR 201101-0001-87

---

Some of the operational characteristics are:

a. Effect of widths of approach lane
b. Effect of composition of traffic
c. Effect of opposing traffic for Capacity reduction
d. Effect of parked vehicles
e. Effect of gradient

### 11.5.4.4  Computation Analysis

The following procedures are used, by calculating the:

- Corrected saturation flow for each approach using factors listed above.
- The capacity ratio for each approach and thus for each signal phase.
- Reserve capacity for the intersection.
- Vehicular delay.
- Queue length.

When the reserve capacity of the intersection or the LOS is not acceptable, the signal timing phasing or layout may have to be modified. If the reserve capacity or LOS is still unacceptable, then grade separation for the major road should be considered.

### 11.5.4.5  Signal Timings

The optimum cycle time, $C_O$, is usually calculated from Webster's Formula:

$$C_O = (1.5 + 5)/(1 - Y)$$

where
   $L$ = total lost time (sec)
      = nl + Rwhere
   $n$ = no. of phases/cycles
   $l$ = lost time/phase
   $R$ = all Red time
   $Y$ = capacity ratio of intersection
   The effective green time for each phase "$g_i$", is given by the following:

$$g_i = Yi \times (C_O - L)/Y$$

In general, the minimum cycle time should not be less than 45 sec and the maximum cycle time should not exceed 120 sec.

### 11.5.5  CAPACITY OF ROUNDABOUTS

### 11.5.5.1  Size of Roundabout

For large variation in various sizes, the movements of traffic in a roundabout cannot be analyzed by one universal method. In large conventional roundabouts, weaving

**TABLE 11.20**
**Size of Roundabout**

| | Diameter of Inscribed Circle (m) | Diameter of Center Circle (m) |
|---|---|---|
| Conventional | $D_I > 50$ | $D_C > 25$ |
| Small | $50 > D_I > 20$ | $25 > D_C > 4$ |
| Mini | $20 > D_I$ | $4 > D_C$ |

*Source:* Page No. 77, A Guide to the Design of At-Grade Intersections, JKR 201101-0001-87

motion in the turning road sections between legs may be assumed, but this is not similar in smaller roundabouts. Separate formulas are proposed to determine the capacity of roundabouts of different sizes. The distinction of the sizes are defined below in Table 11.20:

### 11.5.5.2 Capacity Calculations

*11.5.5.2.1 Conventional Roundabout*

$$Q_p = \frac{160W \times \left(1 + e/W\right)}{1 + \left(W/L\right)}$$

where
$Q_p$ = capacity of weaving section (veh/h)
W = width of weaving section (m)
e = 1/2 (e1 + e2): average entry width (m)
L = length of weaving section (m)

*11.5.5.2.2 Small Roundabout*

$$Qp = K \times \left(W - \sqrt{A}\right)$$

where
Qp = the capacity of the whole intersection (veh/h)
W = the sum of basic full widths on all approaches (m)
A = area added to basic intersection by flared approaches (m²)
K = a specific factor may be taken as:
70 for three legs
50 for four legs
45 for five legs

*11.5.5.2.3   Mini Roundabout*

The same formula as in (b) applies. The value "K" should be changed as follows:

60 for three legs
45 for four legs
40 for five legs

### 11.5.5.3   Reserve Capacity

For roundabouts, the concept of LOS is not applicable. The reserve capacity should be calculated, as below:

$$Q_r = (Q_p - Q) \times 100\% / Q$$

where
  $Q_r$ = reserve capacity (veh./hr.)
  $Q_p$ = calculated capacity (veh./hr.)
  $Q$ = Volume of traffic (weaving/total) (veh./hr.)
  In general, the available reserve capacity should not be less than 15%. If found less, then a signalized intersection or grade separation for the major road may be considered.

## 11.6   OTHER RELATED ELEMENTS

### 11.6.1   PEDESTRIAN FACILITIES

#### 11.6.1.1   General

Pedestrian facilities such as crossings, refuge islands, and pedestrian actuated traffic signals are an integral part of intersection design and should be provided where required.

#### 11.6.1.2   Pedestrian Crossing

- The pedestrian crossing should be located to match the flow line of pedestrian traffic. Pedestrian crossing located against the natural flow line of pedestrians would invite jaywalkers outside of it.
- Pedestrian crossings should be placed perpendicular to the direction of the road. This makes the distance to cross and green time to be allotted to pedestrians the shortest. This is desirable to maintain a high capacity for traffic at the intersection.
- Pedestrian crossings should be placed close to the center of the intersection. This makes the area of the intersection smaller and requires less time to pass through it. A smaller intersection has a larger capacity with a shorter clearance time in signal phasing.
- Pedestrian crossings should be placed where drivers approaching have an adequate view of it.

- Pedestrian crossings shorter than 15 m is recommended. If the width of the road to be crossed is more than 15 m, refuge islands should be provided to enable pedestrians to cross it in two green signals.
- The width of pedestrian crossing should be determined for the estimated number of pedestrians and the duration of green time allotted to the pedestrian phase. However, it is not desirable that every pedestrian crossing having different widths. The minimum width should be fixed as 4 m and 2 m for major roads and minor roads, respectively. When necessary, the width should be increased by whole meters.
- Where the pedestrian crossing is used by blind or physically challenged people, a warning sound system should be considered. Although it may be theoretically desirable to place pedestrian crossing at the extension of the sidewalk, it is usually located several meters (3–4 m, or minimum 1m) behind the extension of the boundary line between the carriageway and sidewalk. Barriers should be provided along the rounded corner between both the thresholds to the pedestrian crossing.

This is recommended for the following reasons:

- Pedestrian crossings placed on the elongation of intersecting sidewalks may cross in the intersection. In the cases like this, pedestrians tend to the next intersecting pedestrians tend to wait in the intersection or move to the next intersecting pedestrian crossing without making a complete crossing of the first one.
- Left-turning vehicles must often wait for the pedestrians who are crossing the street in the same phase. The pedestrian crossings should be set back to make some distance for these vehicles so that the pedestrians will not hinder the straight through vehicles.
- Rounded corners of the sidewalk created by setting back two pedestrian crossings give space for installing traffic signs and lighting.
- Rounded corners protected by barriers provide pedestrians a relaxed and safe feeling and invites composed behavior. When space permits, flowerpots may be provided along the corner to enhance aesthetics.
- The pedestrian crossing should be placed 1–2 m back from the nose of the central median.

## 11.6.2 Lighting

Lighting promotes the safety of intersections and the ease and comfort of traffic operations. Intersections that are channelized should include lighting even though it is not warranted. If lighting is not available, the islands should be provided with pavement reflectors.

## 11.6.3 Public Utilities

The location and size of underground and overhead public utility services which are close enough to the intersection to be combined with any planned extension and should be determined in the preliminary stages of the design.

## 11.6.4 Parking

Parking of vehicles near intersections can obstruct the flow of turning traffic. Parking should be prohibited within the specified distance in the Road Traffic Ordinance.

On the approach side of a signalized intersection, parking should be prohibited for a distance large enough to store as many vehicles as can cross the STOP line in one phase from the curb lane. Parking should also be prohibited in the space between vehicle detectors and the intersection. There should also be effective parking restrictions on the exit side of the intersection to enable the curb lane vehicles to disperse.

## 11.6.5 Traffic Signs and Lane Markings

The layout of traffic signs and lane markings should be considered as part of the intersection design. However, the efficiency of a channelized layout should not depend entirely on signs. Once a driver reaches the intersection, the lane markings and layout of the islands should indicate the paths to be followed. Therefore, the signposting and lane marking in advance of the intersection are the most effective which directs the drivers into the appropriate traffic lane before they enter the intersection.

Effective signposting provides the driver with sufficient time to understand and decide on a maneuver before reaching the point where that maneuver must take place. All traffic signs and traffic control devices must follow the relevant standard. Figure 11.27 indicates the typical lane and pavement markings for an urban intersection.

## 11.6.6 Drainage

The provision of proper drainage facilities is important for any road and at intersections. Proper provision responding to the requirements for an effective drainage system and protection of the intersection from flooding should be considered as an integral part of the design of the intersection.

## 11.6.7 Landscaping

Landscaping has been a common neglected feature in the design of roads and at intersections. In the consideration of landscaping for intersections, the element of road safety must neither be neglected nor be compromised. Within the intersection area, only low shrubs/trees should be considered to ensure that sight distances are unaffected.

## 11.6.8 Stop Line

### 11.6.8.1 General

Stop lines show the boundary beyond which no vehicle can stop. They should be located before the entrance to signalized intersections, at the nearside of pedestrian crossings, and on the minor road at the entrance to "Stop Controlled" intersections.

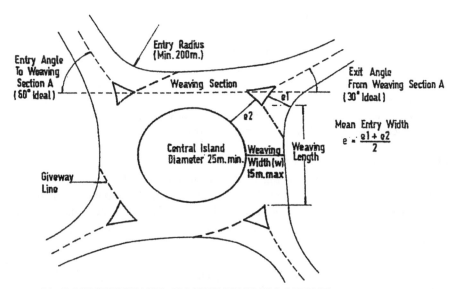

## A) CONVENTIONAL ROUNDABOUT LAYOUT

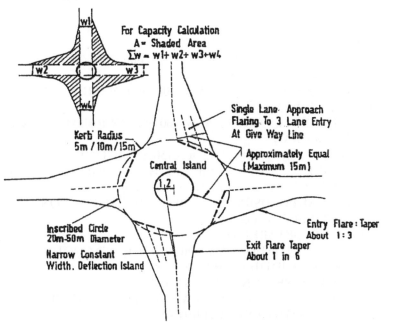

## B) LAYOUT OF SMALL ROUNDABOUT

**FIGURE 11.27**   Notation for Capacity Calculation.
*Source:* Figure 4.5, A Guide to the Design of At-Grade Intersections, JKR 201101-0001-87

Stop line should generally be located:

- Perpendicular to the centerline of the road.
- At 1–2 m before any pedestrian crossing.
- The location from where drivers can have a sufficient sight distance along the intersecting road.
- Where the stopped vehicles will not create any obstacle to the movement of vehicles turning in from the intersecting road.

### 11.6.8.2  Stop Line on Minor Road

The vehicle stopping at the stop line on the narrow minor road would often obstruct the movement of vehicles turning in from the major road. The stop line is usually set back by several meters from the normal position to resolve this issue. This, however, causes a problem in sight distance. In such a case one of the following measures should be taken.

- To make a corner rounding large enough to provide necessary sight distance.
- To signalize the intersection.
- To provide a traffic mirror if traffic from the minor road is light.

1. CONTINUITY LINE (S = 1 m W = 150 Mm)
2. GIVE WAY LINE (S = 300 mm G = 300 mm W = 200 mm) WHERE LENGTHS OF TAPER ARE SHORT.
3. LANE LINE (S = 1 m G = 1.7 m W = 100 mm) IN URBAN AREA WITH CENTRAL MEDIAN
4. LANE LINE (S = 4.5 m G 7.5 m W 100 mm) IN RURAL AREA WITH CENTRAL MEDIAN
5. CHANELISING LINE (CONTINUOUS 150 mm. WIDTH)
6. CENTER LINE (S = 4.5 m 6 = 7.5 m WIDTH = 100 mm) IN RURAL AREA.
7. EDGE LINE AT TRAFFIC ISLAND (W = 150 mm)
8. EDGE LINE AT ROADSIDE (W = 150 mm) WHITE FOR CONTINUOUS GUIDE AND YELLOW FOR
9. PROHIBITION OF PARKING
10. STOP LINE (CONTINUOUS 300 mm WIDTH) AND CONTINUOUS 450 mm OR 600 mm WIDTH) WITH PEDESTRIAN LINE.
11. PEDESTRIAN LINE (CONTINUOUS 300MM WIDTH)
12. LANE LINE (5 = 2.70 m G = 4.5 M W = 100 mm) ON CURVE AND LOW SPEED AREA WITH CENTRAL MEDIAN
13. DOUBLE LINE (W = 100 mm) FOR NO OVERTAKING ON CURVE
14. CENTERLINE FOR SINGLE CARRIAGEWAY IN URBAN AREA (S = 1 m G = 15 m, W = 100MM) FOR DUAL CARRIAGEWAY WITHOUT CENTRAL MEDIAN (S = 1 m G = 15 W = 100MM OR DOUBLE LINE)
15. CENTER LINE (S = 2.70 m) G = 4.5 M W = 100 mm) ON CURVE AND LOW SPEED AREA IN RURAL AREA
16. YELLOW BOX LINE S = STROKE G = GAP W = WIDTH

## 11.7  GENERAL WARRANTS FOR TRAFFIC CONTROL SIGNALS

### 11.7.1  GENERAL

The request to install new traffic signals or upgrading existing signalized locations may originate from various sources. The most usual sources include:

- Responsible agencies (e.g., City Halls, Municipalities, etc.).
- Traffic Enforcement Agencies (e.g., Police).
- Transporters' Association, Industrial or commercial developers and operators.
- Media/General Public.

Upon receiving the request, the responsible agency must determine whether such a request is justified. For this purpose, a set of criteria for verification were developed. These criteria should be viewed as guidelines, not as hard and fast values. The satisfaction of criteria does not guarantee that the signal is needed. Conversely, if any of the criteria are not fully satisfied does not mean that signalization would not serve a useful purpose. Awareness about local conditions and sound engineering judgment helps to understand the guidelines more effectively.

In general, the following steps should be taken before the installation to traffic signal control:

- Determine the function of the intersection in connection with the overall road system. A system of major roads should be designated to channel major flow from one part of the city to another. Intersection controls must be related to the major road system.
- A comprehensive study of traffic data and physical characteristics of the location is essential to determine the need for signal control along with proper design and operation of the control.
- Determine whether the geometric or physical improvement or regulations can provide a better solution to the problem of safety or efficiency than the installation of signal control.
- Use an established warrant to determine whether signal control for the intersection is justified.

### 11.7.2  WARRANT ANALYSIS

Generally, the following warrants should be considered before installing signal control, these are:

- Vehicular Operations.
- Pedestrian Safety.
- Accident Experience.

Traffic control signals should generally not be considered unless one or more of the warrants in this guideline are met.

#### 11.7.2.1  Warrant 1: Vehicular Operations

*11.7.2.1.1  Total Volume*

Vehicular traffic volume affects the efficiency and the LOS of an intersection. High traffic volume on the major road, especially during peak hours, invariably causes considerable delay to the traffic on the minor road. To determine the need for signal control, traffic volumes on both the major and minor roads should be considered. Signal control is warranted if the traffic volume for any 8 hours of an average day meets the minimum requirements in Table 11.20. For the major road, the total volume of both the approaches and for the minor road, the higher volume approach (one direction only) is used. An "Average" day is defined as a weekday representing volumes normally and repeatedly found at the location.

1. The total volume of both approaches
2. Higher volume approach only

*11.7.2.1.2  Peak Hour Volume*

In case for one peak hour of the day, traffic conditions are such that minor road traffic experiences undue delay or hazard in entering or crossing the major road, the peak hour volume may be used to determine the need for signalization. This criterion warrants signalization when the peak hour major road volume (total vehicles per hour (vph) for both approaches) and the higher volume minor street approach (vph for one direction only) fall above the curve for a given combination of approach lanes shown in Figure 11.28.

The warrant becomes lower when either the 85-percentile speed of major street road exceeds 60 km/hr., or when the intersection lies within an urban area. The peak volume requirement is fulfilled when the volumes referred to fall above the curve for the given combination of approach lanes shown in Figure 11.29.

*11.7.2.1.3  Progressive Movements*

In some locations, it may be desirable to install a signal to maintain proper grouping of vehicles and regulate group speed even if the intersection does not satisfy other

---

#### TABLE 11.21
#### Vehicular Volume Requirements for Warrant 1

| | | Minimum Requirements | | | |
|---|---|---|---|---|---|
| | | Major Road Flow (vph1) | | Minor Road Flow (vph1) | |
| Number of Lanes Each Approach | | | | | |
| Major Road | Minor Road | Urban | Rural | Urban | Rural |
| 1 | 1 | 500 | 350 | 150 | 105 |
| 2 or more | 2 or more | 600 | 420 | 150 | 105 |
| 2 or more | 1 | 600 | 420 | 200 | 140 |
| 2 or more | | 500 | 350 | 200 | 140 |

*Source:* Table A.1, A Guide to the Design of At-Grade Intersections, JKR 201101-0001-87

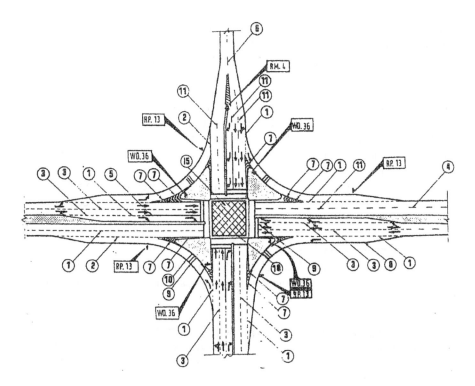

**FIGURE 11.28**   Typical Lane and Pavement Markings.
*Source:* Figure 5.1, A Guide to the Design of At-Grade Intersections, JKR 201101-0001-87

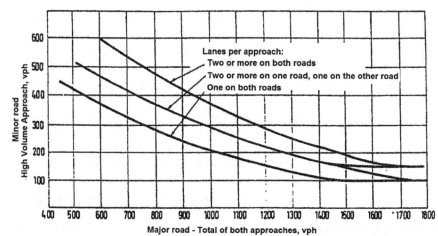

Note: 150 vph applies as the lower thresold volume for a minor road approach with two or more lanes and 100 vph applies as the lower thresold volume for a minor road approach with one lane.

**FIGURE 11.29**   Peak Hour Volume Warrant – Urban or Low Speed.
*Source:* Figure A.1, A Guide to the Design of At-Grade Intersections, JKR 201101-0001-87

warrants for signalization. Several advantages may generate from this type of consideration. Moving the traffic in groups at the desirable speed would reduce the number of stops and delays. Accident reduction may also be expected with the reduction of stops and speeds.

On a one-way road or a road with predominantly unidirectional traffic, this warrant applies when the adjacent signals are so far apart that they do not provide the necessary vehicle grouping and speed control. On a two-way road, the warrant is satisfied when the adjacent signals do not provide the necessary level of grouping and speed control and when the proposed and adjacent signals could constitute a progressive signal system.

A signal installation under this warrant should be based on the 85-percentile speed unless a traffic engineering study indicates that another speed is more appropriate.

### 11.7.2.2  Warrant 2: Pedestrian Safety

A signalized intersection also has higher pedestrian safety. It is warranted for signalization when for each of any 8 hours of an average day the following traffic volume exists:

- On the major road 600 or more vph enter the intersection, which is the total of both approaches or where there is a raised median island 1.2 m or more in width, 1,000 or more vph, which is the total of both approaches, enter the intersection on the major road and
- During the same 8 hours as mentioned, there are 150 or more pedestrians per hour on the highest volume crosswalk crossing the major road.

When the 85-percentile speed of the major road traffic exceeds 60 km/hr in either an urban or a rural area, or when the intersection lies within the built-up area of an isolated community having a population of less than 10,000, the minimum pedestrian volume is considered 70% of the requirements above.

A signal installed at an isolated intersection, under this warrant, should be of the traffic "Actuated Type" with push buttons for pedestrians who want to cross the main road. If such a signal is installed at an intersection within a signal system, it should be equipped and operated with a control device that provides proper coordination.

Special considerations should be given at schools where children in large numbers cross a major road on the way to and from school. The requirement for school children to access the school and bus stop is to be considered either by providing pedestrian Underpass, or Foot Over Bridge or Subway, or Signalized Road Cross.

## 11.8  COMPUTER-AIDED DESIGN OF AT-GRADE INTERSECTIONS

In this session, we shall see the design of an at-grade traffic Intersection for a Four-legged junction with "Tutorial Example Data". The design is based on the alignment of two intersecting main roads and four slip roads for the turning traffic. The design of alignment includes drawing of polylines for horizontal alignment of above six

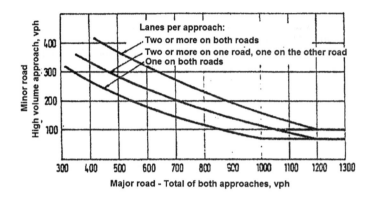

Note: 100 vph applies as teh lower threshold volume for a minor road approach with two or more lanes
and 75 vph applies as the lower threshold volume for a minor approach with one lane.

**FIGURE 11.30**    Peak Hour Volume Warrant – Rural or High-Speed.
*Source:* Figure A.2, A Guide to the Design of At-Grade Intersections, JKR 201101-0001-87

roads, the direction of polyline should be in the direction of the traffic flow in each road. So, the drawing of the polyline is in opposite direction for UK and USA styles of traffic flows. The process essentially includes the steps in the guide as mentioned below.

The computer applications for the "design of at-grade intersections" are explained in Chapter 11 in the "The Guide for Computer Application Tutorials" part of the book available for download from the website of the book.

To download the items under book tutorials, visit the website at: www. roadbridgedesign.com.

For downloading the other items, visit the website at: www.techsoftglobal.com/ download.php.

## BIBLIOGRAPHY

*A Policy on Geometric Design of Highways and Streets 2001*, AASHTO.

*Overseas Road Note 6, A Guide to Geometric Design*, Overseas Unit, Transport and Road Research Laboratory, Department of Transport, Crowthorne, Berkshire, United Kingdom.

*A Guide on Geometric Design of Roads*, Arahan Teknik (Jalan) 8/86, JKR, Malaysia.

*A Guide to the Design of At Grade Intersections*, JKR 201101-0001-87, JKR, Malaysia.

*Design Standards, Interurban Toll Expressway System of Malaysia, Government of Malaysia, Malaysia Highway Authority*, Ministry of Works and Utilities.

*Design Manual for Roads & Bridges*, Ministry of Public Works and Kuwait Ministry, State of Kuwait.

*Geometric Design Manual for Dubai Roads*, Dubai Municipality, Roads Department.

*Geometric Design Standards for Rural (Non-Urban) Highways*, IRC 73-1980, IRC.

*Vertical Curves for Highways*, IRC SP 23-1993, IRC.

*Manual of Specifications & Standards for Four Laning of Highways Through Public Private Partnership*, IRC SP 84-2009, IRC.

*Highway Engineering* by Paul H. Wright and Karen K. Dixon, by John Wiley & Sons, Inc.

# 12 Design of Grade-Separated Interchanges

## 12.1 GENERAL

This section provides an overview of the common types of grade-separated interchanges, describes elements of interchanges, and design requirements for each type. Grade-separations are used to enable two roads to cross each other without interconnection by using a bridge or underpass.

Fully grade-separated (or free flow) interchanges are used to vertically separate some or all of the conflicting streams of traffic by using one or more bridges or underpasses. Conflicts are eliminated, leaving only merging and diverging movements, which occur at ramp terminals.

There are two different traffic systems followed by individual countries in the world, one of these two is "right-hand drive with left-sided traffic," which is known as a UK style of the traffic system, and the other is "left-hand drive with right-sided traffic," which is known as a USA style of the traffic system. UK, India, Malaysia, and Singapore follow the UK style of the traffic system, whereas the USA, Europe, South America, Japan, and Australia follow the USA style of the traffic system. In this chapter the UK style of the traffic system is followed, which may be reversed in the case of the USA style of the traffic system.

## 12.2 GENERAL PRINCIPLES

There are three general types of intersections:

- At-grade intersections.
- Highway grade-separations without ramps.
- Interchanges.

Each of these has a practical usage in respect of the situation, but the applications are not exclusively defined. There is much overlap with the types of intersections for a situation, and the final selection of an intersection type frequently is a compromise after combined consideration of the class of intersecting roads, degree of access control, design traffic volume and pattern, cost topography, and availability of right of way. The general principles in the selection of interchanges are as follows:

### 12.2.1 TRAFFIC AND OPERATION

Through traffic is provided with no difficulty or delays at highway grade-separations. Ramps at interchanges have no severe effect on through traffic as long as its capacity is adequate, the merging or speed-change lanes are of adequate length, or a full complement of turning roadways is provided.

Turning traffic can affect the operation and is handled to the best possible extent on at-grade intersections and interchanges. Ramps are provided at interchanges for the turning movements, where turning movements are light. Some provisions are essential for all turning movements, a one quadrant ramp design may meet the requirements. Ramps are provided in two quadrants where they can be positioned so that the roads crossing and the major highway is free of such interference. An interchange with a separate ramp for every turning movement is suited for heavy volumes of through traffic and for any volume of turning traffic, provided the ramps and terminals are designed with sufficient capacity.

Confusion by some drivers appears unavoidable on interchanges, but such difficulties are minor when compared with the benefits derived from the reduction of delays, stops, and accidents. The confusion is minimized with improved design and increased use of signing and other control devices for the interchange. Interchanges are also adaptable to all kinds of traffic. The presence of a high proportion of heavy trucks in the traffic stream makes the provision of interchanges especially desirable.

### 12.2.2 SITE CONDITIONS

In rolling or hilly topography, interchanges can usually be well fitted to the existing ground, and the through roads can be designed to higher standards.

The design of the interchange is simple in flat terrain, but it may be necessary to introduce grades that may not favor operation. Interchanges in flat terrain generally are not as pleasing in appearance as those applied to rolling terrain. The deficiency in appearance can be improved when it is possible to re-grade the whole of the interchange area with proper landscaping.

### 12.2.3 TYPE OF HIGHWAY AND INTERSECTING FACILITY

As the hazards generated by stoppages and direct turns at an intersection increase with the design-speed, a high design-speed highway or expressway warrants an interchange earlier than low design-speed roads, despite having similar traffic volumes. The ramps provided for high design-speed highways should permit suitably high turning speeds, for which built-in facilities for changing speed are considered more important.

The selection of the interchange type will vary with the terrain, development along the highway, and right of way conditions, but in general, it will be based on ramp layouts that expedite the entrance to or the exit from the major road or expressway. Ramp connections may also require the provision of frontage roads.

Local service is easy to provide for with certain types of at-grade intersections, whereas considerable additional facilities may be required in the case of interchanges.

### 12.2.4 SAFETY

The accidents caused by crossing and turning movements are reduced by separating the grades of the intersecting roadways. But any grade-separation structures are themselves somewhat of a hazard. However, this can be minimized by the use of adequate clear roadway widths and protective devices at bridge abutments and piers.

Depending on the type of interchange used in the UK style of traffic, right-turns may be eliminated or confined to the lower classified crossroad. Left-turning traffic can be accommodated with a high design standard of ramps, which provide operation approaching the equivalent of free flow so conflicts caused by crossing traffic will be eliminated or minimized. In the USA style of traffic, the above mentioned turns will be opposite.

## 12.2.5 STAGE DEVELOPMENT

The projection of traffic for the design should be for a period of 20 years after completion of the road. In areas where traffic estimation is difficult due to the uncertainty of land use or the planning of roadside interference, the design of the formation width shall be based on a period of 20 years, but some of the ramps and pavement construction may be based on 10 years for the initial stage.

Where the development decides to use a single grade-separation structure, staged construction may not be economical unless provisions are made in the original design for future stages of construction. However, ramps are well adapted to staged development.

## 12.2.6 ECONOMIC FACTORS

### 12.2.6.1 Initial Cost

The interchange is an expensive type of intersection. The combined costs of the structure, ramps, through roadways, grading, and landscaping of large areas, possible adjustments in existing roadways and utilities are generally much higher than the cost of an at-grade intersection. The cost of directional interchanges is usually even higher than any simple interchange, as those involve more than one structure.

### 12.2.6.2 Maintenance Cost

Each type of intersection has higher maintenance costs. Interchanges have large pavements and variable slope areas, the maintenance of which, together with that of the structure and landscaping, is much higher than that of an at-grade intersection. Interchanges often involve maintenance and operational costs for the lighting as well.

### 12.2.6.3 Vehicular Operating Cost

Through traffic at an interchange usually follows a direct path on the elevated route, with only minor speed reduction. When grades are steep, the added vehicular costs for the rise and fall due to passing over and under the structure may need to be considered. Left-turning traffic is subject to added vehicular costs of deceleration and acceleration, and may also be subject to the costs of operation on a grade, but travel distance and time is usually shorter when compared to an at-grade intersection. Right-turning traffic is also subject to the added costs of acceleration and deceleration and usually adds to the travel distance compared to direct right-turns at-grade. Directional ramps may eliminate large speed changes, saving travel distance and time compared to at-grade intersections.

For intermediate to heavy traffic, the total vehicle operating costs at an intersection are usually higher than those at an interchange, especially if through movements are dominant.

## 12.3  JUSTIFICATIONS FOR GRADE SEPARATION AND INTERCHANGES

An interchange is a preferred solution for many intersection problems, but because of the high initial cost, it is normal practice to eliminate existing traffic bottlenecks or to correct existing hazardous conditions in the intersection. Replacement of an intersection with an interchange is limited to situations where the cost can be justified, however, due to the widely varying site conditions, traffic volumes, highway types, and interchange layouts, the cost may differ at each location. The conditions that should be considered to reach a decision are as follows:

### 12.3.1  DESIGN DESIGNATION

The decision to develop a highway with full access control between selected terminals becomes the justification for providing highway grade-separations or interchanges for all intersecting highways. Although access control, provision of medians, and elimination of parking and pedestrian traffic are still important measures, the separation of grades on highways provide the greatest increment of safety. An intersection that might justify only traffic signal control when considered as an isolated case, may require a grade-separation or interchange when considered as an integral component of the highway.

### 12.3.2  ELIMINATION OF BOTTLENECKS OR STOP CONGESTION

If the required capacity for an at-grade intersection is unachievable, then there is justification for an interchange where its development and required right of way is possible. Even facilities with partial control of access, the elimination of random signalization contributes greatly to the improvement of free flow characteristics.

### 12.3.3  ELIMINATION OF HAZARDS

At-grade intersections on highways that pass through rural areas, where the traffic volume is low, resulting in high travel speeds, have a disproportionate rate of serious accidents. If there is a lack of inexpensive methods for eliminating hazards, then a highway grade-separation or interchange may be justified.

In rural areas, structures can usually be constructed more cheaply compared to urban areas, and the right of way is equally inexpensive. Due to cost minimizations, the developments can be justified for the elimination of serious accidents. At heavily traveled intersections, serious accidents also justify an interchange. In addition to greater safety, the interchange also facilitates better traffic movement.

### 12.3.4 Site Topography

At some sites, designs for grade-separation are the only option that are economically feasible if the topography at the site does not allow for any other type of intersection which meets the required standards, and is physically impossible to develop at equal cost.

### 12.3.5 Road-User Benefits

Road-user costs at congested at-grade intersections are large due to delays, fuel, tires, oil, repairs, travel time, and accidents that require speed changes, stops, and waiting, which generate significant additional cost compared to an intersection permitting the uninterrupted or continuous operation. In general, interchanges require somewhat more total travel distance than direct crossings at-grade, but this is compensated by the gains due to the reduction in stopping delay. The relation of road-user benefits to the cost of improvement indicates an economic justification for an interchange.

Furthermore, interchanges are usually planned in staged construction, and initial stages may produce incremental benefits that would compare even more favorably with incremental costs.

### 12.3.6 Traffic Volume Warrant

Traffic volumes exceeding the capacity of an at-grade intersection would certainly be a warrant. Although a specific volume of traffic cannot be predetermined, it is nevertheless an important guide, particularly when combined with the traffic distribution pattern and the effect of traffic behavior.

### 12.3.7 Other Justifications

Additional justifications for grade-separation and interchanges that can be considered are as follows:

- Local roads and streets that cannot be feasibly terminated outside the right of way limits of expressways.
- Access to areas not served by frontage roads or other means of access.
- Railway crossings.
- An unusual concentration of pedestrian traffic.
- Cycle tracks and routine pedestrian crossings.
- Access to mass transit stations within the confines of a major arterial road.
- Free flow aspects of certain ramp configurations and completing the geometry of interchanges.

### 12.3.8 Justification for Class of Road

If such justifications for the elimination of bottlenecks and hazards, as well as the site topography, do not arise, and specific additional information on the design

**TABLE 12.1**

**Selection of Intersection Type**

a. Rural Area

| Expressway | Highway | Primary | Secondary | Minor | |
|---|---|---|---|---|---|
| IC. | IC. | IC. | IC. | | **Expressway** |
| | IC. | IC./S.I. | S.I./S.C. | S.C. | **Highway** |
| | | S.I. | S.I./S.C. | S.C. | **Primary** |
| | | | S.C. | S.C. | **Secondary** |
| | | | | S.C. | **Local** |

b. Urban Area

| Expressway | Arterial | Collector | Local Street | |
|---|---|---|---|---|
| IC. | IC. | – | – | Expressway |
| | IC./S.I. | S.I. | S.I./S.C. | Arterial |
| | | S.I. | S.C. | Collector |
| | | | S.C. | Local Street |

*Source:* Table 2.1, A Guide to the Design of Interchanges, Arahan Teknik (Jalan) 12/87, JKR
IC., interchange; S.I., sinalized intersection; S.C., stop control intersection

designation, road-user benefit, or traffic volume is not available, then Table 12.1 may be referred to, to determine whether an interchange is justified or not. The two alternatives, IC (interchange) and S.I. (signalized intersection) are given in the table, provision of an interchange can be considered if the total projected peak-hour traffic volume on the four approaches exceeds 8000 vph.

According to Class of Roads Crossing.

## 12.4   GRADE SEPARATION STRUCTURES

### 12.4.1   Types of Separation Structures

#### 12.4.1.1   General

The separation structure should match the natural lines of the highway approaches in alignment, profile, and cross section. In addition to these geometric considerations, other conditions such as the spans and depths of structures, foundation material at the site, aesthetics, safety, and especially skew may have a significant influence on the engineering and cost feasibility of the structure being considered.

#### 12.4.1.2   Overpass

In the interchange, the bridge deck-type structure is most suitable for the overpass highway. Although there are problems getting both lateral and vertical clearance of the underpass roadway, the supports to the upper roadway are underneath and out of sight.

#### 12.4.1.3   Underpass

For the underpass highway, the most desirable structure concerning vehicular operation is one that will have a single span over the entire highway cross section of the highway and provide lateral clearance from structural supports to the edge of pavement that is consistent with the design requirements.

On divided underpass highways, the center supports should only be used when the median is wide enough to provide sufficient lateral clearance, and with narrow medians, the center piers require protective barriers. Anticipating future widening, the piers or abutment design must provide footings with sufficient space for future widening.

Underpasses are given preference in urban areas and should be used wherever possible. However, careful attention should be given to drainage provisions and a foundation type involving geotechnical considerations.

## 12.4.2 OVERPASS VERSUS UNDERPASS

### 12.4.2.1 General Design Considerations

A detailed study is required for each proposed highway grade-separation to decide whether the main road should be carried over or under the structure. Commonly, the choice is dictated by features such as topography or highway classification.

The design that best fits the existing topography is the most pleasing and economical to construct and maintain. These factors are first considered in the design, the major exception to this is where a major road is primarily designed to suit the topographic and crossroad controls.

The point "(a)" below is common in the case of flat topography, here topography does not govern, and it will be necessary to study the secondary factors. The following general points may be noted:

a. There are certain advantages concerning traffic warning on an undercrossing highway. As a driver approaches, the structure looms ahead, this makes the upper-level crossroad obvious and gives an advance alert of likely interchange connections. This enables the driver to select the right route. However, where an undercrossing highway dips beneath a crossroad, which is at ground level, this advantage is minimized.

b. To give aesthetic preference to the through traffic, a layout is designed where the most important road is the overpass. If a wide overlook is possible from the structure and its approaches, then drivers will feel less restricted.

c. Where turning traffic is significant, the major road is placed at the lower level so that the ramp profiles are best suited. The ramp grades then assist turning vehicles to accelerate as they approach the major road at a lower level and to decelerate as they leave the major highway to reach a higher level of the intersecting road, rather than the reverse.

d. Where there is no pronounced advantage to the selection of either an underpass or an overpass, the type that provides the greatest sight distance on the major road should be preferred. If the road is two-lane, the preferred sight distance is desirable for safe passing distance.

e. An overpass offers the best consideration for staged construction, both for the highway and the structure, with minimum impairment of the original investment. By the lateral extension of both, or construction of, a separate structure and roadway for a divided highway, the ultimate development is therefore reached without changing the initial facility.

f.  Troublesome drainage problems may be reduced by carrying the major highway at the elevated level without altering the grade of the crossroad. In some cases, the drainage problem alone may be sufficient as the reason for choosing to take the major highway to the elevated level rather than under the crossroad.

g.  Where topography control is not essential, a cost analysis, for the bridge type, span length, roadway cross section, angle of skew, and cost of approaches, will determine which of the two intersection roadways should be elevated on the structure.

h.  An underpass may be more advantageous where the major road can be built at the existing ground level, with continuous gradient, and with no pronounced grade changes.

i.  Where a new highway crosses an existing route carrying a large volume of traffic, the crossing by the new highway at an elevated level causes less disturbance to the existing route, and any realignment is usually not required.

j.  The overcrossing structure has no restriction for vertical clearance, which shall be a significant advantage in the case of oversized loads carried by trailers and requiring special permits on a major highway or route.

k.  Desirably, the roadway carrying the highest traffic volume should have the minimum number of bridges for better travel comfort and fewer conflicts when repair and reconstruction are to be carried out.

l.  In some instances, it may be necessary to have the higher volume facility depressed and crossing under the lower volume facility to reduce noise impact to surroundings.

### 12.4.3  Cross Section of Structures

#### 12.4.3.1  Structure Widths

The widths for all structures should follow the requirements as laid down in relevant standards for the geometric design of roads.

#### 12.4.3.2  Clearances

The vertical, as well as horizontal clearances, should also follow the requirements as laid down in relevant standards for the geometric design of roads.

#### 12.4.3.3  Barriers

Protective barriers should be considered for the abutments, piers, and medians when lateral clearances are minimum. The guidelines given in relevant standards for the geometric design of longitudinal traffic barriers should be followed.

### 12.4.4  Grade-separation Without Ramps

There are many situations where grade-separations are made without ramps. Ramps are omitted to prevent interchanges occur close to each other which may result in signing and operation difficult. Ramps are also omitted to eliminate interference with the major road having large traffic volumes and to increase safety and mobility by shifting turning traffic at a few points where it is feasible to provide adequate ramp systems. On the other hand, if such shifting of turning traffic causes the undue concentration of turning movements at one location that should be avoided where it would be better to have a greater number of interchanges.

Where ramp connections are difficult or expensive, it may be practical to omit them at the structure site and shift turning movements elsewhere by way of other intersecting roads.

## 12.5   INTERCHANGE TYPES

### 12.5.1   THREE LEG DESIGN

An interchange with three leg design type is one at an intersection with three intersecting legs consisting of one or more grade-separations and one-way flow for all traffic movements. When all three intersection legs have a through character or the intersection angle with the third intersection leg is small, the interchange may be considered as "Y" type.

Figure 12.1 illustrates the patterns of three leg interchanges with one grade-separation and Figure 12.2 illustrates high-type T and Y interchanges each with more than one structure or with one three-level structure that provides for all of the movements without loops.

Regardless of factors such as the intersection angle and through road character, anyone basic interchange pattern may apply for a wide variant of conditions.

### 12.5.2   FOUR LEG DESIGN

#### 12.5.2.1   General

Interchanges with four intersecting legs may be broadly grouped into the following four types:

- Ramps in one quadrant (ramps in two or three quadrants are known as partial cloverleafs).
- Diamond interchanges.
- Partial cloverleafs.
- Full cloverleafs.

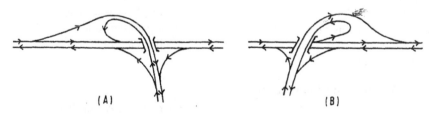

**FIGURE 12.1**   Three Leg Interchange with Simple Structure.
*Source:* Figure 4.1, A Guide to the Design of Interchanges, Arahan Teknik (Jalan) 12/87, JKR

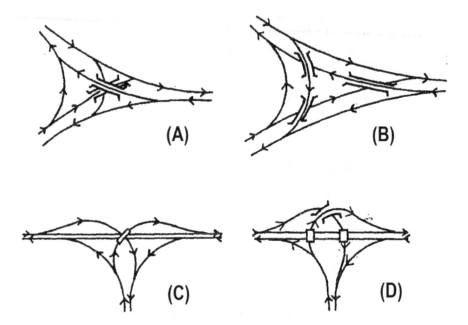

**FIGURE 12.2**    Three Leg Interchange with Multiple Structure.
*Source:* Figure 4.2, A Guide to the Design of Interchanges, Arahan Teknik (Jalan) 12/87, JKR

### 12.5.2.2   Ramps in one Quadrant

Interchanges with a ramp in only one quadrant are applied for intersections of roads with low volumes. Where a grade-separation is provided at an intersection because of topography, even though volumes do not justify the structure, a single 2-way ramp of minimum design shall be adequate for all turning traffic. The ramp terminals may be plain "T" intersections, as illustrated in Figure 12.3.

These types of design are applied at limited locations. At some interchanges, ramp development may be limited to one quadrant because of topography, culture, or other controls, even though the traffic volume demands extensive turning facilities. In such a case, a high degree of channelization at the terminals and the median, together with right-turn lanes for the UK style of traffic or left-turn lanes for USA style traffic, on the through highway is required to control turning movements.

Where a one quadrant interchange is constructed as the initial step in a stage construction program, the initial ramps should be designed as part of the overall development. One or both terminals can be the interchanges of "T" type design. This makes shifting of the interchanges from the place of grade-separation structures. This concept can be applied to situations where ramp connections at-grade-separation structure are difficult.

### 12.5.2.3   Diamond Interchanges

This is the simplest and most common type of interchange. A full diamond interchange is formed when a one-way diagonal type ramp is provided in each quadrant.

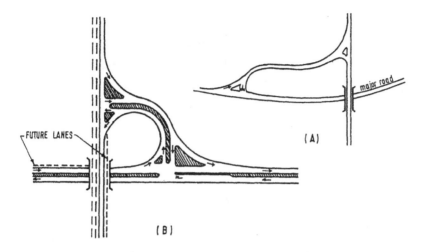

**FIGURE 12.3**  Four Leg Interchanges, Ramps in One Quadrant.
*Source:* Figure 4.3, A Guide to the Design of Interchanges, Arahan Teknik (Jalan) 12/87, JKR

The ramps are aligned with free flow terminals on the major highway and the at-grade right-turns (for UK style of traffic) are confined to the crossroad.

The diamond interchange has several advantages over a comparable partial cloverleaf in that all traffic can enter and leave the major road at relatively high speeds, right-turning movements (for UK style of traffic) have little extra travel and a relatively narrow right of way is required.

Diamond interchanges are applied in both rural and urban areas. These are particularly adaptable to major-minor crossings where right-turns (for UK style of traffic) at graded on the minor road can be handled without hazard or difficulty. The intersection on the crossroad presents a problem in traffic control to prevent wrong way entry from the crossroad. To resolve this issue, a median on the crossroad is important to facilitate proper channelization. The additional signing will also help to prevent improper use of ramps and should be included in the design.

Diamond interchanges usually require signalization when the traffic volume is moderate to heavy in the crossroad. A single-lane ramp usually serves traffic from the through highway, but it may have to be widened to two or three lanes to develop the necessary capacity for the at-grade condition. This would ensure that stored vehicles would not exceed too far along the ramps or onto the highway, thereby causing an undesirable backup effect on the major through road.

Diamond interchanges can be designed with a variety of patterns. Figure 12.4 illustrates examples of conventional types while Figure 12.5 illustrates diamond interchange arrangements to reduce traffic conflicts, by utilizing split diamond or frontage roads and separate turnabout provisions. A serious disadvantage of the split diamond is that traffic leaving the highway cannot return to the major through road at the same interchange and continue in the same direction. Figure 12.6 shows diamond interchanges with more than one structure.

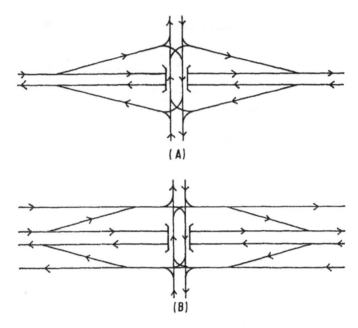

**FIGURE 12.4**   Diamond Interchanges, Conventional Arrangements.
*Source:* Figure 4.4, A Guide to the Design of Interchanges, Arahan Teknik (Jalan) 12/87, JKR

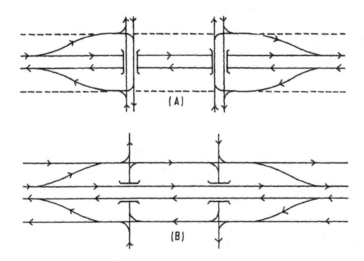

**FIGURE 12.5**   Split Diamond Interchanges.
*Source:* Figure 4.5, A Guide to the Design of Interchanges, Arahan Teknik (Jalan) 12/87, JKR

Traffic signals are used in high-volume situations, and their efficiency is dependent on the relative balance in right-turn volumes for the UK style of traffic. They are normally synchronized to provide continuous movement through a series of right-turns once the area is entered.

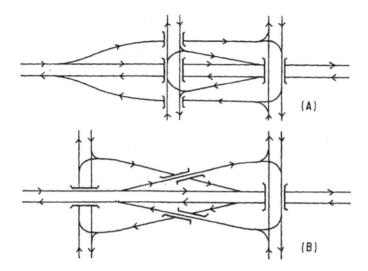

**FIGURE 12.6**   Diamond Interchanges with Additional Structures.
*Source:* Figure 4.6, A Guide to the Design of Interchanges, Arahan Teknik (Jalan) 12/87, JKR

### 12.5.2.4   Cloverleaf

Cloverleafs are four leg interchanges that employ loop ramps to accommodate right-turning (for UK style of traffic) movements. Interchanges with loops in all quadrants are referred to as "full cloverleafs" and all other layouts as "partial cloverleafs."

The principal disadvantages of the cloverleaf are the extra travel distance required for right-turning (for UK style) traffic, the weaving maneuver is generated, the very short weaving length typically available and the relatively large right of way required. When collector-distributor roads are not used, further disadvantages are weaving on the main carriageways, double exits, and problems associated with signing for the second exit.

Because cloverleaf interchanges are expensive considerably more than diamond interchanges, therefore, they are less adaptable in urban areas and are better suited to suburban or rural areas where space is available and there is a need to avoid restrictive at-grade right-turns (for UK style of traffic).

There is an advantage in cloverleaf interchanges for a higher speed, but this must be compared against the disadvantages of extra travel time because of the longer distance and higher cost of a significantly large right of way.

When the sum of traffic on two adjoining loop approaches is about 1,000 vph, interference in weaving occurs rapidly, leading to reduction of capacity of the through traffic. Adequate weaving lengths are thus very important and should be provided. The weaving section should be transferred from the through lanes to a collector-distributor road when the weaving volume exceeds 1,000 vph.

A loop with a single-lane carriageway has a design capacity of 800–1,200 vph. Loop ramp capacity is thus a major control in cloverleaf designs. Loops may be made to operate with two lanes, but this would require careful attention to the design of the terminals. Two-lane loops are considered only in exceptional cases for not being economical from the viewpoint of the right of way, construction cost, and direction of travel.

When a full cloverleaf interchange is made in conjunction with a highway, collector-distributor roads adjacent to the major road should be used. For other highways, where traffic weaving is moderate, full cloverleaf interchanges without collector-distributor roads are acceptable

### 12.5.2.5  Partial Cloverleaf Ramp Arrangements

For partial cloverleafs, ramps should be so arranged that the entrance and exit turns create the least interference to the traffic flow on the major highway. The following points can be considered as the guidelines in the arrangements of ramps at partial cloverleafs:

- The ramp arrangement should enable the major turning movements to be made by left-turn exits and entrances (for UK style of traffic).
- Where the through traffic volume on a major highway is considered greater than that on the intersecting minor road, preference should be given to an arrangement placing the left-turns, either exit or entrance, on the major highway even though it results in a direct right-turn off the crossroad (for UK style of traffic).

At a particular site, topography and adjoining development may be the factors that determine the quadrants in which the ramps and loops can be developed.

Figure 12.7 illustrates various patterns of cloverleaf interchanges.

Figure 12.8 illustrates some patterns of partial cloverleaf ramp arrangements, schematically.

### 12.5.3  Directional and Semi-Directional Design

For important turning movements, direct or semi-direct connections are used to reduce travel distance, increase speed safety and capacity, eliminate weaving, and avoid improper driving on a loop. Higher levels of service by better functioning are also found on direct connections.

Where direct connections are designed with two lanes, the ramp capacity may acquire the capacity of an equivalent number of lanes on the through highway.

In rural areas, there the volume justification is rare for the provision of direct connections in more than one or two quadrants. The remaining right-turning movements are handled satisfactorily by loops or at-grade intersections.

A direct connection is defined as a one-way roadway without deviating greatly from the intended direction of travel. Interchanges that use direct connections for the major right-turn movements are termed as "directional interchanges." When one or more interchange connections are indirect in alignment despite those are more direct than loops, still, the interchange is described as semi-directional. On direct or semi-direct interchanges, the grade-separations more than one are usually involved.

At major interchanges, semi-direct or direct connections are often required in urban areas. Interchanges involving two expressways nearly always require

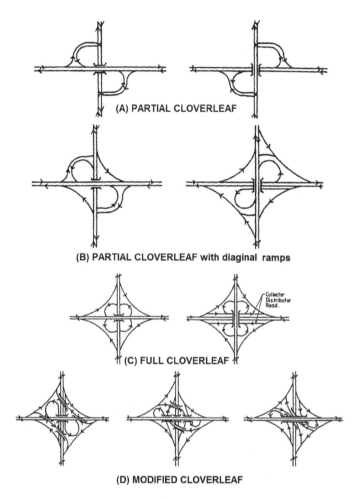

(A) PARTIAL CLOVERLEAF

(B) PARTIAL CLOVERLEAF with diaginal ramps

(C) FULL CLOVERLEAF

(D) MODIFIED CLOVERLEAF

**FIGURE 12.7** Illustrates Various Patterns of Cloverleaf Interchanges.
*Source:* Figure 4.7 (A), (B), (C) & (D), A Guide to the Design of Interchanges, Arahan Teknik (Jalan) 12/87, JKR

directional layouts. There are many layout schemes for directional interchanges that use various combinations of directional, semi-directional, and loop ramps. Any one of them may be appropriate in respect of a certain set of conditions, but only a limited number of layout patterns are used generally. These layouts utilize the least space, have the fewest or least complex structures, minimized internal weaving, and fit the common terrain and traffic conditions. Basic layout patterns of semi-directional interchanges are illustrated in Figure 12.9. These are basically with loops and involving no weaving maneuvers.

The layouts of fully directional interchange have no weaving as this is extremely undesirable. Right-hand exits and entrances (for UK style of traffic) are also undesirable but may be unavoidable because of site restrictions. The most widely used type of directional interchange is the four-level layout system. Layout patterns of full directional interchanges are illustrated in Figure 12.10.

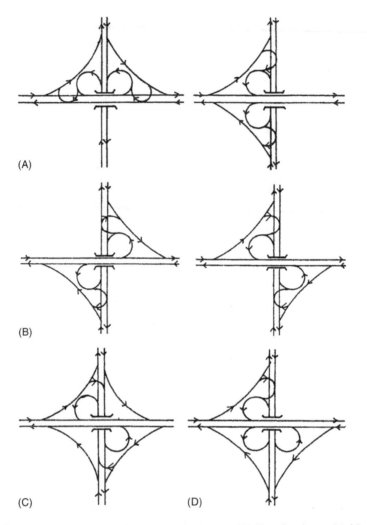

(A)

(B)

(C)                  (D)

**FIGURE 12.8**   Partial Cloverleaf Ramp Arrangements. (A) Two Quadrants Avoid applications if possible. (B) Two Quadrants Diagonally Opposite. (C) Four Quadrants. (D) Three Quadrants.

*Source:* Figure 4.8 (A), (B), (C) & (D), A Guide to the Design of Interchanges, Arahan Teknik (Jalan) 12/87, JKR

## 12.5.4 Rotary Design

A rotary interchange is a roundabout where the major through highway grade-separated. This design eliminates some of the deficiencies of at-grade intersections on the crossroad, and a smoother traffic flow than that of diamond interchange can be expected. Still, this design has the same disadvantages as the roundabout and two grade-separation structures are required in this design.

Through traffic on the minor crossroad has to travel some extra distance on a circularly turning roadway. A vast area is also required to provide sufficient weaving

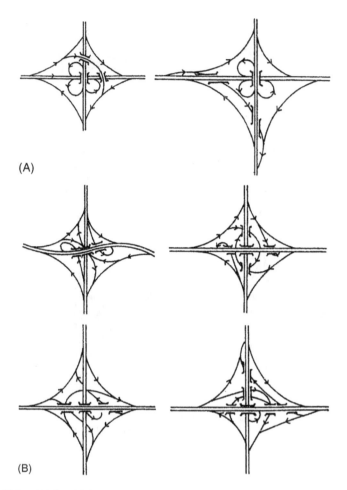

(A)

(B)

**FIGURE 12.9** (A) Semi-Direct Interchanges (with weaving). (B) Semi-Direct Interchanges (with no weaving).
*Source:* Figure 4.9 (A) & (B), A Guide to the Design of Interchanges, Arahan Teknik (Jalan) 12/87, JKR

distances. Therefore, the application of rotary design is limited to extremely exceptional cases, for these shortcomings.

Ramps of cloverleaf can be arranged in the same area required for a rotary interchange. Diamond interchanges can also be a substitute for this design.

Figure 12.11 illustrates some patterns of rotary interchanges.

## 12.5.5 Combination Interchanges

When one or two turning movements have very high traffic volume concerning the other turning movements, an analysis may indicate the need for a combination of two or more of the above discussed types of interchanges. Typical examples are as given below:

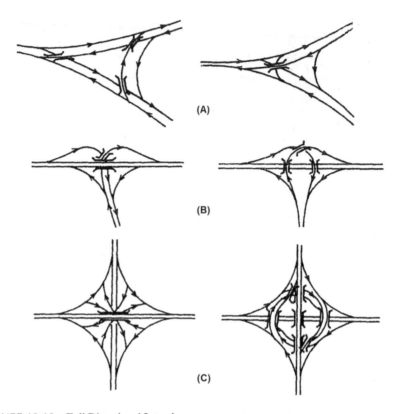

**FIGURE 12.10**   Full Directional Interchanges.
*Source:* Figure 4.10, A Guide to the Design of Interchanges, Arahan Teknik (Jalan) 12/87, JKR

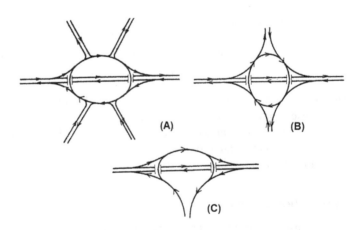

**FIGURE 12.11**   Rotary Interchanges.
*Source:* Figure 4.11, A Guide to the Design of Interchanges, Arahan Teknik (Jalan) 12/87, JKR

- A diamond with a semi-direct connection.
- A cloverleaf with a semi-direct connection.
- A one-half diamond with a one quadrant cloverleaf.

## 12.6 GENERAL DESIGN CONSIDERATIONS

### 12.6.1 INTERCHANGE TYPE DETERMINATION

#### 12.6.1.1.1 Systems Interchanges and Service Interchanges

Interchange types fall into two categories, these are: "systems interchanges" and "service interchange." Systems interchanges connect expressways to expressways while the service interchanges connect expressways to a lesser category.

#### 12.6.1.1.2 Interchanges in Rural Area

The type selection of interchange in rural areas is done based on service demand. When the intersecting roadways are expressways, all directional interchanges may be applicable for high turning volume.

The minimum design is a cloverleaf interchange which can be used at the intersection of two fully "Controlled Access" facilities or where at-grade right-turns are prohibited (for UK traffic type). The cloverleaf interchanges are suitable in rural areas where the right of way is not a constraint and weaving is minimal.

The final application of an interchange may be determined by the need for route continuity, uniformity of exit patterns, single-exit in advance of the reparation structure, elimination of weaving on the major road, signing potential, and availability of right of way.

#### 12.6.1.1.3 Interchanges in Urban Area

Generally, in urban areas, interchanges are closely spaced for which each interchange may be influenced directly by the preceding or following of interchange to the extent that additional traffic lanes may be required to achieve the desired capacity, weaving elimination, and lane balance.

On a continuous urban roadway, all the interchanges should be treated as an integrated system design rather than considering on an individual basis. Cloverleaf interchanges with or without collector-distributor road are not generally, practicable for the urban environment because of the excessive right of way requirements.

#### 12.6.1.1.4 Unbalanced Traffic Distribution

Where turning volumes are high for some movements and low for others, a combination of directional, semi-directional, and loop ramps may be appropriate. When loop ramps are used in combination with direct and semi-direct ramp designs, the loops should be so arranged that weaving sections are avoided.

Application of partial cloverleaf may be appropriate where either the rights-of-way is not available in one or two quadrants or where one or two movements in the interchanges are disproportionate to the others, especially when they require right-turns across traffic (for UK style of traffic), in such situation loop ramps may be utilized to accommodate the heavy right-turn volume.

### 12.6.1.1.5  Cloverleaf Interchanges
In deciding to use cloverleaf interchanges, careful attention should be given to the potential improvement in operational quality by including the collector-distributor roads on the major roadway, in the design.

### 12.6.1.1.6  Diamond Interchanges
The capacity of a diamond interchange is limited by the capacity of the at-grade terminals of the ramps at the crossroad. High through and turning traffic volumes could reject the use of a simple diamond unless signalization is used.

### 12.6.1.1.7  Capacity of Crossroad
The ability of the crossroad to receive and discharge traffic from the major road has a considerable impact on the interchange geometry. For example, loop ramps may be needed to eliminate heavy right-turns (for UK style of traffic) on a conventional diamond interchange.

### 12.6.1.1.8  Factors for Type Determination
After identifying several alternatives for the system design, they can be compared on the following aspects: (1) capacity, (2) route continuity, (3) uniformity of exit patterns, (4) single exits in advance of the separation structures, (5) with or without weaving, (6) potential for signing, (7) cost (8) availability of right of way, (9) potential for stage construction, and (10) compatibility with the existing environment. The most desirable alternatives are to be retained for plan development. Figure 12.12 illustrates the interchanges that are adaptable on expressways as related to classifications of intersecting facilities in a rural and urban environment.

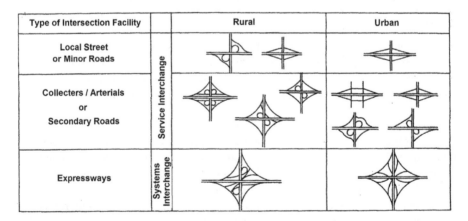

| Type of Intersection Facility | | Rural | Urban |
|---|---|---|---|
| Local Street or Minor Roads | Service Interchange | | |
| Collecters / Arterials or Secondary Roads | | | |
| Expressways | Systems Interchange | | |

**FIGURE 12.12**  Adaptability of Interchanges on Expressways as Related to Types of Interchange Facilities.
*Source:* Figure 5.1, A Guide to the Design of Interchanges, Arahan Teknik (Jalan) 12/87, JKR

## 12.6.2 APPROACHES TO THE STRUCTURES

### 12.6.2.1 Alignment, Profile, and Cross Section

#### 12.6.2.1.1 Major Highways

The geometric standards at the grade-separated interchange should be higher than those for the highway to compensate for the restriction caused by abutments, piers, curbs, and guard rails and to provide better sight distance for diverging and merging maneuvers. It is desired that the alignment and profile of the through highways at an interchange should be relatively flat with high visibility. Any sharp curve would induce the problem in the crossover crown between the super-elevated portion through the traffic lane and the pavement of the ramps located outside of the turning roadway. Sometimes it is possible to design only one of the intersecting roads on a tangent with flat grades. Preferably, the major highway should be so treated. Table 12.2 shows the desirable lowest elements for alignment and profile design of the through highways at interchange area.

#### 12.6.2.1.2 Cross Road

The gradients on the intersecting road at an interchange should be kept to a minimum and in no case should they exceed the maximums established for open-highway conditions. Reduction of vehicle speeds by lane upgrades encourages passing, which is hazardous in the vicinity of ramp terminals. Slow moving through vehicles also encourages abrupt cutting in by vehicles leaving and entering the highways.

#### 12.6.2.1.3 Alignment and Cross Section

In a grade-separated interchange without ramps, the alignment and cross section of the approaches does not invite any special problems except where a change in width is made to include a middle pier or where the median is narrowed for structure economy. At interchanges with ramps, the changes in alignment and cross section may be required to ensure proper operation and to develop the necessary capacity at the ramp

---

### TABLE 12.2
### Desirable Lowest Design Elements of Through Highways at Interchange Locations

| Highway Design Speed (km/h) | Minimum Radius (m) | Maximum Gradient (%) | Minimum Vertical Curve Length in K-Value | |
|---|---|---|---|---|
| | | | Crest | Sag |
| 120 | 2000 | 2 | 450 | 160 |
| 100 | 1500 | 2 | 250 | 120 |
| 80 | 1000 | 3 | 120 | 80 |
| 60 | 500 | 4.5 | 60 | 40 |
| 50 | 300 | 5 | 40 | 30 |

*Source:* Table 5.1, A Guide to the Design of Interchanges, Arahan Teknik (Jalan) 12/87, JKR

---

terminals, particularly where there is no full complement of ramps. Where a two-lane highway is carried through an interchange, provision of medians should be considered for high-speed or high-volume traffic conditions to prevent wrong way right-turns (for UK style of traffic).

### 12.6.2.2 Sight Distance

Sight distance on the highways through a grade-separation should preferably be at least equal or longer than for stopping. The passing sight distance is preferred, where exits are involved.

The horizontal sight distance limitations by piers and abutments at curves are a more difficult problem than that of vertical limitations. At curvature of the maximum degree for a given design-speed, the normal lateral clearance at piers and abutments of underpasses does not provide the minimum stopping sight distance. Similarly, on overpasses with the sharpest curvature for the designs speed, sight deficiencies result from the usual practical offset to bridge railings. This factor emphasis the need for use of below maximum curvature on highways through interchanges. At an interchange area, the minimum design elements of through highways are described in paragraph 5.3.1 and should be used. If there are still some problems in sight distance, the clearances to abutments, piers, or rails should be increased as necessary to obtain the proper sight distance by increasing the spans or widths of the bridge structure.

### 12.6.3 INTERCHANGE SPACINGS

In areas of congested urban development, proper spacing usually is difficult to achieve because of a traffic demand for frequent access. Minimum spacing of arterial interchanges i.e., the distance between intersecting streets with ramps, is determined by weaving volumes, signs, signal progression, and required lengths of speed-change lanes. A generalized rule of thumb for minimum interchange spacing is 1.5 km in urban areas and 3.0 km in rural areas. In urban areas, if the spacing is less than 1.5 km then in place of interchange development may be done by grade-separated ramps or by adding collector-distributor roads. On urban expressways, connection to the arterial is commonly made with a single independent ramp rather than with a full interchange. On the expressways where independent ramps from different streets are closely and irregularly arranged, the distance between ramp terminals must also be examined.

### 12.6.4 UNIFORMITY OF INTERCHANGE PATTERNS

In the design of a series of interchanges, attention must be given not only to each individual but more to the group as well, as it is desirable to provide uniformity in exit and entrance patterns. As the interchanges are closely spaced in urban areas, the available distances are shorter, in which the drivers are to be informed of the course to be followed in leaving an expressway. A dissimilar arrangement of exits between successive interchanges, such as an irregular sequence of the near and far side of structures exit ramp locations or some right off movements may cause confusion resulting in slowing down on high-speed lanes and unexpected maneuvers.

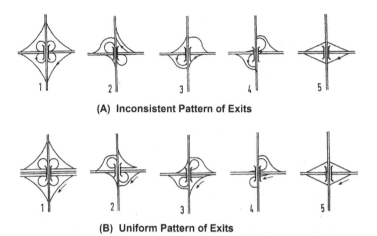

**(A) Inconsistent Pattern of Exits**

**(B) Uniform Pattern of Exits**

**FIGURE 12.13** (a) Inconsistent pattern of exits. (b) Uniform pattern of exits.
*Source:* Figure 5.2, A Guide to the Design of Interchanges, Arahan Teknik (Jalan) 12/87, JKR

The difficulty of right entrance merging (for UK style of traffic) with high-speed through traffic and the requisite lane changing for right exit ramps make these layouts undesirable. In highly special cases all entrance and exit ramps should be on the left, as an exception.

Figure 12.13, gives an example of a consistent and inconsistent pattern of exits.

## 12.6.5 ROUTE CONTINUITY

The provision of a directional path along and throughout the length of a designated route is the "route continuity." The destination of the route pertains to a route number or a name of a major highway. Route continuity is a theoretical extension of the principle of operational uniformity coupled with the application of proper lane balance and the principle of maintaining a basic number of lanes.

The principle of route continuity is to simplify the driving task by which it reduces lane changes, simplifies signing, delineates the through route, and reduces the driver's search for directional signing. Location and design of interchanges, individually and as a group, should be verified for proper signing. Signs used should conform to the relevant specifications on traffic control devices of the country.

Pavement markings, delineators, and other markings also are essential elements of driver communication at interchanges. These should be uniform and consistent by conforming to the relevant standards and guidelines.

Figure 12.14 illustrates the principle of route continuity by applying to a hypothetical route that intersects other major high-volume routes.

## 12.6.6 SIGNING AND MARKINGS

Ease of operation at interchanges, that is, clarity of paths, safety, and efficiency, depends largely on their relative spacing, the geometric layout, and effective signing.

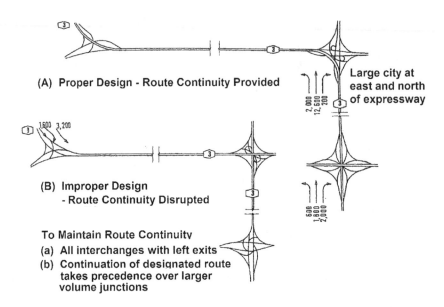

**FIGURE 12.14** (a) Proper design of – route continuity provided. (b) Improper design.
*Source:* Figure 5.3, A Guide to the Design of Interchanges, Arahan Teknik (Jalan) 12/87, JKR

The location of minimum distances is to be checked for, whether providing effective signing can be possible to inform, warn and control drivers. Location and design of interchanges, individually and as a group, should be examined for proper signing. Signs used should conform to the relevant specifications on traffic control devices.

Pavement markings, delineators, and other markings are also essential components of communication to drivers at interchanges. These should also be uniform and consistent following the relevant standards and guidelines.

### 12.6.7 BASIC NUMBER OF LANES

Any route of arterial character should maintain certain uniformity in the number of lanes provided along with it. The basic number of lanes is defined as a minimum number of lanes designated and maintained over a substantial length of a route, irrespective of changes in traffic volume and requirements for the lane balance.

The number of lanes is decided based on the general volume of traffic over a substantial length of the facility. The volume considered here is the "Daily Hourly Vehicles or DHV," which is normally, representative traffic volume in the morning or evening of weekdays at peak hours.

The basic number of lanes may be increased if required, where traffic is developed sufficiently to justify an extra lane and such traffic development raises the volume level over a substantial length of the existing road facility.

The basic number of lanes may be decreased, where traffic is reduced significantly to justify dropping a basic lane, provided there is a general lowering of the volume level on the expressway route as a whole.

## 12.6.8   COORDINATION OF LANE BALANCE OR BASIC NUMBER OF LANES

A proper balance in the number of traffic lanes on the expressway and ramps ensures an efficient traffic operation through and beyond an interchange.

Once the basic number of lanes is determined for each roadway, the balance in the number of lanes should be checked based on the following principles:

- At entrances, the number of lanes beyond the merging of two traffic streams should be not less than the sum of all traffic lanes on the two merging roadways minus one.
- At exits, the number of approach lanes on the highway must be equal to the number of lanes on the highway beyond the exit plus the number of lanes on the exit, minus one.
- At cloverleaf loop ramp exits, an exception to this principle may occur, which follows a loop ramp entrance or at exits between closely spaced interchanges, while a continuous auxiliary lane between the terminals is used. In these cases, the auxiliary lane which is to be dropped as a single-lane may be dropped in a single-lane exit.
- The traveled way of the highway should be reduced by not more than one traffic lane at a time.

Figure 12.15 (12.6.4) describes the application of the principles of lane balance. However, the principles of lane balance may conflict with the concept of continuity in the base number of lanes.

Figure 12.16, describes three different arrangements which illustrate this conflict:

- In arrangement A, lane balance is maintained, but there is no compliance to the basic number of lanes. This pattern may cause confusion and improper operations for through traffic.
- In arrangement B, continuity in the basic number of lanes is provided but the layout pattern does not conform to the principles of lane balance. With this

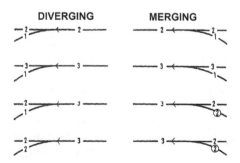

**FIGURE 12.15**   Examples of Lane Balance.
*Source:* Figure 5.4, A Guide to the Design of Interchanges, Arahan Teknik (Jalan) 12/87, JKR

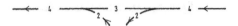

**(A) Lane Balance but no compliance with basic number of Lanes**

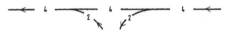

**(B) No Lane Balance but compliance with basic number of Lanes**

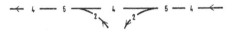

**(C) Compliance with both Lane Balance and basic number of Lanes**

**FIGURE 12.16**   Coordination of Lane Balance and Basic Number of Lanes.
*Source:* Figure 5.5, A Guide to the Design of Interchanges, Arahan Teknik (Jalan) 12/87, JKR

layout pattern, the large exiting and entering traffic volume requiring two lanes would have difficulty in diverging or merging with the mainline flow.

Arrangement C is a layout pattern where the concepts of lane balance and basic number of lanes are brought into harmony by building on the basic number of lanes, that is, by adding auxiliary lanes or removing them from the basic width of the traveled way. Auxiliary lanes may be added to fulfill the capacity and weaving requirements between interchanges, to accommodate traffic pattern variations at interchanges, and for simplifying operations, by reducing lane changing. The principles of lane balance must always be applied in the use of auxiliary lanes.

### 12.6.9   Auxiliary Lanes

An auxiliary lane of the roadway is adjoining the traveled way and is the provision for parking, speed-change, storage for turning, weaving, truck climbing, and other purposes supplementary to traffic movement. The width of an auxiliary lane should equal to that of the through lanes. Adding or removing auxiliary lanes is not counted as the change in the basic number of lanes. The usage of auxiliary lanes should follow the general principle outlined below:

- Where interchanges are closely spaced, i.e., the distance between the ends of the taper on the exit terminal to the next entry should be used to improve operational efficiency.
- The termination at the end of the auxiliary lane may be accomplished by several methods, such as they may be dropped in a two-lane exit; a single-lane exit, or at the physical nose before tapering the through roadway. For these methods, the exit mass of traffic should be visible throughout the length of the auxiliary lane. Figure 12.17(A, B, and C) illustrates these methods.
- Where the single-exit design indicates problems with the haphazard traffic flow caused by vehicles attempting to recover and proceed on the through

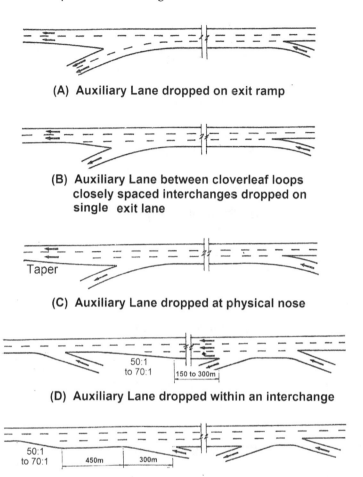

(A) Auxiliary Lane dropped on exit ramp

(B) Auxiliary Lane between cloverleaf loops
closely spaced interchanges dropped on
single exit lane

(C) Auxiliary Lane dropped at physical nose

(D) Auxiliary Lane dropped within an interchange

(E) Auxiliary Lane dropped beyond an interchange

**FIGURE 12.17**   Alternatives in Dropping Auxiliary Lanes.
*Source:* Figure 5.6, A Guide to the Design of Interchanges, Arahan Teknik (Jalan) 12/87, JKR

lanes, the recovery lane should be extended 150–300 m before being tapered
into the through lanes. Within large interchanges, this distance should be
increased to 450 m. When the auxiliary lane is carried through one or more
interchanges, it may be dropped as indicated above or it may be merged into
the through roadway approximately 750 m beyond the influence of the last
interchange. Figure 12.17D and E illustrates the two methods above.

• When successive interchanges are widely spaced, the auxiliary lane originating
at a two-lane entrance should be carried along the expressway for an effective
distance beyond the merging point as shown in Figure 12.18A and B. An auxiliary lane introduced for a two-lane exit should be carried along the expressway for an effective distance in advance of the exit and extended into the ramp
as shown in Figure 12.18C and D.

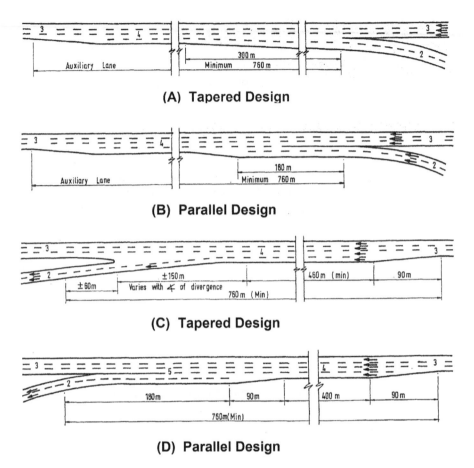

**(A) Tapered Design**

**(B) Parallel Design**

**(C) Tapered Design**

**(D) Parallel Design**

**FIGURE 12.18**    Application of Auxiliary Lanes After Exits and in Advance of Exits.
*Source:* Figure 5.7, A Guide to the Design of Interchanges, Arahan Teknik (Jalan) 12/87, JKR

- Lengths lane taper should not be less than those set up for single-lane ramps with adjustments for grades. A minimum distance of about 760 m is required by traffic for the necessary operational effect to develop the full capacity of two-lane entrances and exits.

## 12.6.10  LANE REDUCTION

The reduction in the basic number of lanes may be applied beyond a principal interchange involving a major fork or a point downstream from the interchange with another expressway. This reduction may be made when the exit volume is sufficiently large to change the basic number of lanes beyond this point on the expressway route as a whole. The basic number of lanes may also be reduced where, a series of exits, as in outlying areas of a city, causes the traffic load on the expressway to drop significantly to justify the lesser number of basic lanes. The lane reduction of a basic lane or an auxiliary lane may be applied at a two-lane exit ramp or between interchanges.

In between successive interchanges, if a basic lane or auxiliary lane is to be dropped, it must be accomplished at 600–900 m from the previous interchange to allow adequate distance for signing.

The lane drop transition should be provided on a tangent horizontal alignment and the approach side of any crest vertical curve. A sag vertical curve is also a good location for a lane drop because it provides good visibility. A left-side lane reduction has advantages in that speeds are generally lower and the merging maneuver from the left is more familiar to most motorists (for UK style of traffic).

The end of the lane drop should be tapered into the highway like that at a ramp entrance. The desirable minimum taper rate should be 50:1 and 70:1.

## 12.7 DESIGN ELEMENTS

### 12.7.1 WEAVING SECTIONS

#### 12.7.1.1 General

Weaving sections are highway segments where vehicle paths crossing each other due to the patterns of traffic entering and leaving at various points of access. They occur where one-way traffic streams cross by merging and diverging maneuvers, and within an interchange are also between entrance ramps followed by exit ramps of successive interchanges and on segments of overlapping roadways.

Weaving sections are formed as simple or multiple. In a multiple weaving section there occur two or more overlapping weaving sections. Multiple weaving sections occur frequently where there is a need for the collection and distribution of high concentrations of traffic, in the urban areas.

As significant indiscipline occurs in traffic movements throughout weaving sections, it is desirable to eliminate or reduce the weaving from the main facility of interchange by designs.

Interchanges that do not involve weaving, are expensive but operate better than those interchanges with weaving areas which are less expensive. Designs to avoid weaving movements require a greater number of larger and more complex structures, with some direct connections. Joint evaluation of the cost of the total interchange and the specific volumes of traffic to be handled is required to reach a sound decision between design alternatives.

In the design of the cloverleaf interchange, consideration should be given to the inclusion of collector-distributor roads on the main facility. Both the facilities are essential where warranted.

Weaving sections in the design should be identified, checked, and adjusted so that the level of service is consistent with the remaining expressway. The desired level of service in the design of a weaving section is dependent on its length, the number of lanes, acceptable degree of congestion, and relative volumes of individual traffic movements.

### 12.7.1.2   Design Considerations

The weaving section should have a length which shall enable to achieve the appropriate level of service same as that of the through expressway. The relationship between the volume of weaving traffic and the required length of the weaving section is discussed in paragraph 12.7.0.

Where weaving takes place directly on the expressway lanes, the weaving length should not be less than 300 m.

## 12.7.2   Collector – Distributor Roads

### 12.7.2.1   General

The purpose of a collector-distributor road is to eliminate weaving and reduce the number of entrance and exit points directly on the through roadways but satisfying the requirements for access to and from the expressway. They may be provided within an individual interchange, in between two adjacent interchanges, or continuously through a series of interchanges. Continuous collector-distributor roads are similar to continuous service roads except that access to the adjacent property is not permitted.

The continuous collector-distributor roads should be incorporated into the basic design to develop an integrated system. The capacity analysis and basic lane determination should be done for the overall system rather than for the separate roadways.

The "Transfer Roads" are the connections between the through roadways and collector-distributor roads. Those may be either one-lane or two lanes roadways and the principle of lane balance applies to the design of transfer roads on both the through roadways and the collector-distributor roadways.

### 12.7.2.2   Design Considerations

Collector-distributor roads may have the width of one or two lanes, which is determined by capacity requirements. Lane balance is required to be maintained at entrances and exits to and from the through roadways, but strict adherence is not mandatory on the collector-distributor road because weaving is handled at a reduced speed. The lane width should be maintained as the same as that of through roadways.

The usual design-speed is considered in the range of 60–80 km/hr, and should not be less than 20 km/hr slower than the design-speed of the main through roadways. The same design standard as that of the through roadways, and with a reduced design-speed should be applied.

The transfer roads and collector-distributor roads should have shoulders of widths equal to those defined for the same design-speed. The outer separation verge between the through roadway and the collector-distributor roads should be as wide as practicable. The minimum width should be great enough for shoulder widths on the collector-distributor roads, and through roadways, providing space for a suitable barrier to prevent indiscriminate crossovers.

Terminals of transfer roads should be designed following the requirements for ramp terminals.

Operational problems will occur in the collector-distributor roads if these are not properly signed, especially those servicing more than one interchange.

## 12.7.3 Exits

### 12.7.3.1 Exit Type 1

In general, interchanges that are designed with single exits are superior to those with two exits. In interchanges with two exits, if one of the exits is a loop ramp and the second exit is a loop ramp preceded by a loop entrance ramp may experience various operational issues. The purposes for providing single exits, where applicable, are as follows:

- To remove weaving from the main roadway and transfer it to a slower speed road.
- To provide a high-speed exit from the main roadway for all exiting traffic.
- To simplify signing and the decision process.
- To fulfill driver expectations by placing the exit in advance of the grade-separation structure.
- To provide uniformity in exit layout patterns.
- To provide an adequate sight distance for all traffic exiting from the main roadway.

The provision for single exits is more expensive for providing added roadway, longer bridges, and in some cases, additional separation structures. However, in urban areas if the required right of way is available, cost alone should not be the main consideration in the decision to omit or provide single exits where practicable. The single-exit design is placed at the exit from the main roadway in advance of the structure and is conducive to a uniform pattern in all exits. Where the through roadway overpasses the crossroad in a vertical curve, it is practically impossible to achieve an adequate sight distance for the loop ramp exit of a conventional cloverleaf interchange. By using the single-exit design, decision sight distance would already be developed owing to the exit occurring on the upgrade.

### 12.7.3.2 Exit Type 2

In some situations, a single-exit does not work as well as two exits. This situation could occur at high-volume, high-speed directional interchanges and the fork following the single-exit from the expressway, particularly when the traffic volume is high enough to warrant a two-lane exit and the distance from the exit terminal to the fork is insufficient for weaving and proper signing. Some confusion is usual at this second decision point, resulting in poor operation and high possibilities of accidents. Because of this, it may be advantageous on some directional interchanges to provide two exits on each expressway leg.

### 12.7.4  RAMPS

#### 12.7.4.1  Ramp Types

The "ramp" generally is a one-way roadway that includes all types, arrangements, and sizes of turning roadways connecting two or more legs at an interchange. The components of a ramp area are a terminal at each leg and a connecting road, usually with some curvature, and on a grade.

Figure 12.19 illustrates several types of ramp layouts and their characteristic shapes. Numerous shape variations are in use but each of those belongs to a class broadly as one of the types shown.

##### 12.7.4.1.1  Diagonal

Diagonal ramps are illustrated in Figure 12.19(A), these are almost always one-way but usually have both left and right-turning movements at the terminal on the minor intersecting road. Although shown as a continuous curve, but the diagonal ramps are most commonly tangent or wishbone in shape with a reverse curve. Diamond-type interchanges generally have four diagonal ramps.

##### 12.7.4.1.2  Loop

The loop may have single turning movements (left or right) or double turning movements (left and right) at either or both ends. Figure 12.19(B) illustrates the case where only single turns are made at both ends. In this loop layout, a right-turning

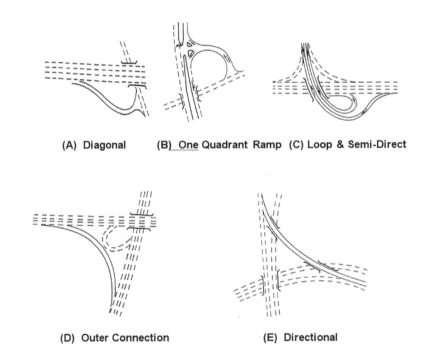

(A) Diagonal     (B) One Quadrant Ramp  (C) Loop & Semi-Direct

(D) Outer Connection              (E) Directional

**FIGURE 12.19**  General Types of Ramps.
*Source:* Figure 6.1, A Guide to the Design of Interchanges, Arahan Teknik (Jalan) 12/87, JKR

movement is made without an at-grade crossing of the opposing through traffic. The loop usually involves more indirect travel distance than that for other types of ramps.

### 12.7.4.1.3   Semi-direct

The semi-direct movement is illustrated in Figure 12.19(C), here, the driver makes a left-turn first, swinging away from the intended direction, gradually turning toward the right, and then completing the movement by following directly around enters the other road. This connection is termed as a "jug-handle," by following its shape. The comparable travel distance on this ramp is less than that by a loop and more than that for a direct connection. This for the UK style of traffic and a mirror image may be assumed for the USA style of traffic.

### 12.7.4.1.4   Direct

The travel distance is shortest in a direct connection. However, it needs at least three structures or a three-level structure. Figure 12.19(D) illustrates an outer connection, while Figure 12.19(E) illustrates a direct connection.

The different types of interchanges are made up of various combinations of these different types of ramp layouts. For example, the trumpet interchange has one loop, one semi-directional ramp, and two right directional or diagonal ramps (for UK style of traffic).

## 12.7.4.2   Design Considerations

### 12.7.4.2.1   Design-Speeds

Desirably, design-speeds at ramps should approximate the lower running speed of the low-volume traffic on the intersecting highways. This design-speed is not always practicable and lower design-speeds, not be less than half of the design-speed for the intersecting highways, may be applied. Table 12.3 mentions guiding values for ramp design-speeds.

The values in Table 12.3 apply to the sharpest ramp curve, usually on the ramp proper. These speeds do not pertain to the ramp terminals, which should be properly transitioned and provided with adequate speed-change facilities to match the highway speed involved.

The highway with the greater design-speed should be the guide in selecting the design-speed for the whole of a ramp. However, the ramp design-speed may vary, the part of the ramp closer to the lower speed highway is designed for the lower speed. This consideration is particularly applicable where the ramp is on an upgrade from the higher speed highway to the lower speed highway.

**Ramps for Left-Turns** Upper range design-speed is often attainable in ramps for a left-turn and a value between the upper and lower range is usually practicable for UK style of traffic.

**Loops** Upper range values of design-speed generally are not attainable in loops. Ramp design-speeds above 50 km/hr for long loops require large areas, which are rarely available in urban areas. These are expensive and require right-turning drivers to travel considerable extra distance, for UK

## TABLE 12.3
### Guide Values for Ramp Design-Speeds

| Highway Design | Ramp Design Speed (km/h) | | |
|---|---|---|---|
| Speed (km/h) | Upper | Middle | Lower |
| 120 | 90,80 | 70 | 60 (50) |
| 100 | 80,70 | 60 | 50 |
| 80 | 70,60 | 50 | 40 |
| 60 | ,50 | 40 | 30 |
| 50 | ,40 | 30 | 25 |

*Source:* Table 6.1, A Guide to the Design of Interchanges, Arahan Teknik
        (Jalan) 12/87, JKR
Value in '(,)' is applicable only for loops.

style of traffic. Usually preferred minimum values should be not less than 40 km/hr (i.e., for 50 m radius).

**Semi-Direct Connections** Design-speeds between the middle and upper ranges shown in Table 12.3 (12.6.1) should be used. A design-speed less than 50 km/hr should not be used for highway design-speeds of more than 80 km/hr.

**Direct Connections** Design-speeds between the middle and upper ranges shown in Table 12.3 should be used. The preferred minimum speed should be 70 km/hr for highway design-speeds of more than 80 km/hr.

**At-Grade Terminals** Where a ramp joins a major crossroad forming an at-grade intersection, speeds given in Table 12.3 do not apply to that portion of the ramp near the intersection because a stop sign or signal control is normally provided. The requirements as specified in the relevant standard for the design of at-grade intersection should be followed.

### 12.7.4.2.2 Curvature

The factor and assumptions of minimum-radius roadway curves for various speeds are described in the relevant standard for the geometric design of roads. They apply directly to the design of ramp curves. To meet the site conditions and other controls, and to fit the natural paths of vehicles, compound curves and spiral transitions are desirable to develop the desired shape of ramps. Proper care should be exercised in the use of compound curvature to prevent unexpected and abrupt speed adjustments.

### 12.7.4.2.3 Ramp Shapes

The general shape of a ramp evolves from the type of ramp selected, as previously discussed, and shown in Figure 12.19. The selection of the specific shape of a ramp may be influenced by factors such as traffic pattern, traffic volume, design-speeds, topography, intersection angle, and type of ramp terminal.

a. **Loop and Outer Connection**

Several forms are available and may be used for the loop and outer connection of a semi-directional interchange, as shown diagrammatically in Figure 12.20. The loop, except for its terminal, is usually applied as a circular arc or, maybe some other type of symmetrical or asymmetrical curve. The asymmetrical arrangement may be adopted where the intersecting roads are not of the same importance and the ramp terminals are designed for different speeds, the ramp in part functioning as a speed-change area. Such shapes should be supported by right of way controls, profile and sight distance conditions, and terminal location. The usually adopted radii of loops are approximately 30–45 m for minor movements on highways with design-speeds of 80 km/hr or less and from 45 to 75 m for more important movements on highways with higher design-speeds. The alignment on a continuous curve (line A) is most desirable for an outer connection. This arrangement, however, may require extensive right of way. Another desirable alignment arrangement has central tangents and terminal curves (lines B-B and C-C). Where the loop is more important than the outer connection, reverse alignment on the outer connection may be used to reduce the area of the right of way, as shown by line D-D. To achieve the desired shape any combination of lines B, C, and D may be used. As generally desirable the loop and the outer connection are separated as described in Figure 12.20(A). However, where the movements are minor and the economy is desired, a portion of the two ramps may be combined into a single two-way road. Where this design is used, a central barrier should be provided to separate the traffic in the two directions. This design is, however, discouraged.

b. **Diagonal Ramps**

Diagonal ramps may be designed for a variety of shapes, depending on the pattern of turning traffic and right of way limitations. The ramp may be a diagonal tangent with connecting curves shown by the solid line in Figure 12.20(B). To favor a left-turning movement the ramp may be on a continuous curve to the left with a spur to the right for right-turns, in terms of UK style of traffic. In the case of restricted right of way along the major highway, it may be necessary to use reverse alignment with a portion of the ramp being parallel to the through roadway.

Diagonal ramps of a certain type usually called slip ramps or roads and maybe more properly termed as cross connections to connect with a parallel service road, as shown in Figure 12.20(C). Where this design is used, it is desirable to have a connection to one-way service roads, connection to two-way service roads introduces the possibility of wrong way entry onto the through lanes and should be avoided.

c. **Semi-direct Connection**

The layout shape of a semi-direct connection is influenced by the location of the terminals concerning the structures, the extent to which the structure

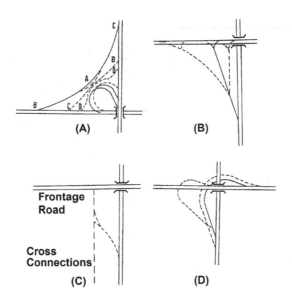

**FIGURE 12.20**   Ramp Shapes.
*Source:* Figure 6.2, A Guide to the Design of Interchanges, Arahan Teknik (Jalan) 12/87, JKR

pavements are widened, and the curve radii necessary to maintain a desired turning speed for an important left-turning speed for an important left-turning movement, this is illustrated in Figure 12.20(D). The angular position or the curvature may be dictated to some extent by the relative design-speeds of the intersection legs and by the proximity of other roadways.

### 12.7.4.2.4   Sight Distance

Sight distance along a ramp should be at least the safe stopping sight distance. Sight distance for passing is not required. There should be a clear view of the whole exit terminal, including the exit nose and a section of the ramp pavement beyond the gore.

The sight distance on an expressway preceding the approach nose of an exit ramp should be higher, desirably by 25% or more, than the minimum for the through traffic design-speed. The ranges in design values for stopping sight distance on horizontal and vertical curves for turning roadways and open road conditions are to be as per the relevant standard for the geometric design of roads.

### 12.7.4.2.5   Grade and Profile Design

The profile of a typical ramp usually consists of a central portion on an acceptable grade having terminal vertical curves and connections to the profiles of the elevated interchange to the profiles of the at-grade legs. The following references to ramp gradient pertained largely to the central portion of the ramp profile. Profiles at the terminals seldom are tangent grades and are largely determined by the through road profiles.

Ramp grades should be as flat as to minimize the driving effort required in maneuvering from one road to another. Most ramps are curved and steep grades on them adversely affect the flow of traffic. The slowing down of vehicles on an ascending ramp is not as serious as on a through road, provided the speed is not decreased sufficiently to result in a peak-hour backup onto the through road. Ramps in most of the diamond interchanges are only 120–360 m long, and the short central portion with the steepest gradient has only a moderate operational effect. Accordingly, gradients on ramps may be steeper than those on the intersecting highways, but a precise relation cannot be set. General values of limiting gradient can be shown, but for anyone ramp, the gradient to be used depends on several factors peculiar to that site and quadrant alone. The flatter the gradient on a ramp, the longer it will be, but the effect of the gradient on the length of the ramp is less than generally thought.

On one-way ramps, a distinction can be made for ascending and descending gradients. Short ramps with upgrades of as much as 5% do not cause discomfort with truck and bus operation. On one-way down ramps, gradients up to 8% do not cause hazards due to excessive acceleration.

As a general consideration, it is desirable that ascending gradients on ramps should not be steeper than that shown in Table 12.4. Where topographic conditions dictate, grades steeper than desirable may be used. One-way descending gradients on ramps should be held to the same general maximums, but in special cases, they may be 2% greater.

The cases in which gradient is a determining factor in the length of the ramp are as follows:

- For intersection angles of 70° or less, it may be essential to place the ramp away from the structure than required for minimum alignment to provide a ramp of sufficient length for a reasonable gradient.
- Where the intersection legs are on acceptable grade, with the upper road ascending and the lower road descending from the structure, the ramp will have to take care of a large difference in elevation that increases with the distance from the structure.
- Where a ramp leaves the lower road on a down grade and meets the higher road on a down grade, the longer-than-usual vertical curves at the terminals may make a lane ramp necessary to meet gradient limitations. The alignment and grade of a ramp must be determined jointly.

## TABLE 12.4
## Allowable Maximum Ascending Gradient

| Ramp Design Speed (km/h) | 90 | 80 | 70 | 60 | 50 | 40 | 30 | 25 |
|---|---|---|---|---|---|---|---|---|
| Allowable maximum Ascending gradient (%) | 3 | 4 | 4.5 | 5 | 5.5 | 6 | 7 | 7.5 |

*Source:* Table 6.2, A Guide to the Design of Interchanges, Arahan Teknik (Jalan) 12/87, JKR

## TABLE 12.5
## Minimum Vertical Curve Lengths and "K" Values

| Ramp Design Speed (km/h) | 90 | 80 | 70 | 60 | 50 | 40 | 30 | 25 |
|---|---|---|---|---|---|---|---|---|
| Crest vertical curve | 50 | 30 | 20 | 15 | 10 | 8 | 5 | 3 |
| Sag vertical curve | 35 | 20 | 15 | 10 | 7 | 5 | 4 | 3 |

*Source:* Table 6.3, A Guide to the Design of Interchanges, Arahan Teknik (Jalan) 12/87, JKR

### 12.7.4.2.6   Vertical Curves

Usually, ramp profiles assume the shape of the letter S consisting of a sag vertical curve at the lower end and a crest vertical curve at the upper end in its simplest form. Additional vertical curves may be necessary, particularly on ramps that overpass or underpass other roadways. Where a crest or sag vertical curve extends onto the ramp terminal, the length of the curve should be decided by using a design-speed between those on the ramp and the highway. Table 12.5 mentions the minimum vertical curve lengths for each ramp design-speed.

### 12.7.4.2.7   Super Elevation and Cross Slope

The general guidelines for values of cross slope design on ramps are as follows:

As related to curvature and design-speed on ramps, the superelevation rates are given in Table 12.6. The highest rate practicable should be adopted, preferably in the upper half or third of the indicated range and particularly on descending ramps.

The cross slope on portions of ramps on tangent normally are slopped one-way at a practical rate ranging from 1.5% to 2% for high-type pavements. Superelevation runoff, or the change in superelevation rate per unit length of the ramp, should not be more than that as mentioned in Table 12.7. The superelevation development starts or ends along the auxiliary pavement of the ramp terminal.

Another important control in developing superelevation along the ramp terminal is that of the crossover crown line at the edge of the through traffic lane. The maximum allowable algebraic difference in cross slope between the auxiliary pavement and the adjacent through lane is given in Table 12.8.

Three segments of a ramp should be analyzed to determine superelevation rates that would be compatible with the design-speed and the configuration of the ramp. The three segments are the exit terminal, the ramp proper, and the entrance terminal which should be studied in combination to ascertain the design-speed and superelevation rate. Three ramp configurations are considered below for this discussion.

a. **Diamond Ramp**

Deceleration to controlling curve speed should occur on the auxiliary lane of the exit, and deceleration is continued to stop or yield conditions that would occur on the ramp proper. Superelevation comparable to open road conditions would not be appropriate on the ramp proper or the forward terminal.

## TABLE 12.6
## Superelevation Rates for Curves

| Radius (m) | Range in Superelevation Rate For Interchange Ramps with Design Speed (km/h) of | | | | | | | |
|---|---|---|---|---|---|---|---|---|
| | 90 | 80 | 70 | 60 | 50 | 40 | 30 | 25 |
| 2000 | .02 | .02 | .02 | .02 | .02 | .02 | .02 | .02 |
| 1,500 | .03 | .02 | .02 | .02 | .02 | .02 | .02 | .02 |
| 1,200 | .04 | .03 | .03 | .02 | .02 | .02 | .02 | .02 |
| 1,000 | .04 | .04 | .04 | .02 | .02 | .02 | .02 | .02 |
| 800 | .05 | .04 | .04 | .02–.03 | .02 | .02 | .02 | .02 |
| 700 | .06 | .05 | .05 | .02–.03 | .02 | .02 | .02 | .02 |
| 600 | .07 | .06 | .05 | .03–.04 | .02–.03 | .02 | .02 | .02 |
| 500 | .08 | .07 | .06 | .03–.04 | .02–.03 | .02 | .02 | .02 |
| 400 | .09 | .08 | .07 | .04–.05 | .03–.04 | .02–.03 | .02 | .02 |
| 350 | .10 | .09 | .08 | .05–.06 | .03–.04 | .02–.03 | .02 | .02 |
| 350 | R min | .09 | .09 | .05–.06 | .03–.05 | .02–.03 | .02 | .02 |
| 250 | = 310 | .10 | .10 | .06–.07 | .04–.06 | .02–.04 | .02–.03 | .02 |
| 200 | | R min | .10 | .08–.09 | .04–.07 | .02–.05 | .02–.03 | .02 |
| 180 | | = 230 | R min | .08–.09 | .04–.07 | .02–.05 | .02–.04 | .02–.03 |
| 160 | | | = 175 | .09–.10 | .05–.08 | .02–.06 | .02–.04 | .02–.03 |
| 140 | | | | .09–.10 | .05–.08 | .02–.06 | .02–.04 | .02–.03 |
| 130 | | | | .09–.10 | .05–.09 | .02–.07 | .02–.05 | .02–.03 |
| 120 | | | | R min | .05–.09 | .02–.07 | .02–.05 | .02–.04 |
| 110 | | | | = 125 | .06–.10 | .02–.07 | .02–.05 | .02–.04 |
| 100 | | | | | .06–.10 | .02–.08 | .02–.06 | .02–.04 |
| 90 | | | | | .06–.10 | .02–.08 | .02–.06 | .02–.05 |
| 80 | | | | | R min | .02–.09 | .02–.07 | .02–.05 |
| 70 | | | | | = 85 | .02–.09 | .02–.07 | .02–.06 |
| 60 | | | | | | .02–.10 | .02–.08 | .02–.06 |
| 50 | | | | | | .02–.10 | .02–.09 | .02–.07 |
| 40 | | | | | | R min | .02–.10 | .02–.08 |
| 30 | | | | | | = 50 | .02–.10 | .02–.09 |
| 20 | | | | | | | | |

*Source:* Table 6.4, A Guide to the Design of Interchanges, Arahan Teknik (Jalan) 12/87, JKR

## TABLE 12.7
## Rate of Change and Pavement Edge Elevation

| Design Speed (Km/hr) | 25 and 30 | 40 | 50 | 60 or more |
|---|---|---|---|---|
| Change in relative rate between centerline and pavement edge per station (percent) | 0.75 | 0.71 | 0.67 | 0.65 |

*Source:* Table 6.5, A Guide to the Design of Interchanges, Arahan Teknik (Jalan) 12/87, JKR

**TABLE 12.8**
**Algebraic Difference in Cross Slope**

| Design Speed of Exit or Entrance Curve (km/hr) | Maximum Algebraic Difference in Cross Slope Crossover (percent) |
|---|---|
| 25 and 30 | 5–8 |
| 40 and 50 | 5–6 |
| 60 and Over | 4–5 |

*Source:* Table 6.6, A Guide to the Design of Interchanges, Arahan Teknik (Jalan) 12/87, JKR

b. **Loop Ramp**

The curvature of the ramp proper could be a simple circular curve or a combination of circular curves and transitions. The design-speed and rate of superelevation would be determined by the curvature of the ramp proper. Superelevation would have to be gradually developed inside and outside the curve for the ramp proper.

c. **Direct and Semi-direct Ramps**

Superelevation rates comparable to open road conditions are appropriate for high-speed direct and semi-direct ramps.

d. **Developing Superelevation**

The method of developing superelevation at free flow ramp terminals is demonstrated diagrammatically in Figure 12.21.

Figure 12.21(A) shows a tapered exit from a tangent section with the first ramp curve falling beyond the required deceleration length. The normal crown is projected onto the auxiliary pavement, and no superelevation is required until the first ramp proper curve is reached. Figure 12.21(B) shows a parallel-type exit from a tangent section that leads into a flat exiting curve. At point b, the normal crown of the through roadway is projected onto the auxiliary pavement. At point c, the crown line can be gradually changed to start the development of superelevation for the exiting curve. At point d, two breaks in the crossover crown line would be conducive for developing a full superelevation in the vicinity of the physical nose. Figure 12.21(C) and 12.6.3(D) show ramp terminals on which the superelevation of the through roadway would be projected onto the auxiliary pavement. Figure 12.21(E) shows a parallel entrance on the high side of a curve. At point d, the ramp would possibly be flat, and full superelevation would be attained at point c. Figure 12.21(F) shows a parallel exit from a tangent section with sharp curvature developing in advance of the physical nose. This is a typical design for cloverleaf loops. Part of the cross slope change can be accomplished over the length of the parallel superelevation being developed at point b. Full superelevation of the ramp proper is reached beyond the physical nose.

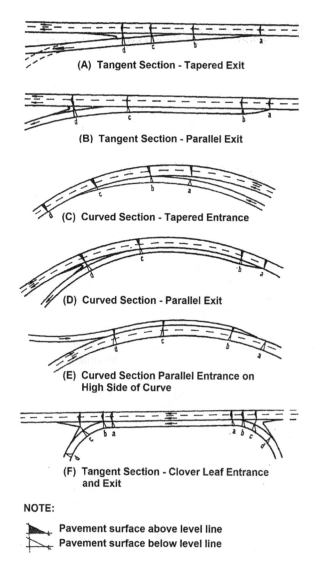

(A) Tangent Section - Tapered Exit

(B) Tangent Section - Parallel Exit

(C) Curved Section - Tapered Entrance

(D) Curved Section - Parallel Exit

(E) Curved Section Parallel Entrance on High Side of Curve

(F) Tangent Section - Clover Leaf Entrance and Exit

NOTE:

Pavement surface above level line
Pavement surface below level line

**FIGURE 12.21** Development of Superelevation at Free Flow Ramp Terminals.
*Source:* Figure 6.3, A Guide to the Design of Interchanges, Arahan Teknik (Jalan) 12/87, JKR

### 12.7.4.2.8 Gores

The term "gore" indicates the area immediately beyond the divergence of two roadways and bounded by the edges of those roadways. The approach nose or a ramp nose is the end of an island or area between diverging roadways. The approach nose faces approaching traffic passing to one or both sides. In common use, the term "gore" refers to both (1) the paved triangular area upstream from the approach nose and (2) the graded area for about 30 m downstream from the approach nose. The geometric layout of the first of these is a very important part of the design for the exit ramp terminal. It is the decision point area for the drivers and must be seen and understood by approaching drivers. The separating ramp roadway not only must be visible but also must be of a geometric shape to fit the likely speeds of vehicles at that point.

In situations of occurrences of a series of interchanges along an expressway, the gores should be uniform and have the same appearance to drivers. The paved triangular area for a left-hand ramp includes the space that is (1) the continuation of the through lane shoulder, which is interrupted by the ramp pavement, (2) the beginning of the right edge shoulder of the ramp pavement, and (3) the elongated triangular area between theoretical edges of the spaces (1) and (2). The whole of this area should be paved to provide an emergency maneuver area that can be utilized safely by drivers who deviate from their proper path. The physical approach nose at the base of the paved area usually rounded for of limited width is located with some offset and wedge taper on each side to the separate shoulders downstream from it. The ground-mounted exit sign should be centered in the gore, downstream from the physical approach nose in the typical exit ramp case.

The width at the base of this triangular paved area, which is at the physical nose, will vary depending on the type of speed-change lane, whether taper-type or parallel-type and on the alignment of the ramp roadway beyond the gore. The speed of traffic using the highway is also a factor. As a general guide, the width at the physical nose should be between 6 and 9 m including the paved shoulders, measured between the traveled way of the mainline and that of the ramp. This dimension should be increased by setting the physical nose further back if the ramp roadway curves away from the expressway immediately beyond the physical nose, it is more applicable if the speed-change lane is of the parallel-type. The dimension should also be increased if running speeds above 100 km/hr are expected to be common.

a. **Marking**

The entire paved triangular area should be conspicuously marked with stripes to delineate the proper paths on each side and to assist the driver in identifying the gore area. This is done by providing chevron markings.

Rumble strips may be placed in the gore area but should not be located too close to the gore nose because such placement renders them ineffectual for warning high-speed vehicles.

b. **Physical Nose**

The physical nose, which constitutes the beginning of the unpaved area within the gore, is generally rounded to provide wedges of paved areas on either side extending downstream in a manner to complete the recovery area and afford space for a vehicle that has made a wrong move to re-enter the traffic stream.

c. **Unpaved Area**

The unpaved area beyond the nose should be graded as nearly level with the roadway as is practicable so that vehicles inadvertently enter will not be upset or abruptly stopped by steep slopes. Heavy signposts, street light poles, and roadway structure supports should be kept well out of the graded gore area.

Yielding or breakaway type supports should be employed for the standard exit sign, and concrete footings, if used, should be kept flush with adjoining ground level.

There may be situations where the placement of a major obstruction in a gore is unavoidable. Cushioning or energy dissipating devices for use in front of hazardous fixed objects should be considered and adequate space should be provided for their installation whenever it is found necessary to construct obstructions in a gore on a high-speed highway.

### d. Merging End

In an entrance terminal the point of convergence, which is the beginning of all paved areas, is defined as the "Merging end." In shape, layout, and extent, the triangular maneuver area at an entrance terminal is like that at an exit. However, it points downstream and separates traffic streams already in lanes, thereby being less of a decision area. The width at the base of the paved triangular area is narrower, usually being limited to the sum of the shoulder widths on the ramp and expressway plus a narrow physical nose 1–2 m wide.

Figure 12.22 diagrammatically shows the details of typical gore designs for expressway exit ramps. Figures 12.6.4(A) and 12.6.4(B) describes a full 4 m recovery lane adjacent to the outside through lane and moderate right offset of the ramp pavement. Figure 12.22(C) shows a major fork, with neither diverging roadway having priority. The offset is equal for each roadway, and striping or rumble strips are placed upstream from the physical nose.

Table 12.9 gives minimum lengths for tapers beyond the offset nose.

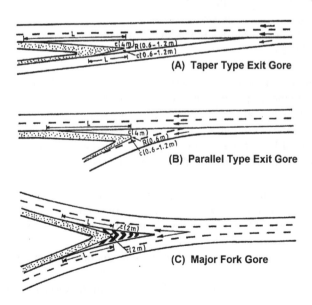

**FIGURE 12.22** Typical Gore Details.
*Source:* Figure 6.4, A Guide to the Design of Interchanges, Arahan Teknik (Jalan) 12/87, JKR

**TABLE 12.9**
**Minimum Length of Taper Beyond and Offset Nose**

| Design Speed of Approach Highway (km/hr) | L = Length in Meters of Nose Taper per Meter of Nose Offset (C) as in Figure 6.4 |
|---|---|
| 50 | 5 |
| 60 | 6 |
| 80 | 8 } × C |
| 100 | 9 |
| 120 | 10 |

*Source:* Table 6.7, A Guide to the Design of Interchanges, Arahan Teknik (Jalan) 12/87, JKR

### 12.7.4.3   Pavement Widths

*12.7.4.3.1   The Width and Cross Section*

The type of operation, curvature, volume, and type of traffic governs the ramp pavement widths. Design widths of ramp pavements for various design traffic conditions are given in Table 12.10. The three general design traffic conditions are:

Traffic condition A – predominantly P vehicles, but some consideration for SU trucks.
Traffic Condition B – sufficient SU vehicles to govern the design, but some consideration for semitrailer vehicles.
Traffic Condition C – Sufficient bus and combination types of vehicles to govern the design

Traffic conditions A, B, and C are described in broad terms as the design data regarding traffic in volume, or percentage of the total, for each type of vehicle may not be available to define the traffic condition with accuracy concerning pavement width.

In general, traffic condition A can be assumed as having a small volume of trucks or only occasional large trucks, traffic condition B is having a moderate volume of trucks, say in the range of 5–10% of the total traffic; and traffic condition C is having more and larger trucks.

For the design of an interchange ramp, when specific information is not available, traffic conditions should be taken for the class of road where the interchange is located as shown in Table 12.11 (12.7.9). If the roads connected by the interchange are of different classes, the higher class should be used.

The selection of the design case is based on the operation of the ramp as below:

• Case I – Light to moderate traffic volume within the service volume of a one-lane ramp.
• Case II – Moderate to near service volume condition.

**TABLE 12.10**
**Ramp Pavement Widths**

| | Pavement Width (m) | | | | | | | | |
|---|---|---|---|---|---|---|---|---|---|
| | Case. I<br>One Lane, One-Way Operation – No Provision for Passing a Stalled Vehicle | | | Case. II<br>One Lane, One-Way Operation-with Pro-Vision for Passing a Stalled Vehicle | | | Case. III<br>Two Lane Operation – Either One-Way or Two-Way | | |
| | | | | Design Traffic Condition | | | | | |
| Radius on Inner Edge of Pavementr (m) | A | B | C | A | B | C | A | B | C |
| 15 | 5.5 | 5.5 | 7.0 | 7.0 | 7.6 | 8.8 | 9.5 | 10.7 | 12.7 |
| 25 | 5.0 | 5.2 | 5.8 | 6.4 | 7.0 | 8.2 | 8.8 | 10.0 | 11.3 |
| 30 | 4.6 | 5.0 | 5.5 | 6.1 | 6.7 | 7.6 | 8.5 | 9.5 | 10.7 |
| 45 | 4.3 | 5.0 | 5.2 | 5.8 | 6.4 | 7.3 | 8.2 | 9.2 | 10.0 |
| 60 | 4.0 | 5.0 | 5.0 | 5.8 | 6.4 | 7.0 | 8.2 | 8.8 | 9.5 |
| 90 | 4.0 | 4.6 | 5.0 | 5.5 | 6.1 | 6.7 | 7.9 | 8.5 | 9.2 |
| 120 | 4.0 | 4.6 | 5.0 | 5.3 | 6.1 | 6.7 | 7.9 | 8.5 | 8.8 |
| 150 | 3.7 | 4.6 | 4.6 | 5.5 | 6.1 | 6.7 | 7.4 | 8.5 | 8.8 |
| Tangent | 3.7 | 4.6 | 4.6 | 5.2 | 5.8 | 6.4 | 7.0 | 8.2 | 8.2 |

*Source:* Table 6.8, A Guide to the Design of Interchanges, Arahan Teknik (Jalan) 12/87, JKR

**TABLE 12.11**
**Traffic Conditions for Ramp Design**

| Area | Class of Road | Traffic Condition |
|------|---------------|-------------------|
| Rural | Expressway | C |
| | Highway | C/B |
| | Primary | B/A |
| Urban | Expressway | C |
| | Arterial | C/B |
| | Collector | B/A |

*Source:* Table 6.9, A Guide to the Design of Interchanges, Arahan Teknik (Jalan) 12/87, JKR

- Case III – Traffic volumes exceed the service volume of a one-lane ramp or two-way operation.

### 12.7.4.3.2   Shoulders and Lateral Clearances

Shoulders and lateral clearances on the ramps should have the following design standards, for UK style of traffic and is opposite for USA style traffic:

For one-way ramps, a paved shoulder width of 1.5 m is desirable on the left and 0.5 m on the right.

Directional ramps with a design-speed of over 60 km/hr should have a paved left and right shoulder of 2.5 m and 1.0 m respectively.

For expressway ramp terminals where the ramp shoulder is narrower than that on the expressway, the paved shoulder width of the through lane should be carried into the exit terminal and should begin within the entrance terminal, with the transition to the narrower ramp shoulder applied gracefully on the ramp end of the terminal. Abrupt changes should be avoided.

Ramps should have a lateral clearance on the left of the outside edge of the paved shoulder of at least 1.8 m (preferably 3 m) and the right a lateral clearance of at least 1.2 m beyond the edge of the paved shoulder.

Where ramps pass under structures, the total width including the paved shoulders should be continued without any change. There should be a lateral clearance of at least 1.2 m from the edge of the paved shoulder, to the piers or abutments.

Ramps on overpasses should also have the full approach width of roadway and shoulders carried over the structure.

Providing edge lines and/or some type of color or texture difference between the traveled way and shoulder is desirable.

### 12.7.4.3.3   Median

Ramps of two-way operation should have a median to separate opposing traffic and to provide the space for the installation of some traffic safety devices. The minimum widths of median required are:

- 3 m for rural areas.
- 2 m for urban areas.

### 12.7.4.3.4 Curbs

Curbs should be considered only to facilitate drainage situations where it is particularly difficult. As in urban areas enclosed drainage is required because of the restricted right of way. Curbs should be located at the edge of the paved shoulders.

In some cases, curbs are used at the ramp terminals but are omitted along the central ramp portions. The use of curbs is not recommended on facilities designed for intermediate or higher design-speeds except in special cases. Mountable curbs should desirably be placed at the outer edge of the paved shoulder. Barrier curbs are seldom used in combination with shoulders, except where pedestrian protection is required. Because of fewer restrictions and more liberal designs, the need for curbs seldom arises in rural areas and is strongly discouraged.

### 12.7.4.4 Ramp Terminals

The terminal of a ramp is that portion adjacent to the through roadway, it includes speed-change lanes, tapers, and islands. Ramp terminals may be at-grade type, as ramp traffic merges with or diverges from high-speed through traffic at flat angles, at the crossroad terminal of diamond or partial cloverleaf interchanges, or the free flow type.

Ramp Terminals are further classified according to the number of lanes on the ramp at the terminal, either single or multilane, and according to the configuration of the speed-change lane, either taper or parallel-type.

#### 12.7.4.4.1 Right-Hand Entrances and Exits

Right-hand entrances and exits are contrary to the concept of driver expectancy when intermixed with left-hand entrances and exits. Right-hand entrances and exits in the design of interchanges should be avoided. Even in the case of major forks and branch connections, the less significant roadway should exit and enter on the left. This is for the UK style of traffic and is to be considered opposite for USA style traffic.

#### 12.7.4.4.2 Terminal Locations

Where diamond ramps and partial cloverleaf arrangements intersect the crossroad at-grade, an at-grade intersection is formed. Desirably, this intersection should be located at an adequate distance away from the separation structure to provide sight distances that permit safe entry of exit on the crossroad.

Diamond ramp terminals create two neighboring intersections on the crossroad within a short distance. Turning traffic at these intersections may be of significant volume in urban areas. Queuing lengths of right-turning vehicles must be considered in determining the distance. Coordination of the traffic control at the two intersections may also be necessary.

Drivers prefer and expect to exit in advance of the separation structure. The use of collector-distributor roads, single exits on partial cloverleafs, and other types of interchanges, automatically locates the mainline exit in advance of the separation structures and is considered a good design.

Designs of exits concealed behind crest vertical curves should be avoided especially on high-speed facilities. The design should place the high-speed entrance ramp terminals on descending grades to facilitate the acceleration. Adequate sight distance at entrance terminals should be available so that merging traffic on the ramp can adjust speed to merge into gaps on the main through road traffic.

### 12.7.4.4.3  Ramp Terminal Design

Vertical profiles of highway ramp terminals are to be designed with a platform on the ramp side of the approach nose or merging end. This platform should have a length of about 60 m or more on which the profile does not greatly differ from that of the adjacent through traffic lane. A platform area should also be provided at the at-grade terminal of a ramp. The length of this platform should be determined by the type of traffic control involved at the terminal and the capacity requirements.

### 12.7.4.4.4  Traffic Control on Minor Crossroads

On minor highways, the terminals would be designed and operated in the same manner as at-grade intersections. The right-turning movements leaving the crossing highway preferably should have median lanes. For low-volume crossroads, the right-turning movements from the ramps normally should be controlled by stop signs. The left-turning movements from the ramps normally should be controlled by stop signs. The left-turning movements from the ramps to multilane crossroads should be provided with an acceleration lane or generous taper or should be controlled by stop or yield signs. Ramps approaching a stop sign should have close to 90° alignment and be nearly level for storage of several vehicles. This is applicable for the UK style of traffic.

Signal controls are usually avoided on express-type highways and limited to the minor crossroads on which other intersections are at-grade, and some of them include signal controls. In or near urban areas, signal control has substantial application at ramp terminals on roads crossing over or under an expressway. Here, the turning movements usually are of large volume, and the cost of right of way and improvements thereon is high. As a result, significant savings may be achieved using diamond ramps with high-type terminals on the expressway and signalized terminals on the crossroads.

### 12.7.4.4.5  Distance Between Terminal and Structure

The terminal of a ramp should not be near the grade-separation structure. If it is not possible to place the exit terminal in advance of the structure, the existing terminal on the far side of the structure should be well removed so that, when leaving, drivers have some distance, after passing the structure, in which to see the turnout and begin the turnoff maneuver. Passing sight distance is recommended.

Ramp terminals on the near side of a grade-separation need not be as far removed as those beyond the structure. Both the view of the terminal ahead for drivers approaching on the through road and the view back along the road for drivers on an entrance ramp are not affected by the structure. Where an entrance ramp curve on the near side of the structure requires an acceleration lane, the ramp terminal should be located to provide length for it between the terminal and the structure, or the acceleration lane could be continued through or over the structure.

### 12.7.4.4.6 Distance Between Successive Ramp Terminals

On urban expressways, there are frequently two or more ramp terminals in close succession along the through lanes. To provide sufficient maneuvering length and adequate space for signing, a reasonable distance is required between terminals. Spacing between successive outer ramp terminals is decided on the classification of the interchange involved, the function of the ramp pairs (entrance or exit), and weaving potential.

Figure 12.23 shows the minimum distances for the spacing of ramp terminals for the various ramp-pair combinations as they apply to the interchange classifications. Where an entrance ramp is followed by an exit ramp, the minimum distance between the successive noses is governed by weaving requirements but should not be less than about 500 m to avoid overlapping maneuver areas. A notable exception to this length policy for entry-exit ramp combinations is the distance between loop ramps of cloverleaf interchanges for these interchanges the distance between entry-exit ramp noses is primarily dependent on loop ramp radii and roadway and median widths. A recovery lane is desirable beyond the nose of the loop ramp exit.

### 12.7.4.4.7 Speed-Change Lanes

An auxiliary lane, including tapered zones primarily for the acceleration or deceleration of vehicles entering or leaving the through traffic lanes, is termed a speed-change lane.

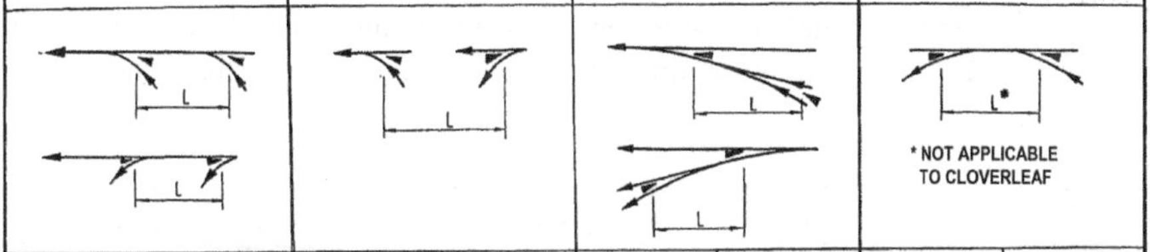

| EN - EN or EX - EX | | EX - EN | | TURNING ROADWAYS | | EN - EX (WEAVING) | | | |
|---|---|---|---|---|---|---|---|---|---|
| FULL EXPRESSWAY | C - D ROAD OR DISTRIBUTOR | FULL EXPRESSWAY | C - D ROAD OR DISTRIBUTOR | SYSTEM INTERCHANGE | SERVICE INTERCHANGE | SYSTEM TO SERVICE INTERCHANGE | | SERVICE TO SERVICE INTERCHANGE | |
| | | | | | | FULL EXP. | C-D Rd OR DIST. | FULL EXP. | C-D Rd OR DIST. |
| MINIMUM LENGTH MEASURED FROM PHYSICAL NOSE TO PHYSICAL NOSE (m) | | | | | | | | | |
| 300 | 250 | 150 | 120 | 240 | 180 | 600 | 480 | 480 | 300 |

*NOT APPLICABLE TO CLOVERLEAF

**FIGURE 12.23** Ramp Terminal Spacing.
*Source:* Figure 6.5, A Guide to the Design of Interchanges, Arahan Teknik (Jalan) 12/87, JKR

A speed-change lane should have sufficient length to enable a driver to make the necessary change between the speed of operation on the highway and the speed on the turning roadway safely and comfortably, as a minimum requirement. In the case of an acceleration lane, there should be additional length sufficient to permit adjustments in speeds of both through vehicles and entering vehicles so that the driver of the entering vehicle can position himself opposite a gap in the through traffic stream and maneuver into it before reaching the end of the acceleration lane.

Speed-change lanes are designed in two general forms, the taper-type, and the parallel-type. The taper-type works on the principle of a direct entry or exit at a flat angle, whereas the parallel-type has an added lane for speed-change. If properly designed, either type will operate satisfactorily, but the same should be applied for all entrances and exits on a route. This consistency would help drivers' expectations. Generally, parallel-type for entrance, taper-type for exit terminal is applied. The exceptions are recommended on relatively sharp curves to the right, where parallel-type may be more suitable.

### 12.7.4.4.8   Single-Lane Free Flow Terminal (Entrances)

a. **Taper-Type**

By relatively minor speed adjustment, the driver entering on a taper-type entrance can see and use an available gap in the through traffic stream. A typical single-lane taper-type entrance terminal is shown in Figure 12.24(A).

The entrance is brought into the expressway by providing a long, uniform tapper. Experience shows a desirable rate of taper of about 50:1–70:1 (longitudinal to lateral) between the outer edge of the acceleration lane and the edge of the through traffic lane. This rate of convergence provides adequate merging length and also defines a proper path for an entering vehicle.

The geometrics of the ramp proper should be such that motorists may attain a speed approximately equal to the average running speed of the expressway less by 10 km/hr within the time they reach the point where the right edge of

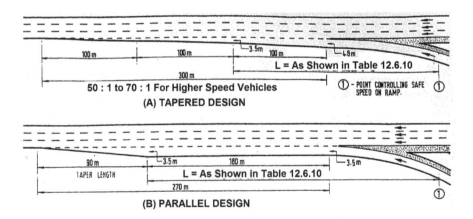

**FIGURE 12.24**   Single-Lane Entrance Ramps.
*Source:* Figure 6.6, A Guide to the Design of Interchanges, Arahan Teknik (Jalan) 12/87, JKR

the ramp joins the traveled way of the expressway. For consistency of application, this point of convergence of the right edge of the ramp and the left edge of the through lane may be assumed to occur where the left edge of the ramp traveled way is 3.5 m from the left edge of the marginal strip of the expressway.

Table 12.12 shows minimum acceleration distances for entrance terminals and their derivation. Where ramps occur in grades, the lengths should be adjusted following Table 12.13.

b. **Parallel-Type**

The parallel-type entrance provides an additional lane of sufficient length to enable the driver to accelerate the vehicle to near expressway speed before merging. A taper is provided at the end of the added lane. The process of entering the expressway is like a lane change to the right under forced conditions, for the UK style of traffic.

A typical design of a parallel-type entrance is shown in Figure 12.24(B). A curve with a radius of 300 m or more and a length of at least 60 m should be provided in advance of the added lane. If this curve has a short radius, motorists tend to drive directly onto the expressway without using the acceleration lane. This behavior results in poor merging operations and increases the accident potential to a high extent.

The length of the taper at the downstream end of a parallel-type acceleration lane should be suitable to guide the vehicle gradually onto the through lane of the expressway. A length of approximately 90 m is suitable for design-speeds up to 120 km/hr.

The length of a parallel-type acceleration lane is generally measured from the point where the right edge of the traveled way of the ramp joins the traveled way of the expressway to the beginning of the taper. The minimum acceleration lengths for entrance terminals are given in Table 12.12, and the adjustments for grades are given in Table 12.13.

The operational and safety benefits are well recognized for long acceleration lanes, particularly when both the expressway and ramp carry high traffic volumes. An acceleration lane of the length of at least 360 m, plus the taper, is desirable wherever it is anticipated that the ramp and expressway will frequently carry traffic volumes approximately equal to the design capacity of the merging area.

## 12.7.4.4.9  Single-Lane Free Flow Terminals (Exits)

a. **Taper-Type**

The taper-type exit is suitable at the direct path preferred by most drivers, permitting them to follow an easy path within the diverging area. The taper-type exit terminal beginning with an outer edge alignment break usually provides a clear indication of the point of departure from the through lane and in general has been found to operate smoothly on a high-volume expressway. The divergence rate of taper should be between 50:1 and 70:1, (longitudinal to lateral).

Figure 12.25(A) shows a typical design for a taper-type exit.

## TABLE 12.12
## Minimum Acceleration Lengths for Entrance Terminals ≤ (Grades 2%)

| Highway Design Speed (km/hr) | Stop Condition | Acceleration Length, L(m) For Entrance Curve Design Speed (km/hr) | | | | | | | |
|---|---|---|---|---|---|---|---|---|---|
| | | 25 | 30 | 40 | 50 | 60 | 70 | 80 | 90 |
| 50 | 60 | – | – | – | – | – | – | – | – |
| 60 | 115 | 100 | 75 | 65 | 40 | – | – | – | – |
| 80 | 235 | 215 | 190 | 180 | 150 | 115 | 50 | – | – |
| 100 | 360 | 325 | 330 | 300 | 275 | 275 | 180 | 120 | 50 |
| 120 | 485 | 470 | 460 | 430 | 405 | 405 | 310 | 250 | 180 |

**TAPE LENGTH OF PARALLEL TYPE ACCELERATION LANE**

| Highway Design Speed (km/hr) | Taper Length (m) |
|---|---|
| 50 | 50 |
| 60 | 60 |
| 80 | 70 |
| 100 | 80 |
| 120 | 90 |

Taper Length | Parallel Type

*Source:* Table 6.10, A Guide to the Design of Interchanges, Arahan Teknik (Jalan) 12/87, JKR

*Note:* "Uniform 50:1–70:1 Tapers" are recommended where "Length of Acceleration Lane" exceeds 400 m or elsewhere if appropriate and space permits

The length available for deceleration may be assumed to start from a point where the left edge of the tapered wedge is about 3–5 m from the left edge of the marginal strip, to the point controlling the safe speed for the ramp. Deceleration may end in a complete stop, as at a crossroad terminal for a diamond interchange, or the critical speed may be governed by the curvature of the ramp roadway. Minimum deceleration lengths for various combinations of design-speeds for the highway and the ramp roadway are given in Table 12.14. Grade adjustments are given in Table 12.13.

b. **Parallel-type**

A parallel-type exit terminal usually begins with a taper, followed by an added full lane of derived length. A typical parallel-type exit terminal is shown in Figure 12.25(C). This type of terminal provides an inviting exit area because the foreshortened view of the taper and the added width are very apparent. However, this design assumes that drivers will exit near the beginning of the added lane and make a speed-change thereafter. It requires a reverse curve maneuver which is somewhat unnatural. Under low-volume conditions, a driver may choose to avoid the reverse curve exit path and turn directly off the

**TABLE 12.13**

**Ratio of Length of Speed-Change on Grade to Length Level Acceleration or Deceleration Lane**

| Design Speed of Highway (km/hr) | Ratio of Length on Grade to Length on Level for a Design Speed of Turning Roadway Curve (km/hr) | | | | |
|---|---|---|---|---|---|
| All speeds | | 3–4% upgrade 0.9 | | | 3–4% downgrade 12 |
| All speeds | | 5–6% upgrade 0.8 | | | 5–6% downgrade 1.35 |

| Design Speed of Highway (km/hr) | Ratio of Length on Grade to Length on Level for a Design Speed of Turning Roadway Curve (km/hr) | | | | |
|---|---|---|---|---|---|
| | 30 | 50 | 60 | 80 | All Speeds |
| | | | 3–4% upgrade | | 3–4% downgrade |
| 60 | 1.3 | | 1.3 | – | 0.7 |
| 80 | 1.3 | | 1.4 | 1.4 | 0.65 |
| 100 | 1.3 | | 1.5 | 1.5 | 0.6 |
| 120 | 1.3 | | 1.6 | 1.7 | 0.6 |
| | | | 5% upgrade | | 3–4% downgrade |
| 60 | 1.5 | | 1.5 | – | 0.6 |
| 80 | 1.5 | | 1.7 | 1.9 | 0.55 |
| 100 | 1.7 | | 1.9 | 2.2 | 0.5 |
| 120 | 2.1 | | 2.3 | 2.6 | 0.5 |

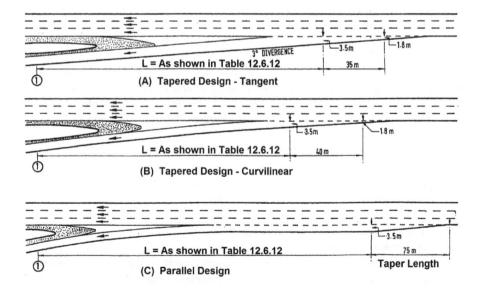

**FIGURE 12.25** Single-Lane Exit Ramps.

*Source:* Figure 6.7, A Guide to the Design of Interchanges, Arahan Teknik (Jalan) 12/87, JKR

**TABLE 12.14**

**Minimum Deceleration Lengths for Exit Terminals (Grades ≤2%)**

| Highway Design Speed (km/hr) | Stop Condition | Deceleration Length, L(m) For Exit Curve Design Speed (km/hr) | | | | | | | |
|---|---|---|---|---|---|---|---|---|---|
| | | 25 | 30 | 40 | 50 | 60 | 70 | 80 | 90 |
| 50 | 70 | 55 | 50 | 45 | – | – | – | – | – |
| 60 | 95 | 90 | 80 | 70 | 60 | 50 | – | – | – |
| 80 | 130 | 125 | 120 | 110 | 95 | 85 | 70 | 55 | – |
| 100 | 160 | 155 | 150 | 140 | 130 | 125 | 105 | 90 | 75 |
| 120 | 190 | 180 | 175 | 170 | 155 | 150 | 130 | 120 | 105 |

*Source:* Table 6.12, A Guide to the Design of Interchanges, Arahan Teknik (Jalan) 12/87, JKR

through lane in the vicinity of the exit nose. Such a maneuver may result in undesirable deceleration at the through lane, in undesirable conflict on the deceleration lane, or excessive speed in the exit nose area.

The ratio from this table multiplies 2 by a length in Table 12.12 or Table 12.14 Gives length of a speed-change lane on grade Taper length need not be corrected for grade.

The length of a parallel-type lane is usually measured from the point where the added lane attains the width of 3.5 m to the point where the alignment of the ramp roadway departs from the alignment of the expressway. It is desirable to provide a transition at the end of the deceleration lane, where the ramp proper is curved. A compound curve may be used with the initial curve desirably having a radius of 300 m or more. If the deceleration lane connects with a relatively straight ramp, a transition or a large radius curve is also desirable. In such cases a portion of the ramp may be considered as a part of the length of deceleration, thus shortening somewhat the required length of a continuous parallel lane. Minimum lengths are given in Table 12.14 and adjustments for grades in Table 12.13.

c. **Free Flow Terminals on Curves**

Where the curves on an expressway are relatively sharp and it is necessary to provide exits and entrances on these curves, some adjustments in design may be needed to avoid operational difficulties.

On an expressway having design-speeds of 100 km/hr or more the curves are sufficiently gentle so that either the parallel-type or the taper-type of speed-change lane fits well. With the parallel-type, the design is the same as that on a tangent and the added lane is usually on the same curvature as the mainline. For the taper-type, the dimensions applicable to terminals located on tangent alignment are also suitable for use on curves. The ramp is tapered relative to the through traffic lanes on the curved section, at the same rate, as the tangent section.

Wherever a part of a tapered speed-change lane falls on a curved alignment, its entire length should be within the limits of the curve. Where the tapper is provided on tangent alignments just upstream from the beginning of the curve, the outer edge of the taper will appear as a kink at the point of curvature.

At ramp terminals on relatively sharp curves as those may occur on expressways having a design-speed of 80 km/hr, the parallel-type of speed-change lanes is advantageous compared to the taper-type. At exits, the parallel-type is less likely to confuse through traffic, and at entrances, this type will usually provide smoother merging operations.

Parallel-type speed-change lanes at ramp terminals on curves are shown by diagrams in Figure 12.26.

Entrances on curved sections of an expressway generally have fewer problems than exits. Figures 12.26(A) and (B) show entrances with the expressway, which are turning to the right and left respectively. The approach curve on the ramp must have a very long radius as it joins the acceleration lane. This aligns the entering vehicle with the acceleration lane and lessens the chances of motorists entering directly onto the through lanes. The taper at the end of the acceleration lane should be long enough, preferably about 90 m to facilitate the merging with the through traffic. When alignment occurs with a reverse curve between the ramp and speed-change lane, an intermediate tangent should be used to aid the attainment of superelevation in transition.

An exit becomes particularly troublesome where the expressway is curved to the right Figure 12.26(C) because traffic on the outside lane tends to follow the ramp, therefore exits on right-turning curves should be avoided, if possible, in the UK style of traffic. Caution must be used in positioning a taper-type deceleration lane on the outside of a right-turning mainline curve. The design should provide a definite break in the left edge of pavement to provide a visual indication to the through driver so that he is not inadvertently led off the through roadway. To make the deceleration lane more apparent to approaching motorists, the taper should be shorter, preferably not more than 30 m in length. The deceleration lane should begin either upstream or downstream from the point of the curve. It should not begin right at the point of the curve as the deceleration lane appears to be an extension of the tangent, and motorists are more likely to be confused. The ramp proper should begin with a section of tangent or a long radius curve to permit a long and gradual reversing of the superelevation.

To avoid operational problems, an alternate design may be developed to locate the exit terminal a considerable distance upstream from the point of the curve. A separate and parallel ramp roadway is provided to connect with the ramp properly.

There is a tendency for vehicles to exit inadvertently, with the highway curved to the left and the exit located on the left (Figure 12.26D). Here also, the taper should be short to provide additional "target" value for the deceleration lane. With this configuration, the superelevation of the deceleration lane is readily applied by continuing the rate from the through pavement and generally increasing it to the rate required on the ramp curve.

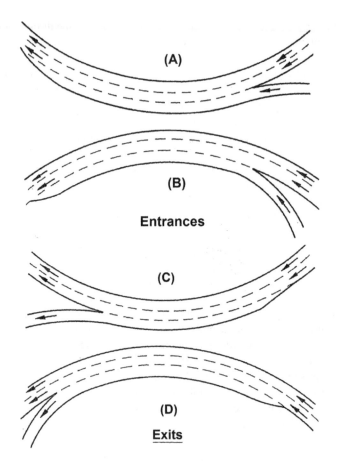

**FIGURE 12.26**   Parallel-Type Ramp Terminals at Curves (Diagrammatic).
*Source:* Figure 6.8, A Guide to the Design of Interchanges, Arahan Teknik (Jalan) 12/87, JKR

### *12.7.4.4.10   Multilane Free Flow Terminals*

Multilane terminals are required where traffic is too high for single-lane operation. Other considerations that may warrant for multilane terminals are through route continuity, queuing on long ramps, lane balance, and design flexibility. The most common multilane terminals consist of two-lane entrances and exits at the expressway. Other multilane terminals are sometimes termed as "major forks" and "branch connections." The latter terms denote a separating and joining of two major routes.

a. **Two-lane Entrances**

Two-lane entrances are warranted in two situations: either as branch connections or because of capacity requirements of the on ramp. To satisfy lane balance requirements, at least one additional lane must be provided downstream. In this addition, a basic lane may also be required for capacity, or an auxiliary lane that may be dropped 750–900 m downstream or at the next interchange. In some instances, two additional lanes may be necessary because of capacity requirements.

When only one additional lane is required either the inside tapered or parallel design will be satisfactory. The choice would be dependent on local usage. Intermixing of the two designs is not recommended within a route system or an urban area system.

If the two-lane entrance is preceded by a two-lane exit, there is probably no need to increase the basic number of lanes on the expressway from a capacity standpoint. In this case, the added lane that results from the two-lane entrance is considered an auxiliary lane, and it may be dropped approximately 750 m or more downstream from the entrance.

Figure 12.27 illustrates simple two-lane entrance terminals where a lane has been added to the expressway. The number of lanes on the expressway has little or no effect on the design of the terminal.

Figure 12.27(A) shows a taper-type entrance and Figure 12.27(B), a parallel-type entrance.

As shown in Figure 12.27(A), the basic layout form of a two-lane taper-type entrance, is that of a single-lane taper, with a second lane added to the left or outer side and continued as an added or auxiliary lane on the expressway. As in the case of a single-lane tapered entrance, an angle of convergence of about 10 degrees or slightly more, corresponding to a rate of convergence of about 1:50 represents a near-optimum design. The length of a two-lane taper-type entrance is approximately the same as that for a single-lane entrance.

As shown in Figure 12.27(B), at the parallel-type of a two-lane entrance, the right lane of the ramp is continued onto the expressway as an added lane.

The left lane of the ramp is made as a parallel lane and terminated with a tapered section of about 90 m long. The length of the left lane should be in the range of 180–360 or more. Major factors in determining the needed length are the traffic volume of the ramp and the volume on the expressway. With traffic from the ramp only slightly above the design capacity of a single-lane ramp, a length of 180 m, plus taper, may fulfill the requirements. Where the combined

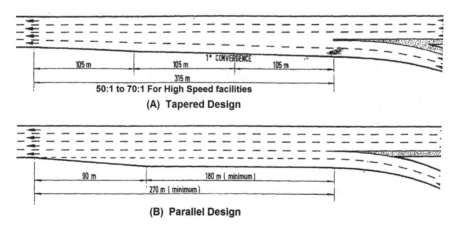

**FIGURE 12.27** Typical Two-Lane Entrance Ramps.
*Source:* Figure 6.9, A Guide to the Design of Interchanges, Arahan Teknik (Jalan) 12/87, JKR

traffic volume for the expressway and ramp is near the design capacity of the expressway downstream from the merging area, a length of up to 900 m may need to blend the traffic smoothly into a free flowing stream.

Traffic operation at a parallel-type with a two-lane entrance is entirely different from that at a taper-type with a two-lane entrance. With the two-lane parallel-type entrance most traffic from the ramp will use the right lane of the ramp, which is continued onto the expressway as an added lane.

With the taper-type of a two-lane entrance, drivers tend to use the left lane rather than the right lane, as they do when using the parallel-type of the entrance. Either form of a two-lane entrance is satisfactory if used exclusively within an area or a region, but they should not be intermixed along a given route. The above descriptions are for the UK style of traffic.

b. **Two-lane Exits**

To fulfill the lane balance requirements and not to reduce the basic number of through lanes, it is usually necessary to add an auxiliary lane upstream from the exit. Approximately 450 m is required to develop the full capacity of a two-lane exit. Typical designs for two-lane exit terminals are shown in Figure 12.28; the taper-type is illustrated in Figure 12.28(A) and the parallel-type in Figure 12.28(B).

In cases where the basic number of lanes is to be reduced beyond a two-lane exit, the basic number of lanes should be carried beyond the exit before the outer lane is dropped. This design provides a recovery area for the through vehicles that remain in that lane.

With the parallel-type of two-lane exit, as shown in Figure 12.28(B) the operation is different from the taper-type, in which the traffic in the outer through lane of the expressway must change lanes to exit.

Lane changing to a significant extent is necessary to enable the "Exit" to operate efficiently. This whole operation requires a sufficient length of the highway, which depends on part of the total traffic volume in the expressway and

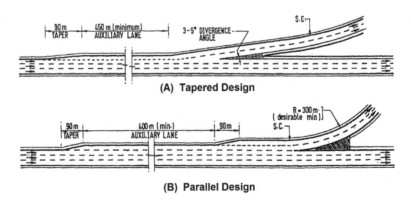

**FIGURE 12.28**   Typical Two-Lane Exit Ramps.
*Source:* Figure 6.10, A Guide to the Design of Interchanges, Arahan Teknik (Jalan) 12/87, JKR

especially on the volume using the exit ramp. The total length from the beginning of the first taper to the point of departure of the ramp traveled way from the left-hand through lane of the expressway will range from 760 m for turning volumes of 1,500 vph or less to 1060 m for turning volume of 3,000 vph.

c. **Major Forks and Branch Connections**
   A major fork is the bifurcation of a directional roadway, of a terminating expressway route into two-directional multilane ramps that connect to another expressway or as the diverging area created by the splitting of an expressway route into two separate expressway routes of about equal importance.

   The design of major forks is subject to the same principles of lane balance as any other diverging area. The total number of lanes in the two roadways beyond the divergence should exceed the number of lanes approaching the diverging area by at least one. It is desired that the number of lanes should be increased by only one. Operational difficulties invariably develop unless traffic in one of the interior lanes has the option to take either of the diverging roadways. The nose should be aligned with the centerline of one of the interior lanes, as illustrated in Figure 12.29.

   The interior lane is continued as a full-width lane both left and right of the gore. Thus, the width of this interior lane will be at least 7.0 m at the painted nose, which is the extension of pavement edge stripes and preferably not over 8.5 m. the length over which the widening from 3.5 to 7 m takes place should be within the range of 300 or 550 m.

   In the case of a two-lane roadway splitting into two, two-lane routes, there is no interior lane. In such cases it is advisable to widen the approach roadway three lanes, thus creating an interior lane. The lane is added on the side of the fork and serves the lesser traffic volume. In the illustration, Figure 12.29(A) the left (lower) fork would be the more lightly traveled of the two. The

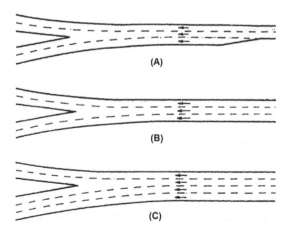

**FIGURE 12.29**  Examples of Major Forks.
*Source:* Figure 6.11, A Guide to the Design of Interchanges, Arahan Teknik (Jalan) 12/87, JKR

widening from 7 m at the approach roadway to about 15 m at the painted nose should be accomplished in a continuous sweeping curve with no reverse curvature in the alignment of the pavement edges.

A branch connection is defined as the beginning of a directional roadway of an expressway, which is formed either by the convergence of two-directional multilane ramps from another expressway or by the convergence of two expressway routes to form a single expressway route.

The number of lanes downstream from the point of convergence may be one-lane less than the combined total number of lanes on the two approach roadways. In some cases, the traffic demand may require that the number of lanes going away from the merging area be equal to the sum of the number of lanes of the two roadways approaching it, and a design of this type will not pose an operational problem. Such a design is illustrated in Figure 12.30(A).

Where a lane is to be dropped, which is a very common case, a means for accomplishing the reduction is discussed in the section "lane reduction." The lane that is terminated will commonly be the exterior lane in the roadway serving the lowest volume per lane. However, some consideration should also be given to the fact that the outer lane from the roadway entering from the left is the slow-speed lane for that roadway, whereas the opposite is true for the roadway entering from the right. If the traffic volumes per lane are nearly equal, it would be proper to terminate the lane on the left, as shown in Figure 12.27(B).

Another consideration is the possibility of a high-speed inside merge, as in Figure 12.30(C). This merge should be treated as any other high-speed merging situation by adding a lane on the left. The above descriptions are for the UK style of traffic.

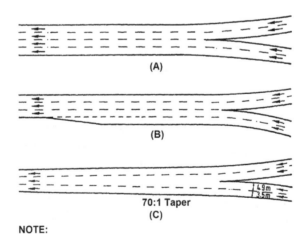

(A)

(B)

70:1 Taper
(C)

NOTE:
**For proper Merge Designs there are Two Lane Entrances**

**FIGURE 12.30**  Examples of Branch Connections.
*Source:* Figure 6.12, A Guide to the Design of Interchanges, Arahan Teknik (Jalan) 12/87, JKR

## 12.8  INTERCHANGE CAPACITY

### 12.8.1  GENERAL

Traffic movement at interchanges is quite different from the uniform flow on an open road. Diverging, merging, and weaving movements frequently take place in the interchange area. Stopping may also occur at the minor crossroad terminals. Every facility provided to meet these various demands of movements must have sufficient capacity to maintain the designated level of service. The different facilities must also provide the same level of service as that assumed for the highway crossing at the interchange. There must not be any bottleneck or spot congestion.

The capacity at the points of those special movements governs the performance of the whole interchange. Checking of the capacity thus must be done at the points of the special movements.

The capacity analysis for the ramp terminals and weaving section should follow, which is detailed in "Highway Capacity Manual" – Special Report 209, Transportation Research Board, 1985. (Chapter 4 and 5). The following sections are taken from the above publication and give a brief outline of the methodology and procedures that are adopted.

### 12.8.2  RAMP TERMINALS

#### 12.8.2.1  Ramp Components

A ramp may consist of up to three geometric elements of interest, those are:

- Ramp terminal at the expressway.
- Ramp roadway proper.
- Ramp terminal at the minor crossroad.

A ramp terminal at the expressway is generally designed to permit high-speed merging or diverging movements to take place with minimum disruption to the adjacent expressway traffic stream. Geometric elements such as provision and length of acceleration/deceleration lanes, angle of convergence and divergence, relative grades, and other aspects may impact ramp operations.

From location to location, the ramp roadway itself may also vary widely. They can vary in the number of lanes, length, design-speed, grades, and horizontal curvature. They seldom create operational difficulties unless a traffic accident disrupts its length.

The ramp terminal at the minor crossroad can be of a type permitting uncontrolled merging or diverging movements or it can be of a type of an at-grade intersection, in which the requirements are to be considered as specified in the relevant standard for the design of at-grade intersections.

#### 12.8.2.2  Operational Characteristics

A ramp terminal at the expressway is an area of competing traffic demands for space.

In the merge area, vehicles at the entrance ramp, try to find openings or "gaps" in the traffic stream at the adjacent lane on the expressway. As most ramps are on the left-side of the facility, the most direct impact is on the shoulder lane of the expressway, which is designated as lane 1.

At exit ramps, the basic maneuver is to diverge. The existing vehicles must occupy the lane adjacent to the ramp so that, there is a net effect of other drivers redistributing themselves among the other lanes.

A ramp operates efficiently only if all its elements, the terminals with expressways and minor crossroads, and the ramp roadway have been properly designed. It is important to note that a breakdown on any one of these elements will adversely affect the operation of the entire ramp, which may also become critical to the facilities it connects.

### 12.8.2.3   Computational Procedure for Ramp Terminals at the Expressway

The step-by-step computational procedure for the capacity analysis of ramp terminals as detailed from the Highway Capacity Manual is as follows:

Step 1 – Establish ramp Geometry and Volumes
The establishment of a geometric configuration includes the type, location, and volumes on the adjacent ramps.
Step 2 – Compute Lane one volume
This is computed by using the nomographs or the approximation procedure. The lane one volume is dependent on the ramp volume, the total expressway volume upstream of the ramp, the distance and volume of the adjacent upstream and /or downstream ramps, and the type of ramp in question.
Step 3 – Convert all volumes to Passenger Car Units Per Hour
Step 4 – Compute Checkpoint Volumes
For each ramp analysis, there are up to three checkpoint volumes for each ramp or pair of ramps, these are merging volume, diverging volume, or total expressway volume.
Step 5 – Convert checkpoint volumes to Peak Flow Rates
Step 6 – Find relevant Levels of Service

The level of service for a given analysis is found by comparing the checkpoint flow rates for merging, diverging, and total expressway volume with the given criteria. In many cases, the various operational elements like, merges, diverges, expressway flows, will not have the same level of services (LOS). In such cases, the worst resultant LOS is assumed to govern and would then be considered for improvement if the resultant LOS is unacceptable.

The LOS of the merging and diverging point locations should be in balance with the expressway as a whole. It is desirable in fact to have the LOS of the merging and diverging points better than the LOS for the total expressway volume.

### 12.8.3   Weaving Sections

### 12.8.3.1   General
Weaving is defined as the crossing of two or more traffic streams traveling in the same general direction along a significant length of the highway, without the aid of

traffic control devices. Weaving areas are formed when a merge area is closely followed by a diverging area, or when an on ramp is closely followed by an off-ramp and the two are joined by an auxiliary lane.

Weaving areas require intense lane changing maneuvers as drivers must access lanes appropriate to their desired exit point. Thus, traffic in a weaving area is subject to turbulence above that is normally present on basic highway sections. This turbulence presents special operational problems and design requirements.

Figure 12.31 shows the formation of a weaving area. If entry and exit roadways are referred to as "legs," vehicles traveling from leg A to leg D must cross the path of vehicles traveling from leg B to leg C. flows A-D and B-C are, therefore, referred to as weaving flows. Flows A-C and B-D may also exist in the section, but these need not cross the path of other flows and are referred to as non-weaving flows. Figure 12.31 shows a simple weaving area, formed by a single merge point followed by a single diverge point. Multiple weaving areas are formed by a single merge followed by two diverges or two merges followed by a single diverge.

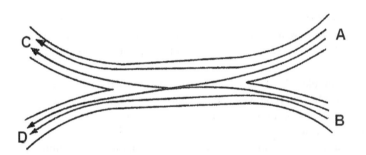

**FIGURE 12.31** Formation of a Weaving Section.
*Source:* Figure 7.1, A Guide to the Design of Interchanges, Arahan Teknik (Jalan) 12/87, JKR

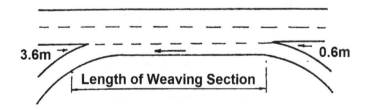

**FIGURE 12.32** Length of Weaving Section.
*Source:* Figure 7.2, A Guide to the Design of Interchanges, Arahan Teknik (Jalan) 12/87, JKR

### 12.8.3.2  Weaving Length

The length measurement of the weaving area is shown in Figure 12.32, it is the length from the merge gore area at a point where the left edge of the expressway shoulder lane and the right edge of the merging lane is 0.6 m apart to a point at the diverge gore area where the two edges are 3.6 m apart.

The various weaving movements, by all lane changes, must be completed within the weaving length. These create constraints for the time and space in which the drivers must make all the required lane changes and therefore cause an increase in the intensity of lane changing and the level of turbulence.

### 12.8.3.3  Configuration

Configuration refers to the relative placement and number of entry and exit lanes for the section and can have a major impact on the operational characteristics of weaving areas.

There are three primary types of weaving configurations and are referred to as types A, B, and C as shown in Figures 12.33, 12.34, and 12.35 respectively. These are defined in terms of the minimum number of lane changes that must be made by weaving vehicles as they travel through the section.

#### 12.8.3.3.1  Type A Weaving Areas

Weaving areas of type A, require that each weaving vehicle make a one-lane change to make the desired movement. Figure 12.33 shows two examples of weaving areas of type A.

Weaving vehicles are usually confined to occupying the two lanes adjacent to the crown line because, in type A, weaving vehicles in the weaving area must cross the crown line. Weaving and non-weaving vehicles, therefore, generally share the lanes adjacent to the crown line.

Configuration significantly has effects on operations and should limit the maximum number of lanes that weaving vehicles may occupy while traversing the section.

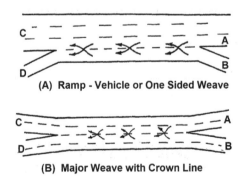

**(A)  Ramp - Vehicle or One Sided Weave**

**(B)  Major Weave with Crown Line**

**FIGURE 12.33**  Type A Weaving Areas.
*Source:* Figure 7.3, A Guide to the Design of Interchanges, Arahan Teknik (Jalan) 12/87, JKR

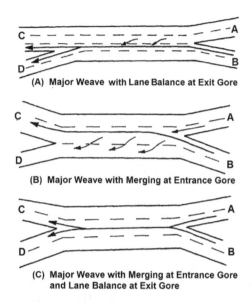

**FIGURE 12.34**  Type B Weaving Areas.
*Source:* Figure 7.4, A Guide to the Design of Interchanges, Arahan Teknik (Jalan) 12/87, JKR

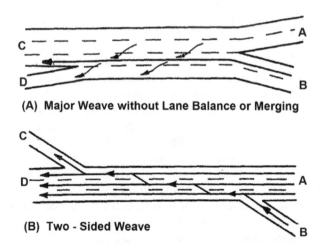

**FIGURE 12.35**  Type C Weaving Areas.
*Source:* Figure 7.5, A Guide to the Design of Interchanges, Arahan Teknik (Jalan) 12/87, JKR

### 12.8.3.3.2   Type B Weaving Areas

Weaving areas of type B, are referred to as major weaving sections, these involve multilane entry and/or exit legs and. They are characterized by:

- One weaving movement may be accomplished without making any lane change.
- The other weaving movement requires a maximum one-lane change.

Type B weaving areas are extremely efficient in carrying large volumes of weaving, mainly for having the provision of a "through lane" for one of the weaving movements. Weaving maneuvers can be accomplished with a single-lane change from the lane or lanes adjacent to this "through lane." Thus, weaving vehicles can occupy an unrestricted number of lanes in the weaving section and are not as in type A sections.

### 12.8.3.3.3   Type C Weaving Areas

Type C weaving areas are like type B sections but, here one or more "Through lanes" are provided for one of the weaving movements. The distinguishing feature between type B and C is the number of lane changes required. A Type C weaving area is characterized by:

- One weaving movement may be accomplished without making a lane change.
- The other weaving movement requires two or more lane changes.

## 12.8.3.4   Weaving Width and Type of Operation

Another geometric characteristic having a significant effect on weaving area operations is the width of the weaving area, expressed as the number of lanes in the section. It is, however, not only the total number of lanes that impact weaving area operations but proportional use of those lanes by weaving and non-weaving vehicles.

The nature of weaving movements creates turbulence in the traffic stream and causes a weaving vehicle to consume more of the available roadway space than a non-weaving vehicle. The exact nature of the relative space use depends on the relative weaving and non-weaving volumes using the weaving area and the number of lane changes weaving vehicles must make. As discussed, the latter depends on the configuration of the weaving section. Thus, the proportional use of space is dependent not only on relative volumes but also on the configuration of the weaving area.

The configuration has a further impact on the proportional use of available lanes, it can limit the ability of weaving vehicles to use outer lanes in the section. This limitation is most severe in sections of type A, in which all weaving vehicles must cross a crown line, and is least severe in sections of type B.

In general, vehicles in a weaving area will make use of available lanes in such a way that all component traffic streams approximately achieve the same average running speed, with weaving flows somewhat slower than non-weaving flows.

Occasionally, the configuration limits the ability of weaving vehicles to occupy the proportion of available lanes required to achieve this equivalent or balanced operation.

In such cases, weaving vehicles occupy a smaller proportion of the available lanes than desired, while non-weaving vehicles occupy a larger proportion of lanes than for balanced operation. The operation of the weaving area is then classified as "constrained" by the configuration. The result of the constrained operation is that non-weaving vehicles will move at significantly higher speeds than weaving vehicles. Where configuration does not restrain weaving vehicles from occupying a balanced proportion of available lanes, the operation is classified as "unconstrained."

### 12.8.3.5   Computational Procedure for Simple Weaving Areas

The step-by-step computational procedure for the evaluation of the LOS in an existing or projected simple weaving area is given in details in the Highway Capacity Manual, which is briefly as follows:

Step 1 – Establish Roadway and Traffic Conditions.

All existing or projected roadway and traffic conditions are to be specified. Roadway conditions include the length, number of lanes, and type of configuration for the weaving area under study as well as the lane widths and the general terrain.

Traffic conditions include the distribution of vehicle types in the traffic stream and the peak-hour factor where the component flows have differing peaking characteristics.

Step 2 – Convert all traffic volumes to Peak Flow under ideal conditions.
Step 3 – Construct Weaving Diagram.
Step 4 – Compute unconstrained weaving and non-weaving speeds.
Step 5 – Check for Constrained Operation.
Step 6 – Check for Weaving Area Limitations.
Step 7 – Determine the LOS.

Levels of service in weaving areas are directly related to the average running speeds of weaving and non-weaving vehicles and for a given analysis, it is found by comparing the calculated average weaving and non-weaving speeds with the given criteria.

The analysis of multiple weaving areas is similar except that it involves the developing of the appropriate weaving diagrams for each sub-segment of the area in question, which is then analyzed as a simple weaving area.

## 12.9   INTERCHANGE SIGNAGE

### 12.9.1   GENERAL

The development of a signing system for an interchange must be approached on the consideration that the signing is primarily for the benefit and guidance of drivers who are not familiar with the route or area. The signing therefore must provide clear instructions with an orderly progress to drivers for their destinations.

The installation of the signs must be taken as an integral part of the interchange and expressway facility and as such must be planned concurrently with the geometric design.

Drivers should be confronted with consistent signing on the approaches to interchanges as they drive from one area to another and when driving through rural or urban areas. Geographical, geometric, and operating factors create significant differences between urban and rural conditions and the signing must take these into account.

### 12.9.2 Types of Interchange Signing

#### 12.9.2.1 Standard Traffic Signs

Standard traffic signs are normally installed at ramp terminals, i.e., exits, entrances, and gore areas. The standard traffic signs must conform to the specifications of the relevant standard. The full complement of standard traffic signs must be installed to ensure the proper operation of the interchange.

Examples of standard traffic signs installed at ramp terminals are:

- At gore areas – a combination double arrow and an obstruction marker sign is to be provided.
- At ramp exit terminals – STOP, GIVE WAY, or traffic signal signs should exist together with NO ENTRY and /or advanced signs.
- At ramp entrance terminals – GIVE WAY signs together with its advanced sign should be provided.

#### 12.9.2.2 Guide Signs

The layout design and application of the various types of guide signs must follow the specified requirements of the relevant standards. The following are the general guidelines that should also be followed:

- Destination Signs – On the expressway approaches a minimum of two destination signs (i.e., 500 m and 1 km away) should be installed. While on the minor crossroad approaches one destination sign (at 500 m away) insufficient.
- Directional Signs – On the expressway, at least two-directional signs, with one at the gore area if feasible, and a complimentary overhead gantry sign at the beginning of the taper section of the exit should be installed. On a minor crossroad, the overhead gantry sign is usually not required.
- Distance signs – these are required on all the roadways leaving away from the interchange area.

Figures 12.35 and 12.36 give the typical guide signing requirements for a diamond interchange and a full cloverleaf interchange respectively, without having the collector-distributor roads. Figures 12.35 and 12.36 also show the signing requirements only for the northbound approach of the main expressway facility and the eastbound

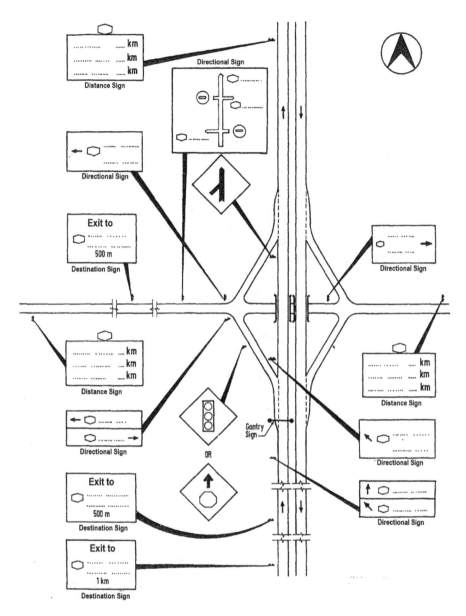

**FIGURE 12.36** Typical Signage Requirements at A Diamond Interchange.
*Source:* Figure 8.1, A Guide to the Design of Interchanges, Arahan Teknik (Jalan) 12/87, JKR

approach of the crossroad. The full signing of the interchange should cover all approaches and ramps.

Figures 12.35 and 12.36 should be taken only as typical examples. At every interchange, the signing arrangements may be different due to physical or operational constraints. This may need the necessary modifications should be made. For

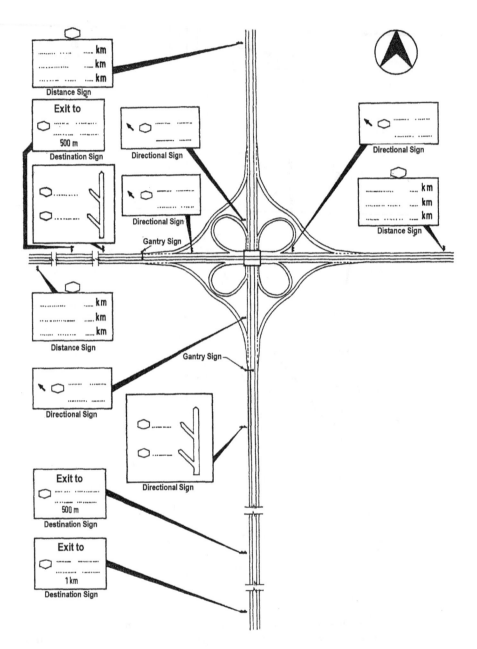

**FIGURE 12.37** Typical Signage Requirements at a Full Cloverleaf Interchange.
*Source:* Figure 8.2, A Guide to the Design of Interchanges, Arahan Teknik (Jalan) 12/87, JKR

example, if the gore area is physically restrictive, the directional sign can be installed as a butterfly sign (with only one post).

### 12.9.2.3 Gantry Signs

Gantry signs are very effective at interchanges and should be provided at all ramp exits on the expressway approaches before the interchange, especially where the physical constraints at the site make it difficult to provide the normal directional signs adequately. This will be more apparent in the layout. Structural design and other requirements of the gantry signs should follow the latest specified requirements of the relevant standards.

### 12.9.3 TYPES AND DETAILS OF GRADE-SEPARATED INTERCHANGES

1 Terminology
    1.1    Complete and incomplete interchanges
2 Between two controlled- or limited-access highways (system interchange)
    2.1    Four-way interchanges
        2.1.1    Cloverleaf interchange
        2.1.2    Stack interchange
        2.1.3    Cloverstack interchange
        2.1.4    Turbine interchange
        2.1.5    Roundabout interchange
        2.1.6    Other/hybrid interchanges
    2.2    Three-way interchanges
        2.2.1    Trumpet interchange
        2.2.2    Directional T interchange (Full Y interchange)
        2.2.3    Semi-directional T interchange
        2.2.4    Other/hybrid interchanges
    2.3    Two-way interchanges
3 Between a controlled- or limited-access highway and a road without access control (service interchange)
    3.1    Diamond interchange
    3.2    Parclo interchange/folded diamond
    3.3    Diverging diamond interchange
    3.4    Single-point urban interchange
    3.5    Other/hybrid interchanges

The descriptions of road junctions in this chapter are for countries like the USA, Europe, Australia, Japan, etc., where vehicles drive on the right side of the road. For countries where driving is on the left, like UK, India, Malaysia, Singapore, etc., the layout of the junctions is the same, only left/right is to be considered as reversed.

- A highway interchange (in the U.S.) or motorway junction (in the UK) is a type of road junction, is a freeway junction linking one motorway to another or sometimes to just a motorway service station. In the UK motorway network, most (but not all) junctions with other roads are numbered sequentially. In the

U.S., interchanges are either numbered according to cardinal interchange numbers or by mileage (typically the latter in most states).

- A highway ramp (as in exit ramp and entrance ramp) or slip road is a short section of road which allows vehicles to enter or exit a controlled or freeway or motorway.
- A directional ramp always tends toward the desired direction of travel. This means that a ramp that makes a left-turn exit from the left-side of the roadway (a left exit). Left directional ramps are relatively uncommon as the left lane is usually reserved for high-speed through traffic. Ramps for a right-turn are almost always right directional ramps. Where traffic drives on the left, these cases are reversed.
- A non-directional ramp goes in a direction opposite to the desired direction of travel. Many loop ramps (as in a cloverleaf) are non-directional.
- A semi-directional ramp exits a road in a direction opposite from the desired direction of travel but then turns toward the desired direction of travel. Many flyover ramps (as in a stack) are semi-directional.
- A U-turn ramp leaves the road in one driving direction, turns over or under it, and rejoins in the opposite direction.

**Off-ramp accessed from collector lanes along the highway**

Weaving is an undesirable situation in which traffic veering right and traffic veering left must cross paths within a limited distance, to merge with traffic on the through lane.

The German Autobahn system splits Autobahn-to-Autobahn interchanges into two types: a four-way interchange, the Autobahn Kreuz (AK), where two motorways cross, and a three-way interchange, the Autobahn Dreieck (AD) where two motorways merge.

Some on ramps have a ramp meter, which is a dedicated ramp only traffic light that throttles the flow of entering vehicles.

### 12.9.4 Complete and Incomplete Interchanges

A complete interchange has enough ramps to provide access from any direction of any road in the junction to any direction of any other road. Barring U-turns, a complete interchange between two freeways requires eight ramps, while a complete interchange between a freeway and another road (not a freeway) requires at least four ramps. Using U-turns, the number of ramps for two freeways can be reduced to six, by making cars that want to turn left either pass by the other road first, then make a U-turn and turn right, or turn right first and then make a U-turn. Depending on the interchange type and the connectivity offered other numbers of ramps may be used. For example, if a highway interchanges with a highway containing a collector/express system, additional ramps can be used to strictly link the interchanging highway with the collector and express lanes respectively. For highways with high-occupancy vehicle lanes, ramps can be used to service these carriageways directly, thereby increasing the number of ramps used.

An incomplete interchange has at least one or more missing ramps that prevent access to at least one direction of another road in the junction from any other road in the junction.

#### 12.9.4.1 Four-Way Cloverleaf Interchange Between Two Controlled Or Limited-Access Highways (System Interchange)

**A typical cloverleaf interchange**

A cloverleaf interchange is typically a two-level, four-way interchange where all left-turns are handled by loop ramps (right-turns if traveling on the left). To go left, vehicles first cross over or under the targeted route, then bear right onto a sharply curved ramp that loops roughly 270 degrees, merging onto the interchanging road from the right, and crossing the route just departed.

The major advantage of a cloverleaf is that they require only one bridge, which makes such junctions inexpensive if the land is plentiful. A major shortcoming of the cloverleaf, however, is weaving (see definition above) and the consequent low capacity of this design.

Cloverleaf also uses a considerable area of land and is more often found along older highways, in rural areas, and within cities with low population densities. A variant design is to separate all turning traffic into a parallel carriageway to minimize the problem of weaving. Collector and distributor roads are similar but are usually separated from the main carriageway by a divider, such as a guard rail or Jersey barrier.

### 12.9.4.2  Stack Interchange

**Four level Stack**

A stack interchange is a four-way interchange whereby left-turns are handled by semi-directional flyover/under ramps. To go left (right in countries with left-hand drive), vehicles first turn slightly right (on a right-turn off-ramp) to exit, then complete the turn via a ramp that crosses both highways, eventually merging with the right-turn on ramp traffic from the opposite quadrant of the interchange. A stack interchange, then, has two pairs of left-turning ramps, of which can be stacked in various configurations above or below the two interchanging highways.

### 12.9.4.3  Cloverstack Interchange (Partial Cloverleaf Interchange)

**Three level Cloverstack**

**Two level Cloverstack**

In the late 1960s, the partial cloverleaf interchange (parclo) designs modified as the freeway traffic emerged, eventually leading to the cloverstack interchange. Its ramps are longer to allow for higher ramp speeds and loop ramp radii are made larger as well. The large loop ramps eliminate the need for a fourth and sometimes a third-level in a typical stack interchange, as only two directions of travel use flyover/under ramps.

As three-level four-way interchanges, cloverstacks are cheaper to build and require less land than four-level stack interchanges. By using the loop ramps in opposite quadrants, weaving is also eliminated. Unfortunately, the loop ramps are not as efficient as a stack interchange's flyover/under ramps in terms of traffic flow. Thus, a cloverstack is oriented in such a way that the third-level semi-directional flyover ramps serve the higher traffic movements, with loop ramps serving lower traffic movements.

The cloverstack design is commonly used to upgrade cloverleaf interchanges to increase their capacity and eliminate weaving.

### 12.9.4.4 Turbine Interchange

 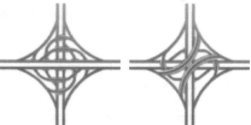

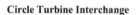

Circle Turbine Interchange        2-level Turbine        3-levelTurbine Interchang

Another alternative to the four-level stack interchange is the turbine interchange (also known as a whirlpool). The turbine/whirlpool interchange requires fewer levels (usually two or three) while retaining semi-directional ramps throughout and has its left-turning ramps sweep around the center of the interchange in a spiral pattern in right-hand driving.

Turbine interchanges offer slightly less vehicle capacity because the ramps typically turn more often and change height quicker. They also require more land to construct than the typical four-level stack interchange.

In areas with rolling or mountainous terrain, turbine interchanges can take the advantage of the natural topography of the land due to the constant change in the height of their ramps, and hence these are commonly used in these areas where conditions apply, by reducing construction costs compared to turbine interchanges built on level ground.

### 12.9.4.5   Roundabout Interchange

**Roundabout Interchange**

A further alternative is found often and is called a roundabout interchange. This is a normal roundabout except one (two-level) or both (three-level) mainlines pass under or over the whole interchange. The ramps of the interchanging highways meet at a roundabout or rotary on a separated level either above or below or in the middle of the two highways.

Roundabout interchanges are much more economical in the use of materials and land than other interchange designs, as the junction does not normally require more than three bridges to be constructed. However, their capacity is limited when compared to other interchanges and can become congested easily with high traffic volumes.

### 12.9.4.6   Other/Hybrid Interchanges

**Other/Hybrid Interchanges**

Hybrid interchanges use a mixture of interchange types and are not uncommon. Their construction can consist of multiple interchange designs such as loop ramps, flyovers, and roundabouts.

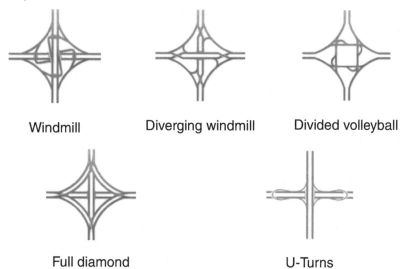

| Windmill | Diverging windmill | Divided volleyball |

| Full diamond | U-Turns |

### 12.9.4.7 Full Diamond U-Turns

A windmill interchange is like a turbine interchange, but it has much sharper turns, reducing its size and capacity. A variation of the windmill, called the diverging windmill, increases capacity by altering the direction of traffic flow of the interchanging highways, making the connecting ramps much more direct. The interchange is named for its similar overhead appearance to the blades of a windmill.

Divided volleyball interchanges create a wide median between the carriageways of the two interchanging highways, using this space for connecting ramps.

Full diamond interchanges are large, multi-level interchanges that use flyover/under ramps to handle both right- and left-turns.

On interchanges with U-turns, to complete a left-turn, traffic must either pass the interchange, make a U-turn, and then exit right or exit right first and then make a U-turn.

### 12.9.4.8 Three-way Interchanges

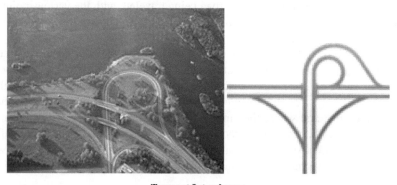

**Trumpet Interchange**

Trumpet interchanges have been used where one highway terminates at another high-way. These involve at least one loop ramp connecting traffic either entering or leav-ing the terminating expressway with the far lanes of the continuous highway. Trumpet interchanges are named as such due to their resemblance to trumpets. The bell of a trumpet can be seen where the terminating highway begins to interchange with the continuous highway and the resemblance to the tubing is seen along the connecting loop ramps. These interchanges are useful for highways as well as toll roads, as they concentrate all entering and exiting traffic into a single stretch of roadway, where toll booths can be installed. A double-trumpet interchange version can be found where a toll road meets another toll road or a free highway. They are also useful when most traffic on the terminating highway is going the same direction. The turn that is not used often would get the slower loop ramp.

## 12.10   OTHER DESIGN FEATURES

### 12.10.1   TESTING FOR EASE OF OPERATION

Each section of the expressway, which includes a series of interchanges with a suc-cession of exits and entrances should have tested for operational characteristics of the route after the preliminary design, including adaptability, capacity, and operational features. The test is an evaluation of the section from a driver's point of view, for the ease of operation and for route continuity, both of which are affected by the location, proximity, and sequence of exits and entrances, the merging, diverging, and weaving movements necessary, practicability of signing, and clarity of paths to be followed.

The testing of a route may be done by isolating that part of the plan, for each path, and examining it only concerning other parts of the layout that will affect a driver on the path being tested. Certain weaknesses of operation not evident on the overall plan may be revealed in testing a single path of travel.

The testing of the plan should be done by drawing of tracing the path separately of each principal origin and destination and studying thereon those physical features that will be encountered by a driver. The test can also be made on an overall plan on which the path to be studied and the truncated parts of connecting roads are colored or shaded.

The peak-hour volumes, number of traffic lanes, and peak-hour and off-peak-hour running speeds are to be shown on the plan. This will enable the designer to visualize exactly what the driver sees, only the road being traveled, with the various points of entrances and exits and the signing along with it and have a sense of the accompanying traffic.

Such an analysis indicates whether or not confusion is likely to occur because of exits and entrances being close together or whether interference is likely to occur because of successive weaving sections. It would also show whether the path is clearly defined if it is feasible to sign the facility properly and if major or overhead signs are required, and where they may be placed. The test may show whether the path is easy to travel, direct in character, and free from sections that might confuse the driver, or it may show that the path is sufficiently complex and confronted with disturbing elements requiring an adjustment in design. As the result, it may indicate to move or eliminate certain ramps. At its best, the test may show whether it is necessary to eliminate an interchange or to introduce collector-distributor roads to prevent interference with through traffic, or to make some other radical changes in design.

## 12.10.2 Grading, Aesthetics and Landscape Development

Grading at an interchange is determined majorly by the alignments, profiles, cross sections, and drainage requirements for the intersecting highways and ramps. Each through roadway or ramp should not be treated as a separate unit and graded to standard cross sections without regard to its relationship with adjacent roads and the surrounding topography. Instead, the whole construction area should be designed as a single unit to keep construction and maintenance costs to a minimum, to obtain maximum visibility, and enhance the appearance of the area.

The aesthetic aspect of all structures is also important and to be taken into consideration to ensure that it harmonizes with the surrounding areas and the subsequent interchanges. Landscape development and design of the entire interchange area should also be considered as part of the overall facility and should be included during the geometric design stage.

## 12.10.3 Alignment Design

The initial bridge control study in which the preliminary alignment and profiles of the intersecting roads are developed to determine the controls for bridge design is an important and early step in the design of interchanges. Requirements for clearances, curbs, and position and extent of walls should be examined concerning general grading before conclusions are drawn for the bridge design, particularly for lengths of wing walls. Minor modifications in alignment and profile, in abutments and walls, and related earthworks may produce a better desirable solution.

At interchanges, and elsewhere as feasible, steep roadside earth slopes should be avoided for all roads and ramps. Flatter slopes should be used wherever feasible, for economical construction and maintenance, to increase safety, and to enhance the appearance of the area. Broad rounded drainage ways or swale-like depressions should be used, where feasible, to encourage good turf and easy mower maintenance. "V" shaped ditches and small ditches with steep side slopes should be avoided. Drainage channels and related structures should be as un-exposed and maintenance-free as feasible. They should not be an eyesore, or hazardous to the traffic.

Where good trees and other desirable landscape features exist, the contour grading and drainage plan should be designed to protect and preserve these features, as far as possible.

## 12.10.4 Treatment of Pedestrian Traffic

Where ramps are connected to a crossroad that is not operated exclusively for motor vehicles and the volume of pedestrians, bicycles, and other slow moving vehicles on the road are considerable, pedestrian walks or cycle tracks should be provided. It should be grade-separated from the ramps or detoured outside of the ramps if possible.

If an at-grade crossing is inevitable, it should be designed so that pedestrian traffic and vehicular traffic will cross nearly at a right angle. Sufficient sight distances should also be provided. With proper arrangement of guard rails, tree planting, and traffic islands, the passage of pedestrians should be channelized, crossings other than those at designated areas are to be discouraged.

### 12.10.5  LIGHTING

Lighting is always desirable, and sometimes necessary at interchanges where a series of serious judgments and decisions are required in diverging, merging, and lane changes. Drivers should be able to see not only the road ahead but also the entire turning roadway area to drive properly the paths to be followed. They should also see all other vehicles which may influence their behavior.

The designer should consider lighting the entire interchange area when the value of information justifies it. This particular point is significant recommending for intensive use of high-mast lighting at interchanges. The principal objective in the application of high-mast lighting to highway interchanges is to provide the visual advantage to the drivers as available by daylight. Illuminating adequately enables the driver to see all things pertinent to the decision-making process in time to assimilate the information and then plan and execute his maneuvers effectively. The driver can also distinguish roadway geometry, obstructions, terrain, and other roadways, each in its proper perspective.

Additional advantages of high-mast lighting are related to safety and aesthetics. Fewer poles lower the occurrences of collision. The masts should be located away from the roadway so that the possibility of a collision with the luminary support is virtually eliminated. Daytime aesthetics are greatly improved by removing many poles generally necessary to light complex interchanges and intersections with continuous lighting.

Without lighting there may be a noticeable decrease in information and the usefulness of the interchange at night, this would cause more cars slowing down and moving with uncertainty than during daylight hours. Consideration should be given in providing visibility at night, the hazardous parts of grade-separation structures such as curbs, piers, and abutments, by roadway lighting or reflectorizing devices. The greater the volume of traffic, particularly turning traffic, the more important is the fixed-source lighting at interchanges.

### 12.10.6  DRAINAGE

The provision of proper drainage facilities is of utmost importance at interchanges and careful attention must be given to the requirements for an adequate drainage system and protection of the interchange facility from flooding. The design of the drainage system must be considered an integral part of the design of the interchange.

The specifications and requirements of the relevant standards are to be followed.

### 12.10.7 PUBLIC UTILITIES

Underground and overhead public utility installations which are close enough to the interchange and likely to be affected, their location and size should be determined in the preliminary stages of the design. The design for its relocation should also be carried out.

Service tunnels/culverts or ducts should also be provided for future services to avoid the digging of the carriageway pavement.

## 12.11 PROCEDURE FOR THE DESIGN OF INTERCHANGES

A brief step-by-step procedure for the design of interchanges is as follows:

Step 1 – Forecasting of traffic.
> For existing roads, the traffic forecast can be based on traffic census by the Highway Planning Unit or from a manual in which classified counts are estimated separately.
> For new roads, the traffic forecast will have to be based on traffic studies carried out by following the approved procedures.
> The service period in the design should be taken as 20 years after completion of the interchange. While stage construction is allowed, it should be restricted only to the ramps and pavement.

Step 2 – Warrants for Grade-separation.
> With the traffic forecast and other relevant information such as site conditions, accident records, road-user benefits, etc., a warrant for the grade-separation must be established.
> If grade-separation is not warranted, the junction or intersection can be designed as an at-grade facility following the relevant standard for intersection design.

Step 3 – Determine the Type of Interchange.
> The required type of interchange is next determined. A more detailed traffic study may be necessary to determine the traffic movements, the composition of traffic, etc. before deciding the configuration of the interchange. Other factors such as physical constraints at the site may also affect the type of configuration that can be selected and may need to be taken into consideration. Where different configurations are possible, the most cost-effective one should be selected.

Step 4 – Preliminary Geometric Design.
> With the selected interchange configuration, a preliminary layout design of the interchange is carried out based on the requirement of traffic volumes and geometric elements.

Step 5 – Capacity Analysis.
> The capacity analysis is carried out for the various checkpoints of the preliminary layout design. If the LOS determined is unacceptable, the layout design is then reviewed and revised until the required LOS is achieved. If the LOS is still unsatisfactory, the type of interchange configuration selected may have to be revised.

Step 6 – Detailed Design.

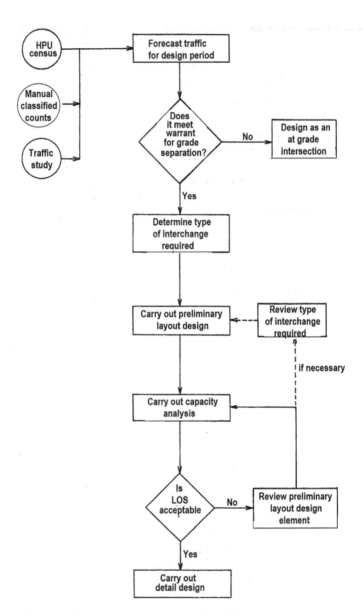

**FIGURE 12.38**   Flow Chart of Procedure for Design of Interchanges.
*Source:* Figure A.1, A Guide to the Design of Interchanges, Arahan Teknik (Jalan) 12/87, JKR

The detailed design of the various elements of the interchange is then carried out i.e.:

- Geometric elements such as horizontal and vertical alignments, etc. should be optimized for an economic design.
- Bridge structures.
- Pavement structure.

- Drainage elements.
- Geotechnical elements.
- Signing and pavement.
- Grading and landscaping.
- Lighting.
- Other miscellaneous elements.

The design of an interchange loop may follow the calculations mentioned here.

If the vertical clearance between the top level underneath the road and the soffit of the bridge is 5.5 m and the structural depth of the girder and deck slab is 2.0 m, then the total grade difference is 5.5 + 2.0 = 7.5 m. A 2.5% grade of the loop connecting the road underneath and bridge top level, will require the length of = 7.5 × 100/2.5 = 300 m.

The length of perimeter of a loop circle = 2 × л × radius, (where, л = 3.1416).

Equating the above two lengths, 2 × л × Radius = 300,

Hence, The Radius of the interchange loop = 300/(2 × 3.1416) = 46.758 = 50 m. (say).

## 12.12  COMPUTER-AIDED DESIGN OF GRADE SEPARATED INTERCHANGES

In this session, we shall see the design of grade-separated interchanges with "Tutorial Example Data." The design is based on the alignment of two intersecting main roads and four slip roads for the turning traffic. The design of alignment includes drawing of polylines for horizontal alignment of above intersecting roads, Loop ramps, and slip roads, the direction of polyline should be in the direction of the traffic flow in each road. So, the drawing of the polyline is in opposite direction for UK and USA styles of traffic flows. The process essentially includes the steps in the guide as mentioned below.

Refer to Tutorial "12.1 Loop Design" inside folder "Chapter 12 Interchange Design", under menu item "Book Tutorials" available on the website at: www.roadbridgedesign.com. The computer applications for the "Design of Grade-separated Interchanges" are explained in Chapter 12 in the "The Guide for Computer Application Tutorials" part of the book available for download from the website www.roadbridgedesign.com.

For downloading the other items visit the websites at: www.techsoftglobal.com/download.php

## BIBLIOGRAPHY

*A Policy on Geometric Design of Highways and Streets 2001*, AASHTO.

*Overseas Road Note 6, A Guide to Geometric Design*, Overseas Unit, Transport and Road Research Laboratory, Department of Transport, Crowthorne, Berkshire, United Kingdom.

*A Guide on Geometric Design of Roads*, Arahan Teknik (Jalan) 8/86, JKR, Malaysia.

*A Guide to the Design of At Grade Intersections*, JKR 201101-0001-87, JKR, Malaysia.

*A Guide to the Design of Interchanges*, Arahan Teknik (Jalan) 12/87, JKR, Malaysia.

*Manual on Traffic Control Devices Road Marking & Delineation*, Arahan Teknik (Jalan) 2D/85, JKR, Malaysia.

*Guidelines for Presentation of Engineering Drawings*, Arahan Teknik (Jalan) 6/85, Pindaan 1/88, JKR, Malaysia.

*Design Standards, Interurban Toll Expressway System of Malaysia*, Government of Malaysia, Malaysia Highway Authority, Ministry of Works and Utilities.

*Design Manual for Roads & Bridges, Ministry of Public Works and Kuwait Ministry*, State of Kuwait.

*Geometric Design Manual for Dubai Roads*, Dubai Municipality, Roads Department.

*Geometric Design Standards for Rural (Non-Urban) Highways*, IRC 73-1980, IRC.

*Vertical Curves for Highways*, IRC SP 23-1993, IRC.

*Guidelines for the Design of Interchanges in Urban Areas*, IRC 92-1985, IRC.

*Lateral and Vertical Clearances at Underpasses for Vehicular Traffic*, IRC 54-1974, IRC.

*Highway Engineering* by Paul H. Wright and Karen K. Dixon, by John Wiley & Sons, Inc.

# 13 Design of Flexible Pavement

## 13.1 GENERAL

**Design Requirements.** The preparation and/or selection of the inputs are required for the design of new pavement construction or reconstruction for the strengthening of an existing pavement. For the design requirements for several types of flexible pavement structures on both highways and low-volume roads, only certain sets of inputs are required for a given structural design combination. The inputs are classified under five separate categories:

**Design Variables.** This category refers to the set of criteria which is considered for each type of pavement design procedure.

**Performance Criteria.** This is the "serviceability" which represents the user-specified set of boundary conditions within which a given pavement design alternative should perform, during the design life.

**Material Properties for Structural Design.** This category covers all the data related to material properties for the pavement and roadbed soil which are required for the design of the pavement structure.

**Structural Characteristics.** This refers to certain physical characteristics of the pavement structure which are considered in the design and have an effect on its performance, during the design life.

Because of the consideration of reliability in the AASHTO Guide, it is strongly recommended that the designer use mean or average values rather than "conservative estimates" for each of the design inputs required by the procedures. This is important since the equations were developed using mean values and actual variations. The designer must use mean values and standard deviations associated with the other conditions.

Table 13.1 identifies all possible design input requirements and indicates the specific types of structural designs for which they are required, which are (i) means that a particular design input (or set of inputs) must be determined for that structural combination, (ii) indicates that the design input should be considered because of its potential impact on the results. Under the "Flexible" heading, AC refers to asphalt concrete surfaces and ST to surface treatments.

**TABLE 13.1**

**Design Requirements for the Different Initial Pavement Types That Can Be Considered**

| Description | Flexible | | Rigid | | Aggr. Surf. |
|---|---|---|---|---|---|
| | AC | ST | JCP/JRCP | CRCP | |
| **2.1 Design Variables** | | | | | |
| 2.1.1 Time Constraints | | | | | |
|     Performance Period | 1 | 1 | I | 1 | 1 |
|     Analysis Period | **1** | **1** | **1** | **1** | I |
| 2.1.2 Traffic | 1 | 1 | 1 | 1 | 1 |
| 2.1.3 Reliability | I | 1 | 1 | 1 | |
| 2.1.4 Environmental Impacts | | | | | |
|     Roadbed Swelling | 2 | 2 | 2 | 2 | |
|     Frost Heave | 2 | 2 | 2 | 2 | |
| **2.2 Performance Criteria** | | | | | |
| 2.2.1 Serviceability | 1 | 1 | 1 | 1 | 1 |
| 2.2.2 Allowable Rutting | | | | | 1 |
| 2.2.3 Aggregate Loss | | | | | 1 |
| **2.3 Material Properties for Structural Design** | | | | | |
| 2.3.1 Effective Roadbed Soil Resilient Modulus | 1 | 1 | | | 1 |
| 2.3.2 Effective Modulus of Subgrade Reaction | | | 1 | 1 | |
| 2.3.3 Pavement Layer Materials Characterization | 2 | 2 | 1 | 1 | 1 |
| 2.3.4 PCC Modulus of Rupture | | | 1 | 1 | |
| 2.3.5 Layer Coefficients | 1 | 1 | | | |
| **2.4 Pavement Structural Characteristics** | | | | | |
| 2.4.1 Drainage | | | | | |
|     Flexible Pavements | 1 | 1 | | | |
|     Rigid Pavements | | **1** | **1** | | |
| 2.4.2 Load Transfer | | | | | |
|     Jointed Pavements | | | 1 | | |
|     Continuous Pavements | | | | | |
|     Tied Shoulders or Widened Outside Lanes | | | 2 | 2 | |
| 2.4.3 Loss of Support | | | 1 | 1 | |
| **2.5 Reinforcement Variables** | | | | | |
| 2.5.1 Jointed Pavements | | | | | |
|     Slab Length | | | 1 | | |
|     Working Stress | | | 1 | | |
|     Friction Factor | | | 1 | | |
| 2.5.2 Continuous Pavements | | | | | |
|     Concrete Tensile Strength | | | | 1 | |
|     Concrete Shrinkage | | | | 1 | |
|     Concrete Thermal Coefficient | | | | 1 | |
|     Bar Diameter | | | | 1 | |
|     Steel Thermal Coefficient | | | | 1 | |
|     Design Temperature Drop | | | | 1 | |
|     Friction Factor | | | | 1 | |

Table 2.1, Part II, of the AASHTO Guide, 1993
AC – Asphalt Concrete      CRCP – Continuously Reinforced Concrete Pavement
ST – Surface Treatment      1 – Design input variable that must be determined
JCP – Jointed Concrete Pavement      2 – Design variable that should be considered
JRCP – Jointed Reinforced Concrete Pavement

## 13.2   DESIGN VARIABLES

### 13.2.1   Time Constraints

**Analysis Period.** The "Analysis Period" for the pavement design is the entire performance period. When for budgetary constraints or other considerations the performance period is considered in parts, then the strategy is to be decided for separate design life for each part of the performance period. These parts are identified as "Strategic Stages." For example, if the Analysis Period is 20 years, the first stage may be of 10 years performance period for which the initial pavement structure is to be designed, which will be less expensive than the pavement structure for 20 years of the entire analysis period. An overlay is to be designed for the second stage of 10 years, which is to be constructed after 10 years of the performance period of the initial pavement structure, as it is over.

**Performance Period.** This is the period that is the design life for the initial pavement structure and will last before it needs rehabilitation by the overlay. In the case of the three-stage Analysis Period, it also refers to the design life for each overlay which is the performance time between rehabilitation operations. The performance period is equivalent to the time elapsed as a new, reconstructed, or rehabilitated structure deteriorates from its initial serviceability to its terminal serviceability. In actual practice, the performance period is significantly affected by the type and level of maintenance applied during the design lives. The predicted performance is based on the maintenance practices as per AASHO Road Test.

The minimum performance period is the shortest design life a given stage should last. For example, the initial pavement structure lasts at least 10 years of its design life before the rehabilitation by an overlay. The limit depends on factors as the public's perception of how long a "new" surface should last, budget for initial construction, life-cycle cost, and other engineering considerations.

The maximum performance period or design life is the maximum practical amount of time that the user can expect from a given stage. The selection of longer design life than its actual lasting may result in unrealistic designs. The consideration of the maximum practical performance period of a given pavement type will help in considering the accurate life-cycle cost.

**Analysis Period.** This refers to the period for which the analysis is to be conducted, i.e., the length of time that any design strategy must cover. This is the total time for the design lives of the initial structure and all the periodic overlays.

The present consideration is given for longer analysis periods, this is better suited to the evaluation of alternative long-term strategies based on life-cycle costs. The analysis period is considered to include one rehabilitation, i.e., One initial structure and at least one overlay. For high volume urban freeways, longer analysis periods may be considered. Following are general guidelines:

| Highway Conditions | Analysis Period (years) |
|---|---|
| High volume urban | 30–50 |
| High volume rural | 20–50 |
| Low-volume paved | 15–25 |
| Low-volume aggregate surface | 10–20 |

## 13.2.2 TRAFFIC

The design for both highways and low-volume roads are done based on cumulative 18-kip (w18) Equivalent Single Axle Loads (ESAL) expected during the analysis period. The procedure for converting mixed traffic into these 18-kip ESAL units is done with different equivalence factors for different vehicles considered in the spectrum plotted from the results of an axle load survey. The procedure is presented in Part I and Appendix D of the AASHTO Guide for pavement design. Detailed equivalency values are given in Appendix D of the AASHTO Guide.

If the initial pavement structure is expected to last for the entire analysis period without any rehabilitation by an overlay, then the total traffic over the analysis period is to be considered.

In case of considering multiple-stage construction, anticipating the rehabilitation or resurfacing due to lack of initial funds, roadbed swelling, frost heave, etc., then the user must prepare a graph of cumulative 18-kip ESAL traffic versus time, as illustrated in Figure 13.1. This will be used to separate the cumulative traffic into the periods (stages) during which it is encountered.

The predicted traffic is generally the cumulative 18-kip ESAL axle applications expected on the highway, whereas the designer requires the axle applications in the design lane. Thus, the design must consider the direction factor to get the design traffic by direction and the lane distribution factor when the number of lanes is more than two.

In single-lane roads, all the wheels pass through the same location and create larger stress on the pavement. On roads with more than one-lane the wheels of vehicles pass through different locations and the stress on the pavement is much less.

The equation below may be used to determine the traffic (w18) in the design lane:

$$w_{18} = D_D \times D_L \times W_{18}$$

where

W$_{18}$ is the cumulative two-directional 18-kip ESAL units predicted for a specific section of highway during the analysis period (from the planning group).

D$_D$ is the directional distribution factor, expressed as a ratio, that accounts for the distribution of ESAL units by direction, e.g., east–west, north–south, etc.

D$_L$ is the lane distribution factor, expressed as a ratio, that accounts for the distribution of traffic when two or more lanes are available in one direction, and

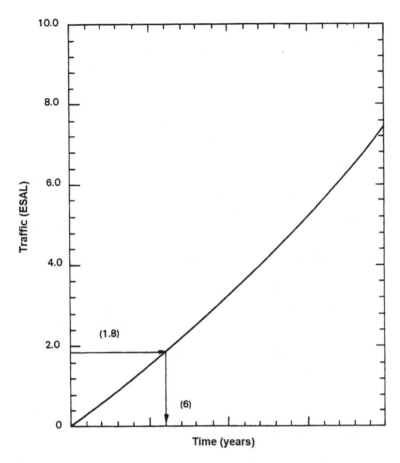

**FIGURE 13.1** Example Plot of Cumulative 18-kip ESAL Traffic Versus Time.
*Source:* Figure 2.1 AASHTO Guide, 1993

Although the $D_D$ factor is generally 0.5 for most roadways, considering equal traffic in either direction, but there are instances where more weight may be moving in one direction than the other. Here for the design, the side with heavier vehicles should be designed for a greater number of ESAL units. Experience has shown that $D_D$ may vary from 0.3 to 0.7, depending on which direction is more loaded and which is less loaded.

For the $D_L$ factor, the following table is given in AASHTO as a guide:

| Number of Lanes in Each Direction | Percent of 18-kip ESAL in Design Lane |
|---|---|
| 1 | 100 |
| 2 | 80–100 |
| 3 | 60–80 |
| 4 | 50–75 |

### 13.2.3  RELIABILITY

Reliability concepts were introduced in AASHTO, it is a provision of incorporating some degree of certainty into the design process to ensure that the various design considerations are applicable over the analysis period. The reliability design factor considers for chance variations in both traffic prediction (w18) and the performance prediction (W18) and provides a predetermined level of assurance (R) that pavement structure will last for the design life.

The risk of not performing to expectations must be minimized by selecting higher levels of reliability. Table 13.2 presents recommended levels of reliability for various functional classifications. Note that the higher levels correspond to the facilities which receive the most used highways having a high volume of traffic, while the lowest level, 50%, corresponds to local roads.

NOTE: Above values are based on a survey of the AASHTO pavement Design Task Force.

The design-performance reliability is controlled using a reliability factor (FR) that is multiplied times the analysis period traffic prediction ($w_{18}$) to produce design applications ($W_{18}$) for the design equation. For a given reliability level (R), the reliability factor (FR) is a function of the overall standard deviation (So) that accounts for both chance variation in the traffic prediction and normal variation in pavement performance prediction for a given $W_{18}$.

Thus, by treating design uncertainty with reliability factor, the designer should no longer use "conservative" values for all the other design input requirements, instead, the designer should use his best estimate of the mean or average value for each input value. The selected level of reliability and overall standard deviation will account for the combined effect of the variation of all the design variables. Application of the reliability concept requires the following steps:

- Define the functional classification of the facility and determine whether the location is rural or urban.
- Select a reliability level from the range given in Table 13.2. The greater the value of reliability, the stronger pavement structure is required.
- A reasonable standard deviation (So) should be selected for justifying the local conditions. Values of So developed at the "AASHO Road Test" did not include traffic error. However, the performance prediction error was decided as.25 for rigid and.35 for flexible pavements. This corresponds to a total standard deviation for the traffic of 0.35 and 0.45 for rigid and flexible pavements, respectively.

### 13.2.4  ENVIRONMENTAL EFFECTS

In several ways, pavement performance can be affected by the environment. Temperature and moisture changes can affect the strength, durability, and load carrying capacity of the pavement and roadbed materials. Another major direct effect is the roadbed swelling, pavement blowups, frost heave, disintegration, etc., which may cause loss of riding quality and serviceability. Other inherent effects on the pavement

**TABLE 13.2**
**Suggested Levels of Reliability for Various Functional Classifications**

| | Recommended Level of Reliability | |
|---|---|---|
| Functional Classification | Urban | Rural |
| Interstate and other freeways | 85–99.9 | 80–99.9 |
| Principal arterials | 80–99 | 75–95 |
| Collectors | 80–95 | 75–95 |
| Local | 50–80 | 50–80 |

*Source:* Table 2.2, Part II, of the AASHTO Guide, 1993

performance are aging, drying, and overall material deterioration due to weathering, these are considered in terms of their prediction models.

Considerations are necessary for quantifying the input requirements for evaluating roadbed swelling and frost heave are explained here. The serviceability of pavement, based on such factors for pavement blowups, may be added to the design procedure.

The objective is to elaborate the use of a graph of serviceability loss versus time, which is presented in Figure 13.2. The serviceability loss due to the environment is additional to that resulting from cumulative axle loads. The environmental loss is a result of the summation of losses from both swelling and frost heave, as obtained from Figure 13.2. The chart is useful to estimate the serviceability loss at intermediate periods, for example, at 13 years the loss is 0.73. In case of either only swelling or only frost heave is considered, then the respective one curve is to be referred to on the graph.

## 13.3 PERFORMANCE CRITERIA

### 13.3.1 SERVICEABILITY

The serviceability of a pavement is its ability to serve the type of traffic which uses the facility. The primary measure of serviceability is the Present Serviceability Index (PSI), which ranges from 0 (unusable surface) to 5 (perfect surface). The basic design philosophy of the AASHTO method is the serviceability-performance concept, which is the designing of a pavement based on a specific total traffic volume and a minimum level of serviceability to be available at the end of the performance period or the design life.

The terminal serviceability index ($p_t$) to be available at the end of the design life, is based on the lowest index that will be tolerated before rehabilitation, resurfacing, or reconstruction becomes necessary. The value of ($p_t$) is used as 2.5 or higher for the design of major highways and 2.0 for highways with lesser traffic volumes. Generally used values for minimum levels of pt. are given below as obtained from studies in connection with the AASHO Road Test.

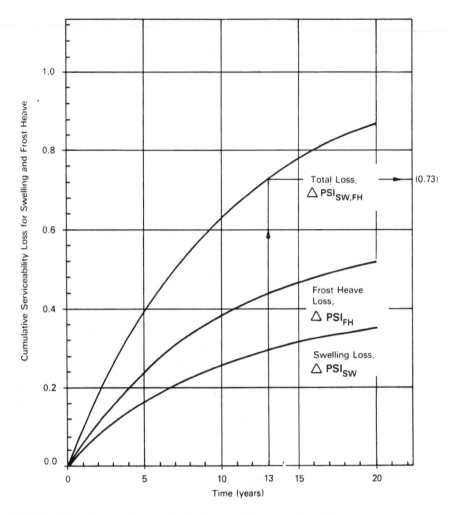

**FIGURE 13.2**   Chart to Select a Reliability Level from the Given Range.
*Source:* Figure 2.2 of the AASHTO Guide, 1993

| Terminal Serviceability Level | Percent of People Stating Unacceptable |
|---|---|
| 3.0 | 12 |
| 2.5 | 55 |
| 2.0 | 85 |

For minor highways where the budget restricts the initial pavement structure, it is suggested to consider a reduced design period or reduced total traffic volume, rather than by designing by considering a terminal serviceability less than 2.0. A proper consideration must also be given for the selection of initial serviceability ($p_o$), which along with traffic volume, governs terminal serviceability for the pavement structure.

The time to terminal serviceability depends on the original or initial serviceability ($p_o$). The $p_o$ values are 4.2 for flexible pavements and 4.5 for rigid pavements as observed at the AASHO Road Test.

Once $p_o$ and $p_t$ are established, the following equation applicable to flexible, rigid, and aggregate-surfaced roads, gives the total change in serviceability index:

$$\Delta PSI = p_o - p_t$$

### 13.3.2 ALLOWABLE RUTTING

As there is no suitable consideration to incorporate rutting in the design of asphalt concrete (AC) surface pavements, therefore, rutting is considered only as a performance criterion for aggregate-surfaced roads. The allowable rut depth with a range from 1.0 to 2.0 inches for an aggregate-surfaced road is dependent on the average daily traffic.

### 13.3.3 AGGREGATE LOSS

For aggregate-surfaced roads, the aggregate loss occurs due to traffic and erosion. The pavement structure becomes thinner and the load carrying capacity is reduced. This reduction in the thickness of the pavement structure increases the rate of surface deterioration.

To resolve this issue, it is necessary to estimate (1) the total thickness of aggregate that will be lost during the performance period, and (2) the minimum thickness of aggregate that is required to be available to keep a serviceable working surface of the pavement structure.

The information available to predict the rate of aggregate loss is not adequate. AASHTO gives an example of a prediction equation developed with limited data on sections experiencing greater than 50% truck traffic:

$$GL = 0.12 + 0.1223(LT)$$

where

GL is the total aggregate loss in inches
LT is the number of loaded trucks in thousands

A second equation, which was developed from a recent study in Brazil on typical rural sections, can be employed by the user to determine the input for gravel loss:

$$GL = (B/25.4)/(.0045LADT + 3380.6/R + 0.467G)$$

where

GL      = aggregate loss, in inches, during the period being considered
B       = number of bladings during the period being considered
LADT  = average daily traffic in design lane (for one-lane road use total traffic
          in both directions)
R       = average radius of curves, in feet
G       = absolute value of grade, in percent

Another equation, developed through a British study done in Kenya, is more applicable to areas where there is very little truck activity and thus the facility is primarily used by cars. Since this equation (below) is for annual gravel loss, the total gravel loss (GL) would be estimated by multiplying by the number of years in the performance period:

$$AGL = \left[ T^{2/}\left(T^2 + 50\right) \right] \times f\left(4.2 + .092T + 0.889R^2 + 1.88VC\right)$$

where

AGL is the annual aggregate loss, in inches
T is the annual traffic volume in both directions, in thousands of vehicles
f   = .037 for lateritic gravels
    = .043 for quartzitic gravels
    = .028 for volcanic gravels
    = .059 for coral gravels
R is the annual rainfall, in inches
VC is the average percentage gradient of the road, and AASHTO Guide mentioned that all the equations have serious shortcomings; therefore, whenever possible, local information about aggregate loss should be referred to as input to the procedure.

## 13.4  MATERIAL PROPERTIES FOR DESIGN OF THE PAVEMENT STRUCTURE

### 13.4.1  EFFECTIVE ROADBED SOIL RESILIENT MODULUS

The characterization of materials in the AASHTO Guide is done as the elastic or resilient modulus. For roadbed materials, laboratory resilient modulus tests (AASHTO T 274) should be performed on representative samples in stress and moisture conditions simulating those of the primary moisture seasons.

Alternatively, the seasonal resilient modulus values may be determined from correlations with clay content, moisture, plasticity index (PI), etc. in the roadbed soil. The purpose of identifying seasonal moduli is to estimate the relative damage a pavement is affected during each season of the year and consider it as part of the overall design. A realistic roadbed soil resilient modulus may be established which is equivalent to the combined effect of all the seasonal modulus values.

The seasonal moisture conditions for which the roadbed soil samples should be tested are those which result in significantly different resilient moduli. For example, in a climate that is not subjected to general sub-freezing temperatures, it would be important to test for differences between the wet and dry seasons. Practical values of resilient moduli of 20,000 to 50,000 psi may be used for frozen conditions, and spring-thaw conditions, the retained modulus maybe 20 to 30% of the normal modulus during the summer and fall periods.

It is necessary to separate the year into the various component time intervals during which the different moduli are effective. In making this breakdown, it is not practicable to specify a time interval of less than one-half-month for any given season.

The length of the seasons and the seasonal roadbed resilient moduli are all that is required of roadbed support for the design of rigid pavements and aggregate-surfaced roads. For the design of flexible pavements, however, the seasonal data must be used in the effective roadbed soil resilient modulus described earlier. This is accomplished with the aid of the chart in Figure 13.3. The effective modulus is a weighted average value that gives the equivalent annual damage obtained by treating each season independently in the performance equation and summing the damage. The effective roadbed soil resilient modulus determined from this chart applies only to flexible pavements designed using the serviceability criteria. It is not necessarily applicable to other resilient modulus-based design procedures.

Since an average value of resilient modulus of different seasons is used, design sections with a coefficient of variations greater than 0.15 (within a season) should be subdivided into smaller sections. For example, if the average value of resilient modulus is 10,000 psi, then approximately 99% of the data of each season should be in a range of 5,500 to 14,500 psi.

The first step of this process is to enter the seasonal moduli in the season's respective periods. If the smallest season is one-half-month, then all seasons must be defined in terms of one-half months. If the smallest season is one month, then all seasons must be defined in terms of whole months.

The next step is to estimate the relative damage ($u_f$) values in respect of each seasonal modulus. This is done using the vertical scale or the corresponding equation shown in Figure 13.3. For example, the relative damage is 0.51, corresponding to a roadbed soil resilient modulus of 4,000 psi.

Next, the $u_f$ values should all be added together and divided by the number of seasonal increments (12 or 24) to determine the average relative damage. The effective roadbed soil resilient modulus ($M_R$), then, is the value corresponding to the average relative damage obtained from the "$M_R - u_f$" scale. Figure 13.4 provides an example of the application of the effective $M_R$ estimation process. Again, it is emphasized that this effective $M_R$ value should be used only for the design of flexible pavements based on serviceability criteria.

## 13.4.2 EFFECTIVE MODULUS OF SUBGRADE REACTION

Like the effective roadbed soil resilient modulus used for flexible pavement design, an effective modulus of subgrade reaction (k-value) is used for rigid pavement design. The k-value is directly proportional to roadbed soil resilient modulus. The duration of seasons and seasonal moduli developed in the previous section are used

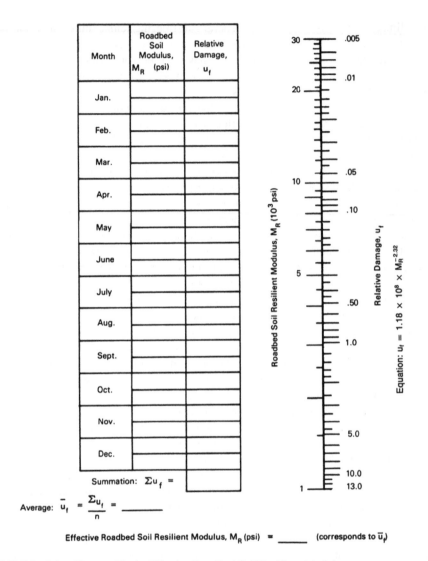

**FIGURE 13.3** Chart to Obtain Effective Roadbed Soil Resilient Modulus.
*Source:* Figure 2.3 of the AASHTO Guide, 1993

as input to the estimation of an effective design k-value. But, because of the effects of subbase characteristics on the effective design k-value, it is determined in an iterative design procedure.

### 13.4.3 PAVEMENT LAYER MATERIALS CHARACTERIZATION

Although there are many types of material properties and laboratory test procedures for assessing the strength of pavement structural materials, the use of "layer coefficients" has been adopted as a basis for design in the AASHTO Guide. With proper

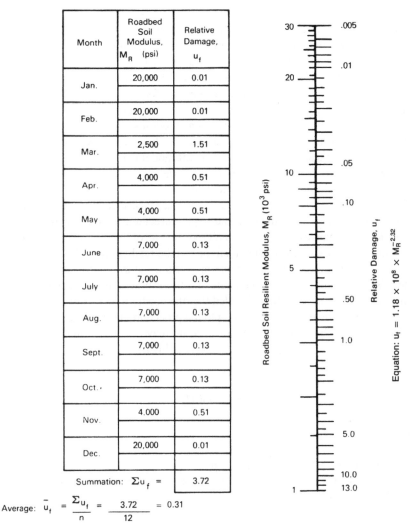

| Month | Roadbed Soil Modulus, $M_R$ (psi) | Relative Damage, $u_f$ |
|---|---|---|
| Jan. | 20,000 | 0.01 |
| Feb. | 20,000 | 0.01 |
| Mar. | 2,500 | 1.51 |
| Apr. | 4,000 | 0.51 |
| May | 4,000 | 0.51 |
| June | 7,000 | 0.13 |
| July | 7,000 | 0.13 |
| Aug. | 7,000 | 0.13 |
| Sept. | 7,000 | 0.13 |
| Oct. | 7,000 | 0.13 |
| Nov. | 4.000 | 0.51 |
| Dec. | 20,000 | 0.01 |
| Summation: $\Sigma u_f$ = | | 3.72 |

Average: $\bar{u}_f = \dfrac{\Sigma u_f}{n} = \dfrac{3.72}{12} = 0.31$

Effective Roadbed Soil Resilient Modulus, $M_R$ (psi) = <u>5,000</u> (corresponds to $\bar{u}_f$)

**FIGURE 13.4**   Example to Obtain Effective Roadbed Soil Resilient Modulus.
*Source:* Figure 2.4 of the AASHTO Guide, 1993

understanding, the layer coefficients have been used in the AASHTO flexible pavement design procedure. Layer coefficients derived from test roads or satellite sections are preferred.

The elastic modulus of any paving or roadbed material is an engineering property. For those material types which undergo significant permanent deformation under load, this property may not reflect the material's inelastic behavior under load. The resilient modulus reflects the material's stress–strain behavior under normal pavement loading conditions. The strength and stiffness of the material are important, the mechanistic-based procedures reflect strength as well as stiffness in the materials characterization procedures. Also, pavement layers with stabilized base materials (WBM or WMM) may be subject to cracking under certain conditions and the stiffness may not be an indicator for this type of distress. The resilient modulus can apply to any type of material, still, it is used as MR applied only to the roadbed soil in the AASHTO procedure. Different notations are used for moduli of subbase ($E_{SB}$), base ($E_{BS}$), asphalt concrete (EAC), and Portland cement concrete (Er).

The procedure for obtaining the resilient modulus for relatively low stiffness materials, such as natural soils, unbound granular layers, and even stabilized layers and AC, should be tested by using the resilient modulus test methods. Although the testing apparatus for each of these types of materials is the same, for unbound materials, there is the need for triaxial confinement.

Alternatively, the bound or higher stiffness materials, such as stabilized bases and AC, may be tested by using the repeated-load indirect tensile test which relies on the use of electronic gauges to measure small movements of the sample under load but is less complex and easier to conduct than the triaxial resilient modulus test.

Because of the small displacements and brittle nature of the stiffest pavement materials, i.e., Portland cement concrete and those base materials stabilized with a high cement content, it is recommended that the elastic modulus of such high-stiffness materials be determined according to the procedure described in ASTM C 469.

The elastic modulus for any type of material may also be estimated using a correlation recommended by the American Concrete Institute for normal weight Portland cement concrete (PCC):

$$E_c = 57,000 \left( f_c \right)^{0.5}$$

where

"Ec" is the PCC elastic modulus (in psi)
"fc" is the PCC compressive strength (in psi) as determined using AASHTO T 22, T 140, or ASTM C 39.

### 13.4.4   Layer Coefficients

Estimating the "a" values, which are the structural layer coefficients in the AASHTO method, and required for standard flexible pavement structural design discussed here. This coefficient value is assigned to each layer material in the pavement structure to convert actual layer thicknesses into the structural number "SN". This layer coefficient has an empirical relationship between SN and thickness. This is a measure of the relative ability of the material to function as a structural component of the pavement. The relative impact of the layer coefficients "a" and thickness "D" on the structural number "SN" is based on the following equation:

$$SN = \Sigma a_i D_i$$

$$i = I$$

Although the elastic or resilient modulus has been adopted as the measure of material quality, still it is necessary to obtain corresponding layer coefficients for the design by the structural number design approach. There are correlations available to determine the modulus from tests such as the R-value, still, it is recommended for direct measurement using AASHTO method T 274 for subbase and unbound granular materials and ASTM D 4123 for asphalt concrete (AC) and other stabilized materials. Research and field studies indicated about many factors govern the values for layer coefficients. The layer coefficient may vary with thickness, underlying support, position in the pavement structure, etc.

The ways for estimating the coefficients by charts are separated into five categories, depending on the type and function of the layer material. These are AC, granular base, granular subbase, cement-treated, and bituminous base. Other materials such as lime, lime flyash, and cement flyash are acceptable materials, and charts may also be developed for them.

**Asphalt Concrete Surface Course.** Figure 13.5 provides a chart to estimate the structural layer coefficient of a dense-graded asphalt concrete (AC) surface course based on its elastic or resilient modulus "$E_{AC}$" at a temperature of 68 °F. Higher modulus above 450,000 psi of ACs are stiffer and more resistant to bending, so they are more susceptible to thermal and fatigue cracking.

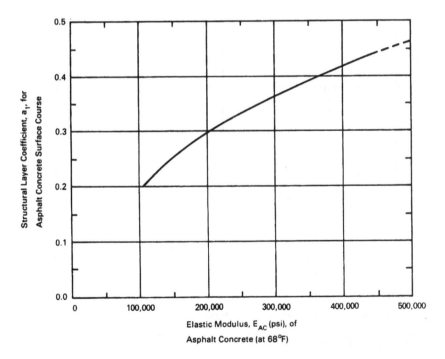

**FIGURE 13.5**    Chart to Estimate the Structural Layer Coefficient of a Dense-Graded Asphalt Concrete Surface Course.
*Source:* Figure 2.5 of the AASHTO Guide, 1993

**Granular Base Layers.** to estimate a structural layer coefficient, $a_2$, a chart is provided in Figure 13.6 from one of four different laboratory AASHO Road Test results on a granular base material, including base resilient modulus, $E_{BS}$. For these correlations, the test values are:

$a_2$ = 0.14
$E_{BS}$ = 30.000 psi
CBR = 100 (approx.)
R- value = 85 (approx.)

As an alternative of Figure 13.6 to estimate the layer coefficient, $a_2$ for a granular base material from its elastic or resilient modulus "$E_{BS}$", the following relationship may be used:

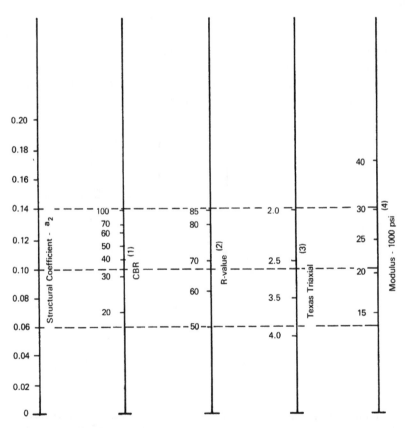

(1)  Scale derived by averaging correlations obtained from Illinois.
(2)  Scale derived by averaging correlations obtained from California, New Mexico and Wyoming.
(3)  Scale derived by averaging correlations obtained from Texas.
(4)  Scale derived on NCHRP project (3).

**FIGURE 13.6**   Chart to Estimate a Structural Layer Coefficient "$a_2$".
*Source:* Figure 2.6 of the AASHTO Guide, 1993

$$a_2 = 0.249\left(\log_{10} E_{BS}\right) - 0.977$$

For aggregate base layers WMM or WBM, EBS is a function of the stress state "θ" within the layer and is given by the relation:

$$E_{BS} = k_{1x}\,\theta \times k_2$$

Where, the stress state or sum of principal stresses is $\theta = \hat{o} + \hat{o} + \hat{o}$ (psi), and $k_1$, $k_2$ are the regression constants that are a function of material type. Typical values of $k1$ and $k_2$ for base materials are: $k_1 = 3,000$ to $8,000$ and $k_2 = 0.5$ to $0.7$, At the AASHO Road Test, modulus values ($E_{BS}$ in psi) for the base were as follows:

| Moisture State | Equation | Stress State (psi) | | | |
| --- | --- | --- | --- | --- | --- |
| | | θ = 5 | θ = 10 | θ = 20 | θ = 30 |
| Dry | $80000^{0.6}$ | 21,012 | 31,848 | 48,273 | 61,569 |
| Damp | $4000^{0.6}$ | 10,506 | 15,924 | 24,136 | 30,784 |
| Wet | $3200^{0.6}$ | 8404 | 12,739 | 19,309 | 24,627 |

Note, $E_{BS}$ is a function of the moisture and the stress state "θ". Values for the stress state within the base course vary with the subgrade modulus and thickness of the surface layer. Typical values for use in the design are:

| | Roadbed Soil Resilient Modulus (psi) | | |
| --- | --- | --- | --- |
| | 3000 | 7500 | 15000 |
| Asphalt Concrete Thickness (inches) | Values for the stress state within the base course | | |
| Less than 2 | 20 | 25 | 30 |
| 2–4 | 10 | 15 | 20 |
| 4–6 | 5 | 10 | 15 |
| Greater than 6 | 5 | 5 | 5 |

Interpolation can be used for intermediate values of roadbed soil resilient modulus. Relationships may be developed for specific base materials (e.g., MR = $k_1 \times \theta \times k_2$) using AASHTO method T 274; Alternatively, values given in Table 13.3 can be used.

**Granular Subbase Layers.** Figure 13.7 provides a chart that may be used to estimate a structural layer coefficient, a3, from one of four different laboratory results on a granular subbase material, including subbase resilient modulus, $E_{SB}$. The AASHO Road Test basis for these correlations is:

a3     = 0.11
ESB   = 15,000 psi

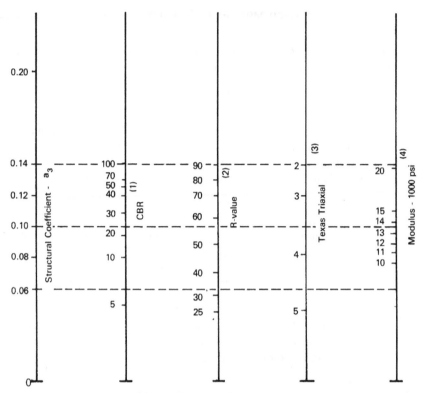

(1)  Scale derived from correlations from Illinois.

(2)  Scale derived from correlations obtained from The Asphalt Institute, California, New Mexico and Wyoming.

(3)  Scale derived from correlations obtained from Texas.

(4)  Scale derived on NCHRP project (3).

**FIGURE 13.7**    Chart to Estimate a Structural Layer Coefficient "$a_3$".
*Source:* Figure 2.7 of the AASHTO Guide, 1993

CBR       = 30 (approx.)
R- value = 60 (approx.)

The relationship between $E_{SB}$ versus "a2", like that for granular base materials is as follows:

$$a_3 = 0.227\left(\log_{10} E_S s\right) - 0.839$$

$E_{SB}$ for aggregate subbase layers are affected by the stress state "$\theta$" like that for the base layer. Typical values for $k_1$ range from 1,500 to 6,000, while $k_2$ varies from 0.4 to 0.6. Values of subbase material from the AASHO Road Test are:

| Moisture State | Developed Relationship | Stress State (psi) | | |
|---|---|---|---|---|
| | | $\theta = 5$ | $\theta = 7.5$ | $\theta = 10$ |
| Damp | $M_R = 5{,}400\theta^{0.6}$ | 14,183 | 18,090 | 21,497 |
| Wet | $M_R = 4{,}600\theta^{0.6}$ | 12,082 | 15,410 | 18312 |

As with the base layers, relationships may be developed for their specific materials; alternatively, the values in Table 13.3 can also be used.

Stress states "$\theta$" which can be used as a guide to select the modulus value for subbase thicknesses between 6 and 12 inches are as follows:

| Asphalt Concrete Thickness (inches) | Stress State (psi) |
|---|---|
| Less than 2 | 10.0 |
| From 2 to 4 | 7.5 |
| Greater than 4 | 5.0 |

**Cement-Treated Bases.** Figure 13.8 provides a chart that may be used to estimate the structural layer coefficient, $a_2$, for a cement-treated base material from either its elastic modulus, $E_{BS}$, or its 7-day unconfined compressive strength (ASTM D 1633).

**Bituminous-Treated Bases.** Figure 13.9 presents a chart that may be used to estimate the structural layer coefficient, a2, for a bituminous-treated base material from either its elastic modulus, $E_{BS}$, or its Marshall stability (AASHTO T 245, ASTM D 1559). This is not shown in Figure 13.9.

## TABLE 13.3
### Typical Values for $k_1$ and $k_2$ for Unbound Base and Subbase

| Materials ($M_R = k_1 \times \theta \times k_2$) | | |
|---|---|---|
| Moisture Condition | $k_1$ | $k_2$ |
| Base | | |
| • Dry | 6000–10,000 | 0.5–0.7 |
| • Damp | 4000–6000 | 0.5–0.7 |
| • Wet | 2000–4000 | 0.5–0.7 |
| Subbase | | |
| • Dry | 6000–8000 | 0.4–0.6 |
| • Damp | 4000–6000 | 0.4–0.6 |
| • Wet | 1500–4000 | 0.4–0.6 |

*Source:* Table 2.3, Part II of the AASHTO Guide, 1993
Range in $k_1$ and $k_2$ is a function of the material quality

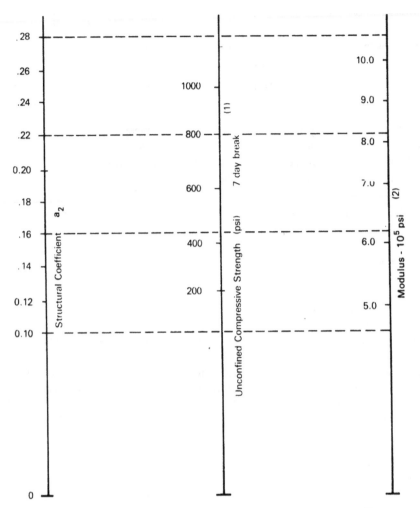

(1)   Scale derived by averaging correlations from Illinois, Louisiana and Texas.
(2)   Scale derived on NCHRP project (3).

**FIGURE 13.8**   Chart to Estimate the Structural Layer coefficient "$a_2$", for a Cement-Treated Base Material.
*Source:* Figure 2.8 of AASHTO Guide, 1993

## 13.5   PAVEMENT STRUCTURAL CHARACTERISTICS

### 13.5.1   DRAINAGE

The selection of inputs is required to treat the effects of certain levels of drainage on predicted pavement performance. The design engineer has to identify what level of drainage is to be achieved under a specific set of drainage conditions. General values corresponding to different drainage levels from the pavement structure are given below:

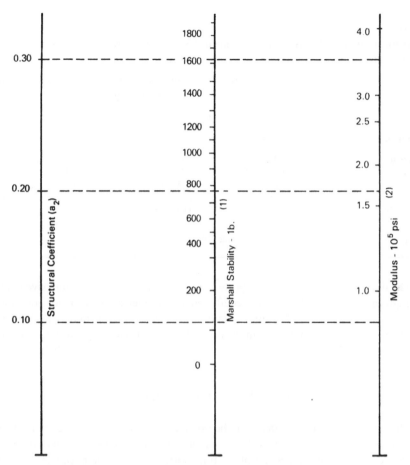

(1)  Scale derived by correlation obtained from Illinois.

(2)  Scale derived on NCHRP project (3).

**FIGURE 13.9**   Chart to Estimate the Structural Layer Coefficient "$a_2$", for a Bituminous-Treated Base Material.

*Source:* Figure 2.9 of AASHTO Guide, 1993

| Quality of Drainage | Water Removed Within |
|---|---|
| Excellent | 2 hours |
| Good | 1 day |
| Fair | 1 week |
| Poor | 1 month |
| Very poor | (water will not drain) |

For comparison purposes, the drainage conditions at the AASHO Road Test are considered to be fair, i.e., free water was removed within 1 week.

**TABLE 13.4**

**Recommended "m" Values for Modifying Structural Layer Coefficients of Untreated Base and Subbase Materials in Flexible Pavements**

| | Percent of Time Pavement Structure Is Exposed to Moisture Levels Approaching Saturation | | | |
| | Less Than | | Greater Than | |
| Quality of Drainage | 1% | 1–5% | 5–25% | 25% |
|---|---|---|---|---|
| Excellent | 1.40–1.35 | 1.35–1.30 | 1.30–1.20 | 1.20 |
| Good | 1.35–1.25 | 1.25–1.15 | 1.15–1.00 | 1.00 |
| Fair | 1.25–1.15 | 1.15–1.05 | 1.00–0.80 | 0.80 |
| Poor | 1.15–1.05 | 1.05–0.80 | 0.80–0.60 | 0.60 |
| Very poor | 1.05–0.95 | 0.95–0.75 | 0.75–0.40 | 0.40 |

*Source:* Table 2.4, Part II of the AASHTO Guide, 1993

The design procedure for the expected level of drainage for a flexible pavement is through the use of modified layer coefficients. A higher effective layer coefficient may be used for improved drainage conditions. The factor for modifying the layer coefficient is referred to as the "m" value and has been integrated into the equation for the structural number "SN" along with layer coefficient "$a_i$" and thickness "$D_i$" thus:

$$SN = a_1D_1 + a_2D_2m_2 + a_3D_3m_3$$

The possible effect of drainage on the asphalt concrete (AC) surface course is not considered. Table 13.4 presents the recommended "m" values as a function of the level of drainage and the percent of the time during the year the pavement structure may normally be exposed to moisture levels approaching saturation, which depends on the average annual rainfall and the existing drainage conditions.

These values apply only to the effects of drainage on untreated base and subbase layers. Although improved drainage is certainly beneficial to stabilized or treated materials, the effects on the performance of flexible pavements are not as profound as those quantified in Table 13.4.

## 13.6   DESIGN PROCEDURE

This chapter describes the design procedure for both asphalt concrete (AC) pavements and overlays by surface treatments (ST) which carry significant levels of traffic, greater than 50,000 18-kip ESAL over the performance period. For both the AC and ST surface types, the design is based on obtaining a flexible pavement structural number (SN) to withstand the projected level of axle load traffic. It is up to the designer to determine whether a single or double overlay by ST or a paved AC surface is required for the specific conditions.

## 13.6.1 Determine Required Structural Number

Figure 13.10 presents the Nomograph provided for determining the design structural number (SN) required for specific conditions, including:

- The estimated future traffic "$W_{18}$", for the performance period.
- The reliability "R", which assumes all input is at average value.
- The overall standard deviation "$S_o$".
- The effective resilient modulus of roadbed material "$M_R$".
- The design serviceability loss, $\Delta PSI = p_o - p_t$.

## 13.6.2 Stage Construction

It has been observed that regardless of the strength or load carrying capacity of a flexible pavement, there may be a maximum performance period associated performance given the initial structure which is subjected to some significant level of truck traffic. If the analysis period is 20 years or more and the performance period is less than 20 years, there may be a need to consider staged construction which is a planned rehabilitation in the design analysis. Here the thickness designs of the initial pavement structure and the subsequent overlays can be evaluated, over their strategic performance periods. It is also important to recognize the need to compound the reliability for each stage of the strategy.

If we consider for each stage of a three-stage strategy with an initial pavement with two overlays, has a 90-percent reliability, the overall reliability of the design strategy is $0.9 \times 0.9 \times 0.9$ is 0.729 or 72.9%. Conversely, if an overall reliability of 95% is desired, the individual reliability for each stage must be $(0.95)^{1/3}$ or 98.3%. The compounding of reliability may be severe for stage construction and may be corrected later on opportunities at the problem areas.

Roadbed swelling and frost heave are both important environmental considerations because of their potential effect on the rate of serviceability loss. Swelling refers to the localized volume changes that occur in expansive roadbed soils as they absorb moisture. A drainage system can be effective in minimizing roadbed swelling if it reduces the availability of moisture for absorption.

If either swelling or frost heave is considered in terms of their effects on serviceability loss and the need for future overlays, then the following procedure should be applied. It requires the plot of serviceability loss versus time that was developed in Section 13.1.4.

The serviceability loss is considered on stage construction strategies because of the planned future need for rehabilitation. In the stage construction approach, the structural number of the initial pavement is decided, and its corresponding performance period or service life is determined. An overlay or series of overlays will serve over the combined performance period which is the analysis period. When swelling and/or frost heave are considered it becomes an iterative process required to determine the duration of the performance period for each stage of the strategy. The iterative process determines when the combined serviceability loss due to traffic and environment reaches the terminal level. It is described in Table 13.5.

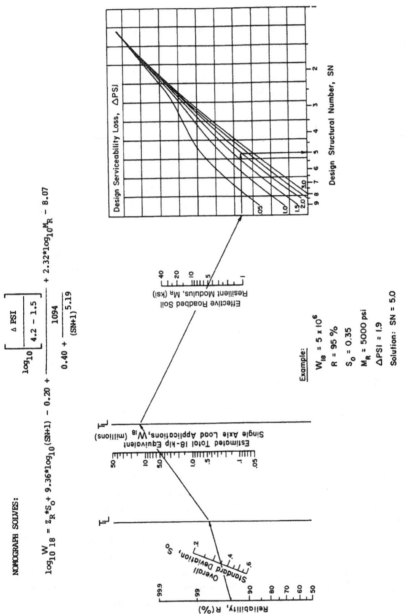

**Figure 3.1.** Design Chart for Flexible Pavements Based on Using Mean Values for Each Input

**FIGURE 13.10** Nomograph Provided for Determining the Design Structural Number (SN).
*Source:* Figure 3.1 of the AASHTO Guide, 1993

**TABLE 13.5**

**Example of a Process Used to Predict the Performance Period of an Initial Pavement Structure Considering Swelling and/or Frost Heave**

| Initial PSI | | | 4.4 | | |
|---|---|---|---|---|---|
| Maximum Possible Performance Period (years) | | | 15 | | |
| Design Serviceability Loss, $\Delta PSI = p_o - p_t = 4.4 - 2.5 = 1.9$ | | | | | |
| (1) | (2) | (3) | (4) | (5) | (6) |
| Iteration No. | Trial Performance Period (years) | Total Serviceability Loss Due to Swelling and Frost Heave $\Delta PSI_{sw,FH}$ | Corresponding Serviceability Loss Due to Traffic $\Delta PSI_{TR}$ | Allowable Cumulative Traffic (18-kip ESAL) | Corresponding Performance Period (years) |
| 1 | 13.0 | 0.73 | 1.17 | $2.0 \times 10^6$ | 6.3 |
| 2 | 9.7 | 0.63 | 1.27 | $2.3 \times 10^6$ | 7.2 |
| 3 | 8.5 | 0.56 | 1.34 | $2.6 \times 10^6$ | 8.2 |

*Source:* Table 3.1, Part II of the AASHTO Guide, 1993

## 13.6.3 ROADBED SWELLING AND FROST HEAVE

| Column No. | Description of Procedure |
|---|---|
| 1 | Estimated by the designer in Step 2. |
| 2 | In Figure 13.2 the total serviceability loss due to swelling and frost heave is calculated. |
| 3 | In step 4, subtract environmental serviceability loss as calculated in Step 3 from total serviceability loss in Column 3, to determine corresponding serviceability loss due to traffic in Column 4. |
| 4 | In step 5, determine from Figure 13.10 keeping all inputs constant, except the use of traffic serviceability loss from Column 4, and applying the chart in reverse. |
| 5 | In step 6, using the traffic from Column 5, estimate the net performance period from Figure 13.1. |

**Step 1** – A realistic structural number "SN" for the initial pavement is to be decided. The maximum initial "SN" is derived for conditions assuming no swelling or frost heave. For example, if the desired overall reliability is 90% by considering an overlay, the design reliability for both the initial pavement and overlay is $0.9^{1/2}$ or 95%, the effective roadbed soil modulus is 5,000 psi, the initial serviceability expected is 4.4, the design terminal serviceability is 2.5, and a 15-year performance period, corresponding to 5 million 18-kip ESAL application, for the initial pavement is assumed, the maximum structural number, by considering for swelling and frost heave conditions, may be determined from Figure 13.10, is 4.4. Any value selected for "SN" less than 4.4 may be appropriate if it does not violate the minimum performance period (Section 13.2.1).

**Step 2** – Select a trial performance period that is expected under the anticipated swelling/frost heave conditions and enter in Column 2. This number
should be less than the maximum possible performance period corresponding to the selected structural number for the initial pavement. In general, for
higher environmental loss, the performance period will be smaller.

**Step 3** – Using the graph of cumulative environmental serviceability loss versus time developed in Section 13.2.4, using Figure 13.2 as an example, estimate the corresponding total serviceability loss due to swelling and frost
heave ($\Delta PSI_{SW,FH}$) that can be expected for the trial period from step 2, and
enter in Column 3.

**Step 4** – Subtract this environmental serviceability loss (Step 3) from the
desired total serviceability loss as, 4.4–2.5 = 1.9, in the example, to establish
the corresponding traffic serviceability loss. Enter the result in Column 4.

$$\Delta PSI_{TR} = \Delta PSI - PSI_{SW,\,FH}$$

**Step 5** – Using Figure 13.10 estimate the allowable cumulative 18-kip ESAL
traffic corresponding to the traffic serviceability loss determined in Step 4
and enter in Column 5. It is important to use the same levels of reliability,
effective roadbed soil resilient modulus, and initial structural number when
applying the flexible pavement chart to estimate the allowable traffic.

**Step 6** – The present traffic count is projected by compound interest, over the
years of the performance period, to reach the final estimated traffic at the
end of the performance period. The corresponding year at which the cumulative 18-kip ESAL traffic that is estimated is determined in Step 5, the
value is entered in Column 6. This may be obtained with the aid of the plot
on cumulative traffic versus time, developed in Section 13.2.2, Figure 13.1
is given as an example.

**Step 7** – Compare the trial performance period with the calculated value in
Step 6. If the difference is greater than 1 year, calculate the average of the
two and use it as the trial value for the start of the next iteration by return-
ing to Step 2. If the difference is less than 1 year, convergence is reached,
and the average is said to be the predicted performance period of the initial
pavement structure corresponding to the selected initial "SN". In the example, after three iterations the convergence was reached, and the predicted
performance period is about 8 years.

The basis of this iterative process is the same, as the estimation of the performance
period of any subsequent overlays. The major differences in the actual application
are that (i) the overlay design methodology presented in AASHTO Part III, is used to
estimate the performance period of the overly and (ii) any swelling and /or frost
heave losses predicted after overlay should restart and then proceed from the point of
time when the overlay was placed.

### 13.6.4  SELECTION OF LAYER THICKNESSES

Once the design structural number "SN" for an initial pavement structure is determined, it is necessary to identify a set of pavement layer thicknesses which, when

combined, will provide the load carrying capacity corresponding to the design "SN". The following equation provides the basis for converting SN into actual thicknesses of surfacing, base, and subbase:

$$SN = a_1 D_1 + a_2 D_2 m_2 + a_3 D_3 m_3$$

where

$a_1, a_2, a_3$ = layer coefficients representative of surface, base, and subbase courses, respectively (see Section 13.4.4)

$D_1, D_2, D_3$ = actual thicknesses (in inches) of surface, base, and subbase courses, respectively

$m_2\ m_3$ = drainage coefficients for base and subbase layers, respectively (see Section 13.5.1)

The "SN" equation does not have a single unique solution, i.e., there are many combinations of layer thicknesses that are satisfactory solutions. The thickness of the flexible pavement layers should be rounded to the nearest $1/2$ inch. When selecting appropriate values for the layer thicknesses, it is necessary to consider their cost effectiveness along with the construction and maintenance constraints to avoid the possibility of producing an impractical design. From a cost-effective view, if the ratio of costs for layer 1 to layer 2 is less than the corresponding ratio of layer coefficients times the drainage coefficient, then the optimum economical design is one where the minimum base thickness is used. Since it is generally impractical and uneconomical to place surface, base, or subbase courses of less than some minimum thickness, the following are provided as minimum practical thicknesses for each pavement course:

| | Minimum Thickness (inches) | |
|---|---|---|
| Traffic, ESAL's | Asphalt Concrete | Aggregate Base |
| Less than 50,000 | 1.0 (or surface treatment) | 4 |
| 50,001 – 150,000 | 2.0 | 4 |
| 150,001–500,000 | 2.5 | 4 |
| 500,001–2,000,000 | 3.0 | 6 |
| 2,000,001–7,000,000 | 3.5 | 6 |
| Greater than 7,000,000 | 4.0 | 6 |

The individual authority should also establish the effective thicknesses and layer coefficients of both single and double ST. The thickness of the surface treatment layer may be negligible in computing "SN", but it has a significant effect on the base and subbase properties due to reductions in surface water percolations.

## 13.6.5  ANALYSIS FOR THE DESIGN OF LAYERED PAVEMENT SYSTEM

For flexible pavements, the structure is a layered system and should be designed accordingly. The structure should be designed by following the procedure shown in Figure 13.11. First, the structural number should be computed over the roadbed soil.

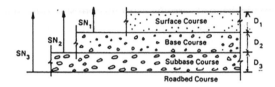

$$D*_1 \geq \frac{SN_1}{a_1}$$

$$SN*_1 = a_1 D*_1 \geq SN_1$$

$$D*_2 \geq \frac{SN_2 - SN*_1}{a_2 m_2}$$

$$SN*_1 + SN*_2 \geq SN_2$$

$$D*_2 \geq \frac{SN_3 - SN*_1 + SN*_2}{a_3 m_3}$$

1)    a, D, m and SN are as defined in the text and are minimum required values.

2)    An asterisk with D or SN indicates that it represents the value actually used, which must be equal to or greater than the required value.

**FIGURE 13.11**   Procedure for Determining Thicknesses of Layers Using a Layered Analysis Approach.
*Source:* Figure 3.2, Part II of the AASHTO Guide, 1993

In the same way, the structural numbers should also be computed over the subbase layer and the base layer, using the applicable strength values for each. The maximum allowable structural number for the subbase material would be equal to the structural number required over the subbase subtracted from the structural number required over the roadbed soil. Similarly, the structural numbers of the other layers may also be computed. The thicknesses for the respective layers may then be determined as indicated in Figure 13.11.

This procedure should not be applied to determine the "SN" required above sub-base or base materials having a modulus greater than 40,000 psi. For such cases, layer thicknesses of materials above the "high" modulus layer should be established based on cost effectiveness and minimum practical thickness considerations.

## 13.7   EXAMPLE FROM APPENDIX H OF THE AASHTO GUIDE

### 13.7.1   Design Analysis

The following example is provided to illiterate the flexible pavement design procure present in Section 3.1 of Part II of the AASHTO Guide. The design requirements for this example are described here in the same order as they are in Part II, Chapter 2 of the AASHTO Guide, and earlier in this chapter.

## 13.7.2 TIME CONSTRAINTS

The analysis period selected for this design example is 20 years. The maximum performance period or service life selected for the initial flexible pavement structure in this example is 15 years. Thus, it will be necessary to consider a stage construction i.e., planned rehabilitation, alternatives to develop design strategies that will last the time analysis period.

## 13.7.3 TRAFFIC

Based on average daily traffic and axle weight data from the planning group, the estimated two-way 18kip ESAL applications during the first year or the pavement's life is $2.5 \times 10^6$ and the projected (compound) growth rate is 3% per year. The directional distribution factor ($D_D$) is assumed to be 50% and the lane distribution factor ($D_L$) for the facility (assume three lanes in one direction) is 30%. Thus, the traffic, during the first year (in the design lane) is $2.5 \times 10^6 \times 0.80 \times 0.50$ or.$0 \times 10^6$ 18-kip ESAL applications. Figure 13.16 provides a plot of the cumulative 18-kip ESAL traffic over the 20-year analysis period. The curve and equation for future traffic ($w_{18}$) are reflective of the assumed exponential growth rate (g) of 3%.

## 13.7.4 RELIABILITY

Although the facility will be in a heavily trafficked state volumes should never exceed half of its capacity. Thus, a 90% overall reliability level was selected for the design. This means that for a two-stage strategy (initial pavement plus one overlay), the design reliability for each stage must be $0.90^{1.2}$ or 95%. Similarly, for a three-stage strategy (initial pavement plus two overlays), the design reliability for all three stages must be $0.90^{1.3}$ or 96.5%.

Another criterion required for the consideration of reliability is the overall standard deviation (S). Although it is possible to estimate this parameter through an analysis of variance of all the design factors (see Volume 2. Appendix EE of the AASHTO Guide), an approximate value of 0.35 will be used for this example problem.

## 13.7.5 ENVIRONMENTAL IMPACTS

Eighty boreholes were obtained along the 16-side length of the project (approximately one every thousand feet). Based on an examination of the borehole samples and subsequent soil classifications, it was determined that the soil at the first twelve borehole sites (approximately 12,000 feet) was basically of the same composition and texture. Significantly *different* results were obtained from examinations at the other borehole sites. Based on this type of unit delineation, this 12,000-foot section of the project will be designed separately from all the rest. The site of this highway construction project is in a location that can be environmentally classified as U.S. Climatic Region II, i.e., wet with freeze–thaw cycling. The soil is considered to be a highly active swelling clay. Because of this and the availability of moisture from high levels of precipitation, a drainage system will be constructed which can remove excess moisture in less than 1 day. The duration of below-freezing temperatures in this environment, however, is not sufficient to result in any problems with frost heaving.

Table H.1 of the AASHTO Guide, 1993 summarized the data used to consider the effects of roadbed swelling on future loss of serviceability. Columns 1 and 2 indicate the borehole number and length of the corresponding section (or segment) of the project. The depth to any rigid foundation at the site is, for all practical purposes, semi-infinite. (Roadbed soil thicknesses greater than 30 feet are considered to be semi-infinite.)

> **Column 4** shows the average PI (P1) of the soil at each borehole location. P1 values above 40 are indicative of a potentially high volume change of the material.
>
> **Column 5** represents the estimated moisture condition of the roadbed material after pavement construction. Because of the plan to construct a "good" drainage system, the future moisture conditions are considered to be "optimum" throughout the project length.
>
> **Column 6** presents the results of applying the chart in Figure 13.14, to estimate the potential vertical rise ($V_R$) at each borehole location.
>
> **Column 7** represents a qualitative estimate of the fabric of the soil or the rate at which it can take on moisture. The natural impermeability of clay materials means that the soil at this site tends toward "tight." This, combined with the relatively low moisture supply (due to the installation of a drainage

**TABLE 13.6**
**Table for Estimating Swell Parameters for Flexible Pavement Design Example**

| Borehole Number | Section Length (ft) | Roadbed Thickness (ft) | Soil Plasticity Index (P1) | Moisture Condition | Potential Vertical Rise (in.) | Soil Fabric | Swell Rate Constant |
|---|---|---|---|---|---|---|---|
| 1 | 900 | >30 | 48 | Optimum | 0.87 | Rel. tight | 0.07 |
| 2 | 1200 | >30 | 56 | ,, | 1.34 | ,, | ,, |
| 3 | 800 | >30 | 67 | ,, | 2.20 | ,, | ,, |
| 4 | 1000 | >30 | 15 | ,, | 0.00 | ,, | 0.10 |
| 5 | 1000 | >30 | 46 | ,, | 0.70 | ,, | 0.07 |
| 6 | 1100 | >30 | 62 | ,, | 1.86 | ,, | ,, |
| 7 | 100 | >30 | 65 | ,, | 2.00 | ,, | ,, |
| 8 | 900 | >30 | 71 | ,, | 2.60 | ,, | ,, |
| 9 | 1200 | >30 | 38 | ,, | 0.28 | ,, | ,, |
| 10 | 800 | >30 | 60 | ,, | 1.80 | ,, | ,, |
| 11 | 900 | >30 | 19 | ,, | 0.00 | ,, | 0.10 |
| 12 | 1200 | >30 | 51 | ,, | 1.04 | ,, | 0.07 |
| **Total** | 12000 | | | | | | |

*Source:* Table H.1 of Appendix H, of the AASHTO Guide, 1993

system), means that the swell rate constant at each location having a "tight" fabric (i.e., PI (P1) greater than about 20) can be estimated at 0.07, see Figure 13.12. For the occasions where P1 was less than 20, a value of 0.10 was used because of the likelihood of greater permeability.

Based on the data in Table H.1 of the AASHTO Guide, 1993, the overall swell rate constant and potential VR are determined by calculating a weighted average; thus:

Swell Rate Constant = 0.075
Potential Vertical Rise (VR) = 1.2 inches

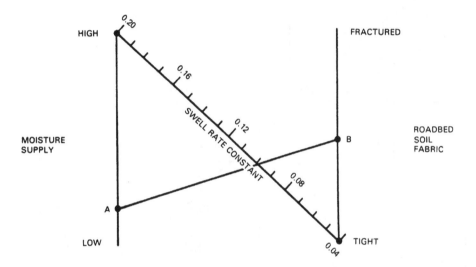

NOTES:     a)  LOW MOISTURE SUPPLY:

                 Low rainfall
                 Good drainage

           b)  HIGH MOISTURE SUPPLY

                 High rainfall
                 Poor drainage
                 Vicinity of culverts, bridge abutments, inlet leads

           c)  SOIL FABRIC CONDITIONS (self explanatory)

           d)  USE OF THE NONOGRAPH

           1)  Select the appropriate moisture supply condition which may be somewhere between
               low and high (such as A).

           2)  Select the appropriate soil fabric (such as B). This scale must be developed by each
               individual agency.

           3)  Draw a straight line between the selected points (A to B).

           4)  Read swell rate constant from the diagonal axis (read 0.10).

**FIGURE 13.12**   Nomograph for Estimating Swell Rate Constant.
*Source:* Figure G.2, Part II, in Appendix G of the AASHTO Guide, 1993

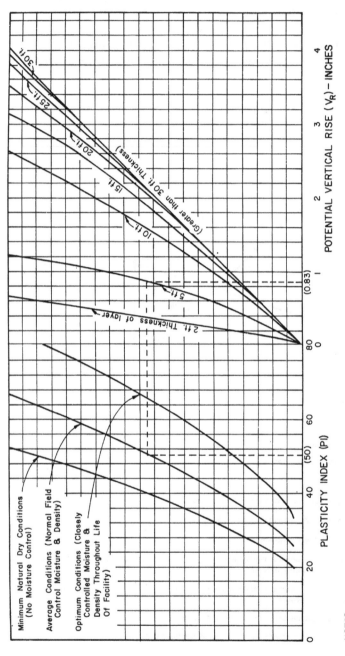

**FIGURE 13.13**  Chart for Estimating the Approximate Potential VR of Natural Soils.

*Source:*  Figure G.3, Part II, of Appendix G of the AASHTO Guide, 1993

The swelling probability is simply the percent of the length of the project which has a potential vertical rise greater than 0.2 inches. 10,000 feet out of the total 12,000 have a $V_R$ greater than 0.2 inches, thus the swelling probability is 84%.

These factors were then used to generate the serviceability loss versus the time curve presented in Figure 13.16. The curve shown was generated using the equation presented in Figure 13.14. This represents a graph of the estimated total environmental serviceability loss versus time since frost heave is not a consideration.

## 13.7.6 SERVICEABILITY

Based on the traffic volume and functional classification of the facility (6-lane state highway), terminal serviceability ($p_t$) of 2.5 was selected. Experience indicates (for this hypothetical example) that the initial serviceability ($P_O$) normally achieved for flexible pavements in the state is significantly higher than that at the AASHO Road

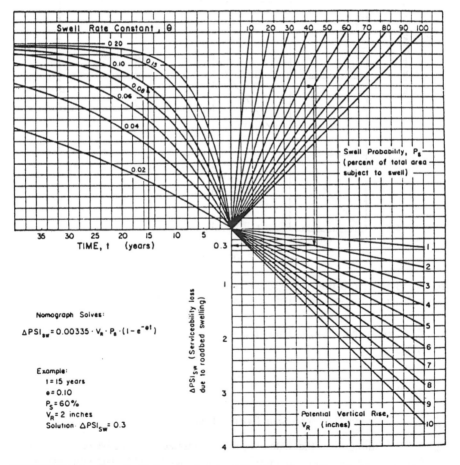

**FIGURE 13.14**   Chart for Estimating Serviceability Loss Due to Roadbed Swelling.
*Source:* Figure G.4 of Appendix G of the AASHTO Guide, 1993

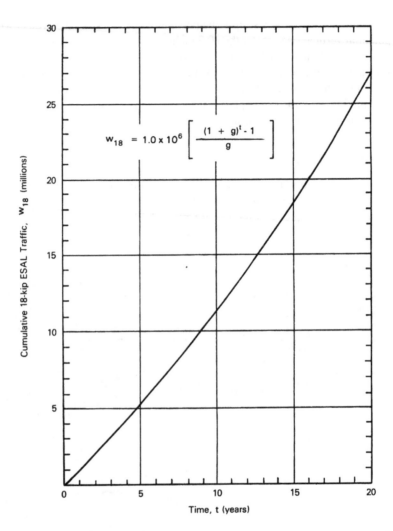

$$w_{18} = 1.0 \times 10^6 \left[ \frac{(1 + g)^t - 1}{g} \right]$$

**FIGURE 13.15**    Plot of Cumulative 18-kip ESAL Traffic Versus Time for Assumed Conditions.
*Source:* Figure H.1 in Appendix H of the AASHTO Guide, 1993

Test (4.6 compared to 4.2). Thus, the overall design serviceability loss for this
problem is:

$$\Delta PSI = p_o - p_t = 4.6 - 2.5 = 2.1$$

### 13.7.7    EFFECTIVE ROADBED SOIL RESILIENT MODULUS

Figure 13.17 summarizes that data used to characterize the effective resilient modulus of the roadbed soil. Individual moduli are specified for 24 half-month intervals to
define the seasonal effects. These values are also reflective of the roadbed support

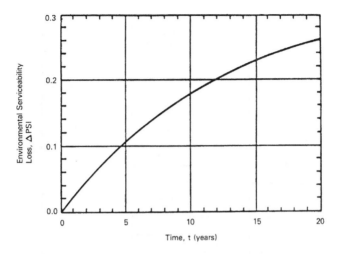

Time, t (years)

**FIGURE 13.16** Graph of Environmental Serviceability Loss Versus Time for Swelling Conditions Considered.
*Source:* Figure H.2 in Appendix H of the AASHTO Guide, 1993

that would be expected under the improved moisture conditions provided by the "good" drainage system:

| Roadbed Moisture Condition | Roadbed Soil Resilient Modulus (psi) |
|---|---|
| Wet | 5,000 |
| Dry | 6,500 |
| Spring-Thaw | 4,000 |
| Frozen | 20,000 |

The frozen season (from mid-January to mid-February) is 1 month long. The spring-thaw season (mid-February to March) is 0.5 months long, the wet periods (March through May and mid-September through mid-November) total 5 months, and the dry periods (June through mid-September and mid-November through mid-January) total 5.5 months. Application of the effective roadbed soil $M_R$ estimation procedure results in a value of 5700 psi.

## 13.7.8 PAVEMENT LAYER MATERIALS CHARACTERIZATION

Three types of pavement materials will constitute the individual layers of the structure. The moduli for each, determined using the recommended laboratory test procedures, are as follows:

| | |
|---|---|
| Asphalt Concrete: | $E_{AC} = 400.000$ psi |
| Granular Base: | $E_{BS} = 30.000$ psi |
| Granular Subbase: | $E_{SB} = 11.000$ psi |

| Month | Roadbed Soil Modulus, M (psi) | Relative Damage u |
|-------|-------------------------------|-------------------|
| Jan. | 6,500 | 0.167 |
| | 20,000 | 0.012 |
| Feb. | 20,000 | 0.012 |
| | 4,000 | 0.515 |
| Mar. | 5,000 | 0.307 |
| | 5,000 | 0.307 |
| Apr. | 5,000 | 0.307 |
| | 5,000 | 0.307 |
| May | 5,000 | 0.307 |
| | 5,000 | 0.307 |
| June | 6,500 | 0.167 |
| | 6,500 | 0.167 |
| July | 6,500 | 0.167 |
| | 6,500 | 0.167 |
| Aug. | 6,500 | 0.167 |
| | 6,500 | 0.167 |
| Sept. | 6,500 | 0.167 |
| | 5,000 | 0.307 |
| Oct. | 5,000 | 0.307 |
| | 5,000 | 0.307 |
| Nov. | 5,000 | 0.307 |
| | 6,500 | 0.167 |
| Dec. | 6,500 | 0.167 |
| | 6,500 | 0.167 |
| Summation: u = | | 5.446 |

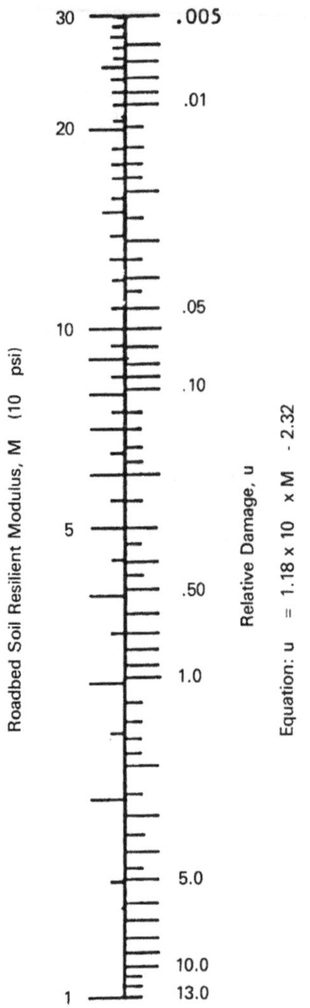

Average: $= \dfrac{\text{}}{n} = \dfrac{5.46}{24} = 0.227$

Roadbed Soil Resilient Modulus, M (10 psi)

Relative Damage, u

Equation: $u = 1.18 \times 10 \times M^{-2.32}$

Effective Roadbed Soil Resilient Modulus, $M_R$ (psi) 5,700
(corresponds to $u_f$ )

FIGURE 13.17   Estimation of Effective Roadbed Soil Resilient Modulus.
*Source:* Figure H.3 in Appendix H of the AASHTO Guide, 1993

These values correspond to the average year-round moisture conditions that would be expected without any type of pavement drainage system. (The effects of positive drainage on material requirements are considered in a letter section.)

### 13.7.9 Layer Coefficients

The structural layer coefficients ($a_1$-values) corresponding to the moduli defined in the previous section are as follows:

Asphalt Concrete:     $a_1 = 0.42$ 2.5, Part II, of the AASHTO Guide, 1993

Granular Base:     $a_2 = 0.14$ 2.6, Part II, of the AASHTO Guide, 1993

Granular Subbase:     $a_3 = 0.08$ 2.7, Part II, of the AASHTO Guide, 1993

### 13.7.10 Drainage Coefficients

The only item that is considered under the heading "Pavement Structural Characteristics" (Part II, Section 2.4 of the AASHTO Guide) in the design of a flexible pavement is the method of drainage. The drainage coefficient (m-value) corresponding to the granular base and subbase materials for a "good" drainage system (i.e., water removed within 1 day) and the balanced wet-dry climate of U.S. Climate Region II is 1.20. (The range in Part II, Table 2.4 of the AASHTO Guide, 1993, for 1 to 5% moisture exposure time is 1.15 to 1.25)

### 13.7.11 Development of Initial Stage of a Design Alternative

Since the estimated maximum performance period (15 years) is less than the design analysis period (20 years), any initial structure selected will require an overlay to last the analysis period. The thickest recommended initial structure (evaluated here) is that corresponding to the maximum 15-year performance period. Thinner initial structures, selected for life-cycle cost analyses, will require thicker overlays (at an earlier date) to last the same analysis period. The strategy with the maximum recommended initial structural number is determined using the effective roadbed soil resilient modulus of 5700 psi, a reliability of 95%, an overall standard deviation of 0.35, a design serviceability loss of 2.1, and the cumulative traffic at the maximum performance period. $18.6 \times 10^4$ 18-Kip ESAL Figure 13.15, for a time of 15 years. Applying Figure 13.10, the result is a maximum initial structural number (SN) of 5.6. Because of serviceability loss due to swelling, however, an overlay will be required before the end of the 15-year design performance period. Using the step-by-step procedure described in Part II (Section 3.1.3 of the AASHTO Guide and 13.1.3 of this book, the service life that can be expected is about 13 years (see Table H.2 of the AASHTO Guide, 1993). Thus, the overlay that must be designed will need to carry the remaining 18-kip ESAL traffic over the last 7 years of the analysis period.

### 13.7.12 Determination of Structural Layer Thicknesses for Initial Structure

The thicknesses of each layer above the roadbed soil or subgrade are determined using the procedure described in Part II (Section 3.1.4 of the AASHTO Guide see

Figure 13.11. For the design SN of 5.6 developed in this example, the determination of the layer thicknesses is demonstrated below:

Solve for the SN required above the base material by applying Figure 13.10 using the resilient modulus of the base material (rather than the effective roadbed soil resilient modulus). Values of $E_{BS}$ equal to 30,000 psi, first stage reliability (R) equal to 95%. $w_{18}$ equal to $16.0 \times 10^6$ and $\Delta PSI_{TR}$ equal to 1.89 (the latter two are from Table H.2 of the AASHTO Guide, 1993) result in an $SN_1$ of 3.2. Thus, the asphalt concrete surface thickness required is:

$$D^*_1 = SN_1/a_1 = 3.2/0.42 = 7.6 \left( or\, 8\, inches \right)$$

$$SN^*_1 = a_1\, D^X_1 = 0.42 \times 8 = 3.36$$

Similarly, using the subbase modulus of 11.000 psi as the effective roadbed soil resilient modulus. $SN_2$ is equal to 4.5 and the thickness of base material required is:

$$D^*_2 = \left( SN_2 - SN^{*,y}_1 \right) = \left( a_2 m_2 \right)$$

$$= \left( 4.5 - 3.36 \right) / \left( 0.14 \times 1.20 \right)$$

$$= 6.8 \left( or\, 7\, inches \right)$$

$$SN*2 = 7 \times 0.14 \times 1.20 = 1.18$$

Finally, the thickness of the subbase required is:

Initial SN 5.6
Maximum Possible performance Period (years) 15
Design Serviceability Loss, PSI = $p_o - p_t$ = 4.6–2.5 = 2.1

*Convergence achieved after only one iteration.

---

**TABLE 13.7**

**Reduction in Performance Period (Service Life) of Initial Pavement Arising from Swelling Considerations**

| (1) | (2) | (3) | (4) | (5) | (6) |
|---|---|---|---|---|---|
| Iteration No. | Trial Performance Period (years) | Serviceability Loss Due to Swelling $\Delta PSI_{SW}$ | Corresponding Serviceability Loss Due to Traffic $\Delta PSI_{TR}$ | Allowable Cumulative Traffic (18-KIP ESAL) | Corresponding Performance Period (years) |
| 1* | 13 | 0.21 | 1.89 | $16.0 \times 10^6$ | 13.2 |

*Source:* Table H.2 of AASHTO Guide, 1993

---

$$D^*_3 = \left(SN_3 - \left(SN^*_1 - SN^*_2\right)\right)/\left(a_3 m_3\right)$$

$$= \left(5.6 - \left(3.36 + 1.18\right)\right)/\left(0.08 \times 1.20\right)$$

$$= 11 \text{ inches}$$

## 13.7.13  OVERLAY DESIGN

Since the initial pavement structure in this design example has a service life of only 13 years, the overlay that should be determined for life-cycle cost analysis must last 7 years and carry the remaining 18-kip ESAL traffic. Thickness design for this overlay is accomplished using an adaptation of the overlay design procedure described in Part III.

The equation used to solve for the flexible overlay required is as follows:

$$SN_{OL} = SN_y - \left(F_{RL} \times SN_{xeff}\right)$$

where

$SN_{OL}$ = structural number of the required asphalt concrete overlay

$SN_y$ = structural number required for a "new" pavement to carry the estimated future traffic for the prevailing roadbed soil support conditions

$F_{RL}$ = effective structural number of the existing pavement at the time the overlay is placed

$SN_{xeff}$ = effective structural number of the existing pavement at the time the overlay is placed

$SN_y$ is determined using Part II, Figure 13.10, with input parameters associated with the design of a new pavement to last the remaining 7 years of the analysis period. The cumulative 18-kip ESAL traffic ($w_{18}$) between years 13 and 20 is $11.3 \times 10^6$ from Figure 13.15, the second stage reliability (R) is 95%, the overall standard deviation ($S_o$) is 0.35, and the effective roadbed soil resilient modulus ($M_R$) is 5700 psi. The serviceability loss due to traffic ($\Delta PSI_{TR}$), which is used as input Figure 13.10, is equal to the design serviceability loss $\Delta PSI$) less the serviceability loss due to environment, in this case, only roadbed swelling ($\Delta PSI_{SW}$).

$$\Delta PSI_{TR} = \Delta PSI - \Delta PSI_{SW}$$

Since,

$$\Delta PSI = P_o - P_t = 4.6 - 2.5 = 2.1$$

and

$$\Delta PSI_{SW} = 0.26$$

Figure 13.16 (year 20)–0.21 (year 13) = 0.05 the serviceability loss due to traffic is:

$$\Delta PSI_{TR} = 2.1 - 0.05 = 2.05$$

Thus, the resulting $SN_y$ from Figure 13.10 is 5.2. The remaining life factor ($F_{RL}$) is established based on the estimated remaining life ($R_{Lx}$) of the original pavement at the time of overlay and the estimated remaining life ($R_{Ly}$) of the overlay when it reaches its design terminal serviceability of 2.5:

- From Figure 13.19, for an $SN_o$ (the structural number of the original pavement) equal to 5.6 and a terminal serviceability (at the time of overlay) of 2.5, the value of $R_{Lx}$ is 42%.
- From Figure 13.10 the estimated future 18-kip ESAL traffic ($N_{fy}$) to ultimate failure (i.e., when the serviceability drops to 2.0 and the remaining life is zero) is $17 \times 10^6$ 18-kip ESAL. This is based on the same input parameters used to estimate $SN_y$, except the serviceability loss used as input to the design chart includes the loss one to roadbed swelling.

$$PSI_{TR} = (4.6 - 2.0) - 0.05 = 2.55$$

Since the estimated future traffic (y) to a terminal serviceability of 2.5 is $11.3 \times 10^6$ 18-kip ESAL, the estimated remaining life ($R_{LY}$) of the overlay when it reaches pt. equal 2.5 is:

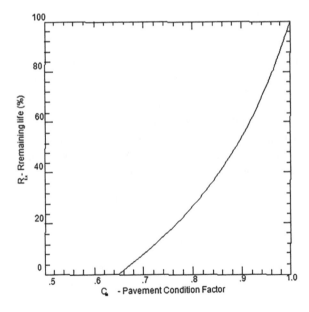

**FIGURE 13.18**   Remaining Life Estimate Predicted from Pavement Condition Factor.
*Source:* Figure 5.13, Part III, of the AASHTO Guide, 1986

$$R_{Ly} = (N_{fy} - y)/N_{fy}$$

$$= (17 \times 10^6 - 11.3 \times 10^6)(17 \times 10^6)$$

$$= 0.335(33.5\%)$$

Thus, by applying the above estimates of $R_{Lx}$ and $R_{Ly}$ in the graph from Figure 13.20 the remaining life factor ($F_{RL}$) is 0.72.

The last factor to be estimated before the required overlay structural number can be determined is the effective structural number ($SN_{xeff}$) of the original pavement at the time of overlay. Since this represents the design for an overlay 13 years after initial pavement construction, $SN_{xeff}$ must be approximated using the following relationship:

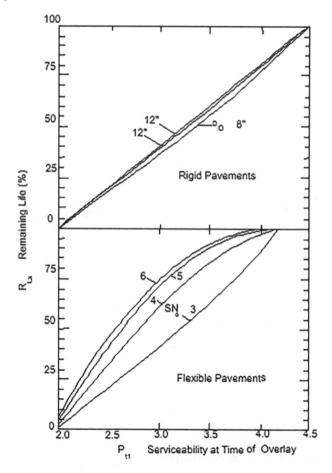

**FIGURE 13.19** Remaining Life Estimate Based on Present Serviceability Value and Pavement Cross Section.
*Source:* Figure 5.15, Part III, of the AASHTO Guide, 1986

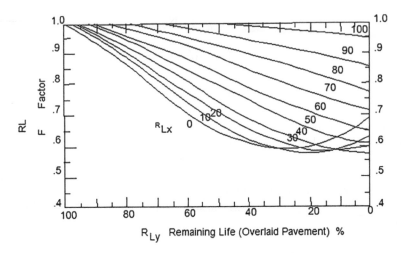

**FIGURE 13.20**   Remaining Life Factor as a Function of the Remaining Life of Existing and Overlaid Pavements.
*Source:* Figure 5.17, Part III, of the AASHTO Guide, 1986

$$SN_{xeff} = C_x \times SN_o$$

where

$SN_o$  = structural number of original pavement (5.6), and
$C_x$   = pavement condition factor.

The letter term, $C_x$, is estimated based on the remaining life $(R_{Lx})$ of the original pavement at the time of the overlay. Referring to Figure 13.18 (with $R_{Lx}$ equal to 42%), $C_x$ is 0.86. Thus, the effective structural number is:

$$SN_{xeff} = 0.86 \times 5.6 = 4.82$$

Inserting the values for $SN_y$, $F_{RL,}$ and $SN_{xeff}$ into the overlay design equation results in a required overlay structural number of:

$$SN_{OL} = 5.2 - (0.72 \times 4.82) = 1.73$$

This translates into a design asphalt concrete overlay thickness of:

$$D_{OL} = SN_{OL}/a_1 = 1.73/0.42 = 4.1 \, inches$$

### 13.7.14   SUMMARY OF DESIGN STRATEGY

The first stage or initial pavement structure has a structural number (SN) of 5.6, which was broken down into the following layer thicknesses:

| Asphalt Concrete | 8 inches |
| Granular Base | 7 inches |
| Granular Subbase | 11 inches |

This structure will last approximately 13 years and carry about 16 million 18-kip ESAL applications. It will require a 4-inch asphalt concrete overlay to last the remaining 7 years of the 20-year analysis period. During those 7 years, the overlaid structure will carry about 11 million 18-kip applications.

## 13.8 COMPUTER-AIDED DESIGN OF FLEXIBLE PAVEMENT IN AASHTO METHOD

The computer applications for the "Design of Flexible Pavement" are explained in Chapter 13 in the "The Guide for Computer Application Tutorials" part of the book available for download from the website www.roadbridgedesign.com.

For downloading the items under book tutorials, visit the website at: www. roadbridgedesign.com.

For downloading the other items visit the website at: www.techsoftglobal.com/download.php.

## BIBLIOGRAPHY

*AASHTO Guide for Design of Pavement Structures 1986*, AASHTO.

*AASHTO Guide for Design of Pavement Structures 1993*, AASHTO.

*Manual on Pavement Design*, Arahan Teknik (Jalan) 5/85, JKR, Malaysia.

*Guidelines for the Design of Flexible Pavements*, IRC 37-2001, IRC.

*Guidelines for the Design of Plain Jointed Rigid Pavements for Highways*, IRC 58-2002, IRC.

*Tentative Guidelines for Strengthening of Flexible Pavements Using Benkelman Beam Deflection Technique*, IRC 81-1997, IRC.

*Guidelines for the Design of Flexible Pavements for Low Volume Rural Roads*, IRC SP 72-2007, IRC.

*Pavements and Surfacings for Highways and Airports* by Michel Sargious, by Applied Science Publishers Ltd.

*Thickness Design, Asphalt Pavement Structures for Highways and Streets*, The Asphalt Institute.

*Asphalt Pavements for Airports*, The Asphalt Institute.

*Highway Engineering* by Paul H. Wright and Karen K. Dixon, by John Wiley & Sons, Inc.

*Principles of Highway Engineering and Traffic Analysis* by Fred L. Mannering, Walter, and Scott, by Wiley India.

# 14 Design of Rigid Concrete Pavement

## 14.1 GENERAL

The design and construction of concrete pavements is a fairly complex subject, and it seems desirable to list the major topics that will be covered in this chapter. Major subjects that will be discussed in subsequent sections include materials, proportioning of concrete mixtures, structural pavement design, and the construction of Portland cement concrete pavements.

## 14.2 TYPES OF RIGID CONCRETE PAVEMENTS

Rigid concrete pavements can be subdivided into three types:

- Jointed plain concrete pavements (JPCP).
- Jointed reinforced concrete pavements (JRCP).
- Continuously reinforced concrete pavements (CRCP).

Jointed plain concrete pavements are characterized by short transverse joint spacing, typically of 15–20 ft. (4.57–6.1 m), with no reinforcing steel in the concrete slab, although some types of reinforcing steel may be used across the joints. Jointed reinforced concrete pavements are characterized by long joint spacing, typically of 30–100 ft. (9.1–30.5 m), with reinforcing steel in the concrete slab to hold the cracks developed in the concrete by shrinkage tightly together. Continuously reinforced concrete pavements theoretically have no joints and contain a greater percentage of steel to control cracking from shrinkage of the concrete.

In the interests of simplification, in the remainder of this text, the term concrete will be taken to be synonymous with Portland cement concrete.

## 14.3 MATERIALS

The materials included in concrete, as generally used in highway construction, are coarse aggregate, fine aggregate, water, cement, and one or more admixtures. These materials will be discussed in turn in the following sections, along with a discussion of typical requirements of specifications related to each.

### 14.3.1 COARSE AGGREGATE

Coarse aggregates most frequently used in Portland cement concrete include crushed stone, gravel, and blast-furnace slag. Other similar inert materials may also be used, and the listed materials may be used singly or in combination with one another. Specific requirements related to coarse aggregations to be used for this purpose may be divided into five groups: deleterious substances, percentage of wear, soundness, weight per cubic foot (slag), and grading.

Highway agencies specify the maximum percentages of various deleterious substances that are permitted to be present in aggregates. Specific requirements related to the cleanliness of the aggregate vary with different agencies. The requirements in Table 13.1 related to deleterious substances are indicative of those in current use.

In certain areas, considerable difficulty has been experienced with aggregates containing deleterious substances that react harmfully with the alkalies present in the cement. Such reactions generally result in an abnormal expansion of the concrete. Methods have been devised as ASTM Methods C227 and C289, for detecting aggregates with these harmful characteristics, and suitable stipulations are included in typical specifications, for example, ASTM C33.

The ability of the coarse aggregate to resist abrasion is generally controlled by the inclusion of the specifications of a maximum permissible percentage of wear in the Los Angeles abrasion test. Maximum permissible percentages of wear in this test (AASHTO Designation T96) range from as low as 30 to as high as 65 in the specifications of the various state highway departments. On average, however, the maximum permissible loss varies from 40 to 50% for all three of the principal types of coarse aggregate, although many specifications do not contain a requirement related to the percentage of wear of slag.

## 14.4   DESIGN REQUIREMENTS

The preparation and/or selection of the inputs are required for the design of new pavement construction or reconstruction for the strengthening of an existing pavement. For the design requirements for several types of pavement structures on both highways and low-volume roads, only certain sets of inputs are required for a given structural design combination.

### 14.4.1   DESIGN INPUTS

The inputs are classified under five separate categories:

**Design Variables.** This category refers to the set of criteria which is considered for each type of pavement design procedure.

**Performance Criteria.** This is the "serviceability" which represents the user-specified set of boundary conditions within which a given pavement design alternative should perform, during the design life.

**Material Properties for Structural Design.** This category covers all the data related to material properties for the pavement and roadbed soil which are required for the design of the pavement structure.

**Structural Characteristics.** This refers to certain physical characteristics of the pavement structure which are considered in the design and have an effect on its performance, during the design life.

**Reinforcement Variables.** This category covers all the reinforcement design variables for longitudinal and transverse bars, tie and dowel bars, needed for the different types of rigid concrete pavements.

Table 13.1 identifies all possible design input requirements and indicates the specific types of structural designs for which they are required, which are (i) means that a particular design input (or set of inputs) must be determined for that structural combination, (ii) indicates that the design input should be considered because of its potential impact on the results. Under "Rigid," the Plain (jointed) Concrete Pavement is referred to as JCP, Jointed Reinforced Concrete Pavement is referred to as JRCP, Continuously Reinforced Concrete Pavement is referred as CRCP, and prestressed concrete pavements is referred to as PCP. PCP is not shown as a column in Table 13.1.

### 14.4.2    EFFECTIVE MODULUS OF SUBGRADE REACTION

Like the effective roadbed soil resilient modulus used for flexible pavement design, an effective modulus of subgrade reaction (k-value) is used for rigid pavement design. The k-value is directly proportional to roadbed soil resilient modulus. The duration of seasons and seasonal moduli developed in the previous section are used as input to the estimation of an effective design k-value. But, because of the effects of subbase characteristics on the effective design k-value, it is determined in an iterative design procedure.

### 14.4.3    PAVEMENT LAYER MATERIALS CHARACTERIZATION

Although there are many types of material properties and laboratory test procedures for assessing the strength of pavement structural materials, the use of "layer coefficients" has been adopted as a basis for design in the AASHTO Guide, 1993. With proper understanding, the layer coefficients have been used in the AASHTO flexible pavement design procedure. Layer coefficients derived from test roads or satellite sections are preferred.

The elastic modulus of any paving or roadbed material is an engineering property. For those material types which undergo significant permanent deformation under load, this property may not reflect the material's inelastic behavior under load. The resilient modulus reflects the material's stress–strain behavior under normal pavement loading conditions. The strength and stiffness of the material are important, the mechanistic-based procedures reflect strength as well as stiffness in the materials characterization procedures. Also, pavement layers with stabilized base materials (WBM or WMM) may be subject to cracking under certain conditions and the stiffness may not be an indicator for this type of distress. The resilient modulus can apply to any type of material, still, it is used as MR applied only to the roadbed soil in the AASHTO procedure. Different notations are used for moduli of subbase ($E_{SB}$), base ($E_{BS}$), asphalt concrete ($E_{AC}$), and Portland cement concrete ($E_C$).

The procedure for obtaining the resilient modulus for relatively low stiffness materials, such as natural soils, unbound granular layers, and even stabilized layers and asphalt concrete, should be tested by using the resilient modulus test methods.

Although the testing apparatus for each of these types of materials is the same, for unbound materials, there is the need for triaxial confinement.

Alternatively, the bound or higher stiffness materials, such as stabilized bases and asphalt concrete, may be tested by using the repeated-load indirect tensile test which relies on the use of electronic gauges to measure small movements of the sample under load but is less complex and easier to conduct than the triaxial resilient modulus test.

Because of the small displacements and brittle nature of the stiffest pavement materials, i.e., Portland cement concrete and those base materials stabilized with a high cement content, it is recommended that the elastic modulus of such high-stiffness materials be determined according to the procedure described in ASTM C 469.

The elastic modulus for any type of material may also be estimated using a correlation recommended by the American Concrete Institute for normal weight Portland cement concrete (PCC):

$$E_c = 57,000 \left( f_c \right)^{0.5}$$

where

"Ec" is the PCC elastic modulus (in psi)
"fc" is the PCC compressive strength (in psi) as determined using AASHTO T
    22, T 140, or ASTM C 39

### 14.4.4 PCC MODULUS OF RUPTURE

The modulus of rupture is the flexural strength of PCC and is required for the design of the rigid pavement. The use of modulus of rupture in the design procedure is the mean value determined after 28 days using third point loading as per AASHTO T 97 and ASTM C 78. If the center-point loading is to be used, then a correlation should be developed between the two tests.

It is recommended in the AASHTO Guide, 1993 that the normal construction specification for modulus of rupture not to be used as input, since it represents a value below which is only for a small percent of the distribution. If it is to be used then some adjustment should be applied, based on the standard deviation of modulus of rupture and the percent (PS) of the strength distribution that normally falls below the specification:

$$S' \left( \text{mean} \right) = S° + z \times \left( SD2 \right)$$

where

S is the estimated mean value for PCC modulus of rupture (psi)
S° is the construction specification on concrete modulus of rupture (psi)
SD is the estimated standard deviation of concrete modulus of rupture (psi)

z  = standard normal variate
   = 0.841, for PS = 20%*
   = 1.037, for PS = 15%
   = 1.282, for PS = 10%
   = 1.645, for PS = 5%
   = 2.327, for PS = 1%

*NOTE: Permissible number of specimens, expressed as a percentage, that may have strengths less than the specification value.

### 14.4.5  LAYER COEFFICIENTS

Estimating the "a" values, which are the structural layer coefficients in the AASHTO method, and required for standard flexible pavement structural design discussed here. This coefficient value is assigned to each layer material in the pavement structure to convert actual layer thicknesses into the structural number "SN". This layer coefficient has an empirical relationship between SN and thickness. This is a measure of the relative ability of the material to function as a structural component of the pavement. The relative impact of the layer coefficients "a" and thickness "D" on the structural number "SN" is based on the following equation:

$$SN = \sum a_i\, D_i$$

$$i = I$$

Although the elastic or resilient modulus has been adopted as the measure of material quality, still it is necessary to obtain corresponding layer coefficients for the design by the structural number design approach. There are correlations available to determine the modulus from tests such as the R-value, still, it is recommended for direct measurement using AASHTO method T 274 for subbase and unbound granular materials and ASTM D 4123 for asphalt concrete and other stabilized materials. Research and field studies indicated about many factors govern the values for layer coefficients. The layer coefficient may vary with thickness, underlying support, position in the pavement structure, etc.

The ways for estimating the coefficients by charts are separated into five categories, depending on the type and function of the layer material. These are asphalt concrete, granular base, granular subbase, cement-treated, and bituminous base. Other materials such as lime, lime fly-ash, and cement fly-ash are acceptable materials, and charts may also be developed for them.

## 14.5  PAVEMENT STRUCTURAL CHARACTERISTICS

### 14.5.1  DRAINAGE

The selection of inputs is required to treat the effects of certain levels of drainage on predicted pavement performance. The design engineer has to identify what level of drainage is to be achieved under a specific set of drainage conditions. General values corresponding to different drainage levels from the pavement structure are given below:

| Quality of Drainage | Water Removed Within |
|---|---|
| Excellent | 2 hours |
| Good | 1 day |
| Fair | 1 week |
| Poor | 1 month |
| Very poor | (water will not drain) |

For comparison purposes, the drainage conditions at the AASHO Road Test are considered to be fair, i.e., free water was removed within 1 week.

The design procedure for the expected level of drainage for a flexible pavement is through the use of modified layer coefficients. A higher effective layer coefficient may be used for improved drainage conditions. The factor for modifying the layer coefficient is referred to as the "m" value and has been integrated into the equation for the structural number "SN" along with layer coefficient "$a_i$" and thickness "$D_i$," thus:

$$SN = a_1 D_1 + a_2 D_2 m_2 + a_3 D_3 m_3$$

The possible effect of drainage on the asphalt concrete surface course is not considered.

In rigid pavements, in the performance equation, the treatment for the expected level of drainage is using a drainage coefficient, "Cd", for a rigid pavement, with an effect similar to that of the load transfer coefficient, "J". As a basis for comparison, the value for "Cd" is 1.0, for conditions at the AASHO Road Test.

Table 14.1 provides the recommended "Cd" values, depending on the quality of drainage and the percent of the time during the year the pavement structure is

## TABLE 14.1
### Recommended Values of Drainage Coefficient, Cd, for Rigid Pavement Design

Percent of Time Pavement Structure Is Exposed to Moisture Levels Approaching Saturation

| Quality of Drainage | Less Than | | Greater Than | |
|---|---|---|---|---|
| | 1% | 1–5% | 5–25% | 25% |
| Excellent | 1.25–1.20 | 1.20–1.15 | 1.15–1.10 | 1.10 |
| Good | 1.20–1.15 | 1.15–1.10 | 1.10–1.00 | 1.00 |
| Fair | 1.15–1.10 | 1.10–1.00 | 1.00–0.90 | 0.90 |
| Poor | 1.10–1.00 | 1.00–0.90 | 0.90–0.80 | 0.80 |
| Very poor | 1.00–0.90 | 0.90–0.80 | 0.80–0.70 | 0.70 |

*Source:* Table 2.5 AASHTO Guide, 1993

normally exposed to moisture levels approaching saturation. The latter depends on the average yearly rainfall and the prevailing drainage conditions.

## 14.5.2 LOAD TRANSFER

The load transfer coefficient, "J", is a factor used in rigid pavement design to account for the ability of a concrete pavement structure to transfer and distribute the wheel load across the joints or cracks. The dowel bars as the load transfer devices, aggregate interlock, and the presence of tied concrete shoulders all effect this value. Generally, the "J" value for the jointed concrete pavement with tied shoulders or else increases as traffic loads increase, because the aggregate load transfer decreases with load repetitions.

Table 14.2 provides ranges of load transfer coefficients for different conditions developed from experience and mechanistic stress analysis. As a general guide, higher "J" values should be used for low k-values, high thermal coefficients, and large variations of temperature. Different project authorities should develop criteria for their aggregates, climatic conditions, etc.

The size and spacing of dowel bars should be determined by the local agency's procedures and/or experience. In general, the dowel diameter should be equal to the slab thickness multiplied by 1/8 inch for example, for a 10-inch pavement, the diameter is 5/4 = 1.25 inch. The dowel spacing and length are normally 12 inches and 18 inches, respectively.

**Jointed Pavements.** The value of "J" recommended for a Plain Jointed Pavement (JCP) or jointed reinforced concrete pavement (JRCP) with a load transfer device, such as dowel bars, at the joints is 3.2. This value is indicative of the load transfer of jointed pavements without tied concrete shoulders.

For jointed pavements without load transfer devices at the joints, a "J" value of 3.8–4.4 is recommended. This is for the higher bending stresses that develop in pavements without the dowel bars but considers the increased potential in developing faults. If the concrete has a high thermal coefficient, then the value of "J" should be increased. On the other hand, if few heavy trucks are anticipated such as a

---

**TABLE 14.2**

**Recommended Load Transfer Coefficient for Various Pavement Types and Design Conditions**

| Shoulder Load Transfer Devices | Asphalt | | Tied P.C.C. | |
|---|---|---|---|---|
| | Yes | No | Yes | No |
| Pavement Type | Load Transfer Coefficient | | | |
| Plain jointed and jointed reinforced | 3.2 | 3.8-4.4 | 2.5-3.1 | 3.6-4.2 |
| CRCP | 2.9-3.2 | N/A | 2.3-2.9 | N/A |

*Source:* Table 2.6 AASHTO Guide, 1993

low-volume road, the "J" value may be lowered since the loss of aggregate interlock will be less. AASHTO Guide, 1993 provides some general criteria for the consideration and/or design of expansion joints, contraction joints, longitudinal joints, load transfer devices, and tie bars in jointed pavements.

**Continuously Reinforced Pavements.** The value of "J" recommended for continuously reinforced concrete pavements (CRCP) without tied concrete shoulders ranges between 2.9 and 3.2, depending on the capability of aggregate interlock against the transverse cracks to transfer load in the future. Earlier a commonly used "J" value was 3.2 for CRCP, but for a better design for crack width control, each authority should determine proper values based on local aggregates and temperature ranges.

**Tied Shoulders or Widened Outside Lanes.** The use of tied PCC shoulders or widened outside lanes results in the reduction of slab stress and increased service life, this is one of the major advantages. Significantly lower "J" values may be used for the design of both jointed and continuous pavements.

For continuously reinforced concrete pavements with tied concrete shoulders, the range of "J" is between 2.3 and 2.9, with a recommended value of 2.6. This value is considerably lower than that for the design of concrete pavements without tied shoulders because of the significantly increased load distribution capability of concrete pavements with tied shoulders. The minimum bar size and maximum tie bar spacing should be the same as that for tie bars between lanes.

The recommended value of "J" should be between 2.5 and 3.1, for jointed concrete pavements with dowels and tied shoulders. A lower "J" value for tied shoulders may be considered when traffic is not permitted to run on the shoulder, by providing curbstones or otherwise.

NOTE: A concrete shoulder of width 3 feet should be a tied shoulder. Pavements with monolithic or tied curb and gutter that provide additional stiffness and keeps traffic away from the edge and may be considered as a tied shoulder.

### 14.5.3 LOSS OF SUPPORT

The factor "LS", in the design of rigid pavements considers the potential loss of support arising from subbase erosion and/or differential vertical soil movements. It is treated in the actual design procedure by lowering the effective k-value based on the size of the void that may develop beneath the slab. Table 14.3 provides the ranges of "LS" depending on the type of material for its stiffness or elastic modulus. For different types of base or subbase, the corresponding values of "LS" should be determined for each type.

The "LS" factor is considered for differential vertical soil movements, that may result in voids beneath the pavement. Thus, even though a non-erosive subbase is used, a void may still develop, which shortens the pavement life. Generally, for active swelling clays or excessive frost heave, LS values of 2.0–3.0 "E" is the elastic or resilient modulus of the material. Experience should be the key element in the selection of an appropriate "LS" value. The effect of "LS" on reducing the effective k-value of the roadbed soil Figure 14.4 may also help select an appropriate value.

**TABLE 14.3**
**Typical Ranges of Loss of Support (LS) Factors for Various Types of Materials**

| Type of Material | Loss of Support (LS) |
|---|---|
| Cement treated granular base (E = 1,000,000–2,000,000 psi) | 0.0–1.0 |
| Cement aggregate mixtures (E = 500,000–1,000,000 psi) | 0.0–1.0 |
| Asphalt treated base (E = 350,000–1,000,000 psi) | 0.0–1.0 |
| Bituminous stabilized mixtures (E = 40,000–300,000 psi) | 0.0–1.0 |
| Lime stabilized (E = 20,000–70,000 psi) | 1.0–3.0 |
| Unbound granular materials (E = 15,000–45,000 psi) | 1.0–3.0 |
| Fine grained or natural subgrade materials (E = 3,000–40,000 psi) | 2.0–3.0 |

*Source:* Table 2.7 AASHTO Guide, 1993

## 14.6   REINFORCEMENT VARIABLES

The reinforcement design procedures are different in jointed and continuous pavements, therefore, the design requirements for each are separated. Information is also provided here for the design of prestressed concrete pavement. In addition to dimensions, consideration is also given to corrosion resistance of reinforcement, especially in areas where pavements are exposed to variable moisture contents and salt applications.

### 14.6.1   JOINTED REINFORCED CONCRETE PAVEMENTS

In the "Jointed" category there are further two types of rigid concrete pavements, plain Jointed Concrete Pavement (JCP), which is designed without any steel reinforcement, and Jointed Reinforced Concrete Pavement (JRCP), which is designed to have significant steel reinforcement, in terms of either steel bars or welded steel mats. If the probability of transverse cracking during pavement life is high due to soil movement and/or temperature/moisture change stresses, additional steel reinforcement is required.

For the case of plain jointed concrete pavements (JCP), the joint spacing is so selected that temperature and moisture change stresses do not produce intermediate cracking between joints. The maximum joint spacing varies, depending on local conditions, subbase types, coarse aggregate types, etc., and by considering the minimum joint movement and maximum load transfer.

The criteria needed for the design of Jointed Reinforced Concrete Pavements (JRCP) are discussed here. These criteria apply to the design of both longitudinal and transverse steel reinforcement.

**Slab Length:** This refers to the joint spacing or length of pavement slab panel, "L" in feet, between free transverse joints. It is an important design consideration since it has a large impact on the maximum concrete tensile stresses and the amount of steel reinforcement required. Because of this effect, the slab length, which is the spacing between the joints, is considered in the design of reinforced or unreinforced jointed concrete pavement.

## TABLE 14.4
### Recommended Friction Factors

| Type of Material | Friction Factor (F) |
|---|---|
| Beneath Slab | |
| • Surface treatment | 2.2 |
| • Lime stabilization | 1.8 |
| • Asphalt stabilization | 1.8 |
| • Cement stabilization | 1.8 |
| • River gravel | 1.5 |
| • Crushed stone | 1.5 |
| • Sandstone | 1.2 |
| • Natura subgrade | 0.9 |

*Source:* Table 2.8 AASHTO Guide, 1993

**Steel Working Stress:** This refers to the allowable working stress, "fs" in psi, in the steel reinforcement. Typically, an allowable value of 75% of the steel yield strength is used for working stress. Therefore, for Grade 40 and Grade 60 steel, the allowable working stresses are 30,000 and 45,000 psi, respectively. For Welded Wire Fabric (WWF) and Deformed Wire Fabric (DWF), the steel yield strength is 65,000 psi and the allowable working stress is 48,750 psi. The minimum wire size should be adequate to prevent the significant impact of potential corrosion on the cross-sectional area.

**Friction Factor:** The friction factor "F" represents the frictional resistance between the bottom of the slab and with the underlying subbase or subgrade layer and is a coefficient of friction. The recommended values for natural subgrade and a variety of subbase materials are given in Table 14.4.

### 14.6.2 Continuously Reinforced Concrete Pavements (CRCP)

The principal steel reinforcement in Continuously Reinforced concrete pavements (CRCP) is the longitudinal steel which is provided "Continuous" throughout the length of the pavement. This longitudinal reinforcement is used to control cracks in the pavement due to volume change in the concrete.

The reinforcement is provided either by steel reinforcing bars or deformed steel wire fabric. It is the restraint of the concrete to fracture, due to the steel reinforcement and subbase friction. A balance between the properties of the concrete and the reinforcement must be maintained for the pavement to perform satisfactorily. The evaluation of this interaction is the basis for longitudinal reinforcement design.

The transverse reinforcement in a continuously reinforced concrete (CRCP) pavement is to control the width of any longitudinal cracks which may develop. Transverse reinforcement may not be required for CRC pavements in which no longitudinal cracking is likely to occur based on soil types, aggregate types, etc. However, if longitudinal cracking is expected, transverse reinforcement will restrain lateral movement and minimize the undesirable effects of a free edge. Transverse reinforcement should be designed based on the criteria and methodology same as for jointed pavements.

The following are the considerations for the design of longitudinal steel reinforcement in CRC pavements.

**Concrete Tensile Strength:** Two measures of concrete tensile strength are used in the design procedure. The modulus of rupture or flexural strength which is derived from a flexural beam test with third point loading is used to determine the required slab thickness. Steel reinforcement design is based on the tensile strength derived from the indirect tensile test which is covered under AASHTO T 198 and ASTM C 496 test specifications. The strength at 28 days should be used for both of these values. These two strengths should be consistent with each other. For this design procedure, the indirect tensile strength should be about 86% of concrete modulus of rupture.

**Concrete Shrinkage:** Drying shrinkage in the concrete caused by water loss is a significant factor in the reinforcement design. Other factors affecting shrinkage are cement content, chemical admixtures, curing method, aggregates, and curing conditions. The value of shrinkage at 28 days is the design shrinkage value.

Both shrinkage and strength of the concrete are strongly dependent upon the water-cement ratio. More the quantity of water added to a mix will increase the shrinkage and decrease the strength. Since shrinkage is inversely proportional to the strength, Table 2.9 provides the values corresponding to the indirect tensile strength as determined above.

**Concrete Thermal Coefficient:** The coefficient of thermal expansion for portland cement concrete varies with factors as water-cement ratio, concrete age, the richness of the mix, relative humidity, and the type of aggregate in the mix. The type of coarse aggregate has the most significant influence. Recommended values of PCC thermal coefficient, which is a function of aggregate type, are presented in Table 14.6.

---

**TABLE 14.5**

**Approximate Relationship between Shrinkage and Indirect Tensile Strength of Portland Cement Concrete**

| Indirect Tensile Strength (psi) | Shrinkage (in./in.) |
| --- | --- |
| 300 (or less) | 0.0008 |
| 400 | 0.0006 |
| 500 | 0.00045 |
| 600 | 0.0003 |
| 700 (or greater) | 0.0002 |

*Source:* Table 2.9 AASHTO Guide, 1993

---

## TABLE 14.6
## Recommended Value of the Thermal Coefficient of PCC as a Function of Aggregate Types

| Type of Coarse Aggregate | Concrete Thermal Coefficient (10-6/°F) |
| --- | --- |
| Quartz | 6.6 |
| Sandstone | 6.5 |
| Gravel | 6.0 |
| Granite | 5.3 |
| Basalt | 4.8 |
| Limestone | 3.8 |

*Source:* Table 2.10 AASHTO Guide, 1993

**Bar or Wire Diameter:** Typically, No. 5 and No. 6 deformed bars are used for longitudinal reinforcement in CRCP. The No. 6 bar is the largest practical size that should be used in CRCP to meet bond requirements and to control crack widths. The design nomographs for reinforcement have limited the bar selection to a range of No. 4–7. The nominal diameter of a reinforcing bar, in inches, is the bar number divided by 8. The wire diameter should be large enough so that any possible corrosion will not significantly reduce the cross section diameter. The relationship between longitudinal and transverse wire should also conform to manufacturers' recommendations.

**Steel Thermal Coefficient:** Unless otherwise specified a value of $5.0 \times 10{-6}$ in./in./°F may be assumed as the coefficient of thermal expansion of the reinforcing steel, for design purposes.

**Design Temperature Drop:** The temperature drop used in the reinforcement design is the difference between the average concrete curing temperature and a minimum temperature. The average concrete curing temperature may be taken as the daily average high temperature during the month of construction of the pavement. This average accounts for the heat of hydration. The design minimum temperature is the daily average low temperature for the coldest month during the pavement life. The temperature drop used in the design for longitudinal reinforcement may be determined by using the formula:

$$DT_D = T_H - T_L$$

where

$DT_D$ is the design temperature drop, °F

$T_H$ is the daily average high temperature during the month the pavement is constructed, °F

$T_L$ is the daily average low temperature during the coldest month of the year, °F

**Friction Factor:** The criteria for the selection of a slab-base friction factor for CRCP is the same as that for jointed pavements as mentioned in Section 2.6.1.

## 14.7   RIGID PAVEMENT DESIGN

This section describes the design for portland cement concrete pavements, which includes the types of Plain Jointed Pavement (JCP), Jointed Reinforced Concrete Pavements (JRCPs), and continuously reinforced concrete pavements (CRCP). As in the design for flexible pavements, it is made on the assumption that these pavements will carry traffic volumes above 50,000 18-kip ESAL during the performance period. An example of the application of this rigid pavement design procedure is presented in Appendix L of AASHTO Guide, 1993.

The AASHTO design procedure is based on the AASHO Road Test pavement performance steps. The use of dowels at transverse joints along the cross section of the roads is inherent in the use of this procedure. Therefore, joint faulting was not a distress consideration at the Road Test. If the design is done without considering dowel bars in joints, then an appropriate J-factor is to be determined (see paragraph 14.5.2, "Load Transfer") and check the design with design procedure adopted by others, such as the PCA procedure.

### 14.7.1   DETERMINING THE EFFECTIVE MODULUS OF SUBGRADE REACTION

Before applying the design chart for determining design slab thickness, it is necessary to estimate the possible levels of slab support that can be provided. This is accomplished by developing an effective modulus of subgrade reaction "k" by using Table 14.7 and Figures 14.1, 14.2, 14.3, and 14.4. An example of this process is demonstrated in Table 3.3 of the AASHTO Guide, 1993. The effective k-value depends upon several different factors other than the roadbed soil resilient modulus.

The first step is to identify the combinations (or levels) that are to be considered and enter them in the heading of Table 14.7 of AASHTO Guide.

- Subbase types are every different type of subbase that has different strength or modulus value. The selection of a subbase type in estimating an effective k-value has a significant effect on its cost-effectiveness as part of the design process.
- Subbase thicknesses (inches) is effective design thickness for each type of subbase should also be identified, in terms of its cost-effectiveness.
- Loss of support (LS) is a factor that is quantified in paragraph 14.5.3, is used to correct the effective k-value based on possible erosion of the subbase material.
- Depth to the rigid foundation (feet) in case of the existence of bedrock within a depth of 10 feet from the surface of the subgrade for a significant length in the project, its effect on the overall k-value and the design slab thickness for that length is to be considered.

## TABLE 14.7
## Table for Estimating Effective Modulus of Subgrade Reaction

| Trial Subbase Type | | | Depth to Rigid Foundation (feet) | | |
| --- | --- | --- | --- | --- | --- |
| Thickness (inches) | | | Projected Slab Thickness (inches) | | |
| Loss of Support, LS | | | | | |
| Month | Roadbed Modulus, $M_R$ (psi) | Subbase Modulus, $E_{SB}$ (psi) | Composite k-Value (pci) (Figure 3.4) | k-Value (pci) on Rigid Foundation (Figure 3.4) | Relative Damage, $u_r$ (Figure 3.5) |
| Jan. | | | | | |
| Feb. | | | | | |
| Mar. | | | | | |
| Apr. | | | | | |
| May | | | | | |
| June | | | | | |
| July | | | | | |
| Aug. | | | | | |
| Sept. | | | | | |
| Oct. | | | | | |
| Nov. | | | | | |
| Dec. | | | | | |

*Source:* Table 3.2 of AASHTO Guide, 1993
Summation: $\Sigma u_r =$
Average: $\bar{u}r = \Sigma u_r =$
Effective Modulus of Subgrade Reaction, k (pci) =
Corrected for Loss of Support: k (pci) =
**Example, $D_{SB}$ = 6 inches, $E_{SB}$ = 20,000 psi, MR = 7,000 psi, Solution: $k_\infty$ = 400 pci**

A separate table needs to be prepared to evaluate a combination of these factors and determine a corresponding effective modulus of subgrade reaction.

The second step of the process is to determine the seasonal roadbed soil resilient modulus values (from paragraph 13.4.1) and enter them in Column 2 of each table. Here also, if the length of the smallest season is one half-month, then all seasons must be defined in terms of consecutive half-month time intervals in the table.

The third step is to assign season wise subbase elastic (resilient) modulus ($E_{SB}$) values. These values, which were discussed in Section 13.4.3, should be entered in

Example:

$D_{SB}$ = 6 inches

$E_{SB}$ = 20,000 psi

$M_R$ = 7,000 psi

Solution: $k_\infty$ = 400 pci

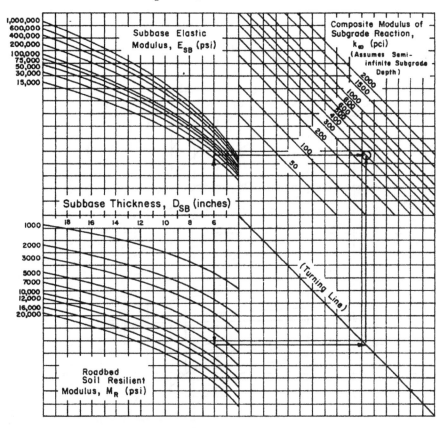

**FIGURES 14.1**  Chart for Estimating Composite Modulus of Subgrade Reaction.
*Source:* Figure 3.3 of AASHTO Guide, 1993

Column 3 of Table 14.7 and should correspond to those for the seasons used to develop the roadbed soil resilient modulus values. For those types of subbase material which are insensitive to season (e.g., cement-treated material), a constant value of subbase modulus may be assigned for each season. For the unbound materials which are sensitive to season values for $E_{SB}$ of 50,000 psi and 15,000 psi may be used for the frozen and spring thaw periods, respectively and the $E_{SB}$ ratio of the subbase to the roadbed soil resilient modulus should not exceed 4 to avoid an artificial condition.

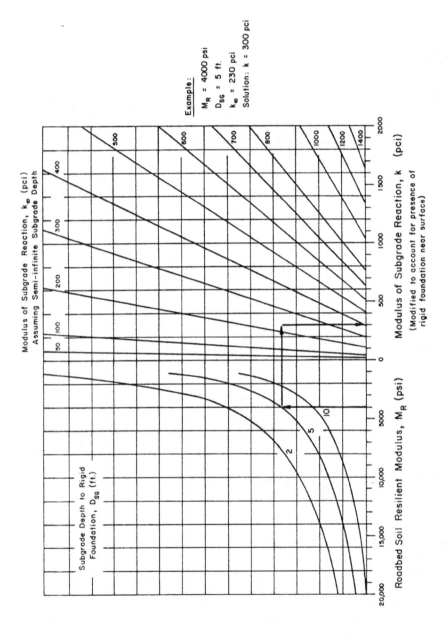

**FIGURES 14.2**   Chart to Modify Modulus of Subgrade Reaction to Consider Effects of Rigid Foundation Near Surface (<10 ft.).
*Source:* Figure 3.4 of AASHTO Guide, 1993

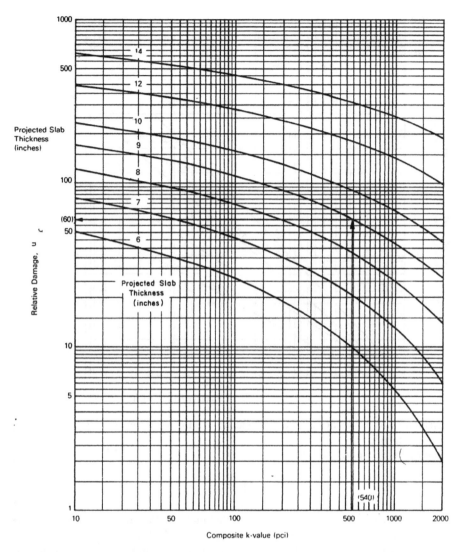

**FIGURES 14.3** Chart of Estimating Relative Damage to Rigid Pavements Based on Slab Thickness and Underlying Support.
*Source:* Figure 3.5 of AASHTO Guide, 1993

The fourth step is to determine the composite modulus of subgrade reaction for each season and enter in Column 4, by assuming a semi-infinite subgrade depth, where the depth of bedrock is greater than 10 feet. This is accomplished with the aid of Figure 14.1. Note that the starting point in this chart is subbase thickness, $D_{SB}$. If the slab is placed directly on the subgrade (i.e., no subbase), the composite modulus of subgrade reaction is determined by using the following theoretical relationship between k-values from a plate bearing test and elastic modulus of the roadbed soil:

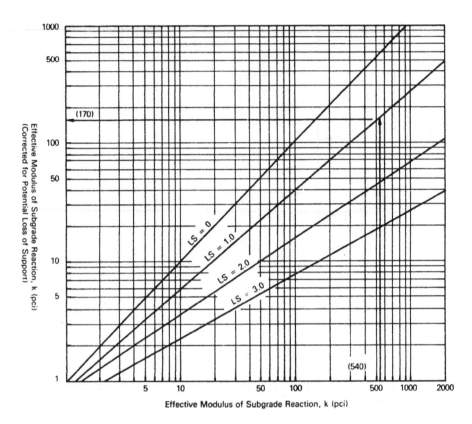

**FIGURES 14.4**   Correction of Effective Modulus of Subgrade Reaction for Potential Loss of Subbase Support.
*Source:* Figure 3.6 of AASHTO Guide, 1993

$$k = M_R / 19.4$$

The fifth step is to determine a k-value that includes the effect of a rigid foundation near the surface. This step is avoidable if the depth to a rigid foundation is greater than 10 feet. Figure 14.2 provides the chart that may be used to estimate this modified k-value for each season. It considers roadbed soil resilient modulus and composite modulus of subgrade reaction, as well as the depth to the rigid foundation. The values for each modified k-value should subsequently be entered in Column 5 of Table 14.7.

The sixth step in the process is to determine the thickness of the slab that will be required, and then use Figure 14.3 to determine the relative damage, $u_r$, in each season and enter the values in Column 6 of Table 3 of the AASHTO Guide, 1993.

The seventh step is to add all the $u_r$ values in Column 6 and divide the total by the number of seasonal increments (12 or 24) to determine the average relative damage, $u_r$. The effective modulus of subgrade reaction, then, is the value corresponding to the average relative damage and projected slab thickness in Figure 14.3.

**TABLE 14.8**

**Example Application of Method for Estimating Effective Modulus of Subgrade Reaction**

Trial Subbase Type: Granular
Depth to Rigid Foundation (feet): 5
Thickness (inches): 6
Projected Slab Thickness (inches): 9
Loss of Support, LS: 1.0

| (1) | (2) | (3) | (4) | (5) | (6) |
|-----|-----|-----|-----|-----|-----|
| Month | Roadbed Modulus, MR (psi) | Subbase Modulus, ESB (psi) | Composite k-Value (pci) (Figure 3.3) | k-Value (pci) on Rigid Foundation (Figure 3.4) | Relative Damage, $u_r$ (Figure 3.5) |
| Jan. | 20,000 | 50,000 | 1,100 | 1,350 | 0.35 |
| Feb. | 20,000 | 50,000 | 1,100 | 1,350 | 0.35 |
| Mar. | 2,500 | 15,000 | 160 | 230 | 0.86 |
| Apr. | 4,000 | 15,000 | 230 | 300 | 0.78 |
| May | 4,000 | 15,000 | 230 | 300 | 0.78 |
| June | 7,000 | 20,000 | 410 | 540 | 0.60 |
| July | 7,000 | 20,000 | 410 | 540 | 0.60 |
| Aug. | 7,000 | 20,000 | 410 | 540 | 0.60 |
| Sept. | 7,000 | 20,000 | 410 | 540 | 0.60 |
| Oct. | 7,000 | 20,000 | 410 | 540 | 0.60 |
| Nov. | 4,000 | 15,000 | 230 | 300 | 0.78 |
| Dec. | 20,000 | 50,000 | 1,100 | 1,350 | 0.35 |

*Source:* Table 3.3 of AASHTO Guide, 1993
Summation: Eu = 7.25
Average: $u_r = \Sigma u_r/n = 7.25/12 = 0.60$
Effective Modulus of Subgrade Reaction, K (pci) = 540
Corrected for Loss of Support: k (pci) = 170

The eighth and final step is to adjust the effective modulus of subgrade reaction for the potential loss of support arising from subbase erosion. Figure 14.4 provides the chart for correcting the effective modulus of subgrade reaction based on the loss of support factor, LS, determined in paragraph 14.5.3. Space is provided in Table 14.7 of the AASHTO Guide, 1993 to enter the final design k-value.

### 14.7.2 Determine Required Slab Thickness

Figures 14.5 and 14.6, in its two parts, present the nomograph which is used for determining the slab thickness for each effective k-value identified in the previous section. The designer may then select the optimum combination of slab and subbase thicknesses based on economics and other requirements. Generally, the layer thickness is rounded to the nearest inch, but the use of controlled grade slip from pavers may permit half-inch increments. In addition to the design k-value, other inputs required by this rigid pavement design nomograph include:

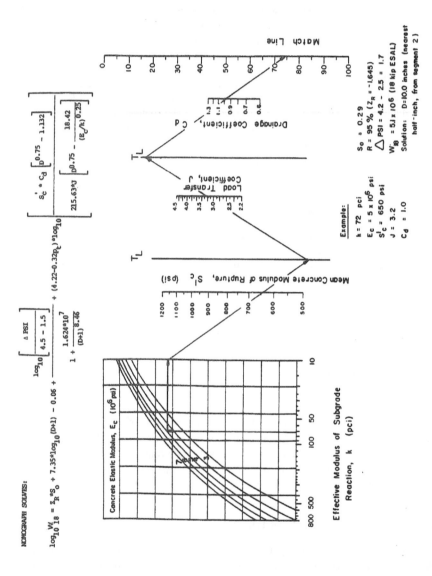

**FIGURE 14.5** (1) Design Chart for Rigid Pavement Based on Using Mean Values for Each Input Variable.
*Source:* Figure 3.7 (Segment 1) of the AASHTO Guide, 1993. (2) Continued – Design Chart for Rigid Pavement Based on Using Mean Values for Each
Input Variable. (*Source:* Figure 3.7 (Segment 2) of the AASHTO Guide, 1993)

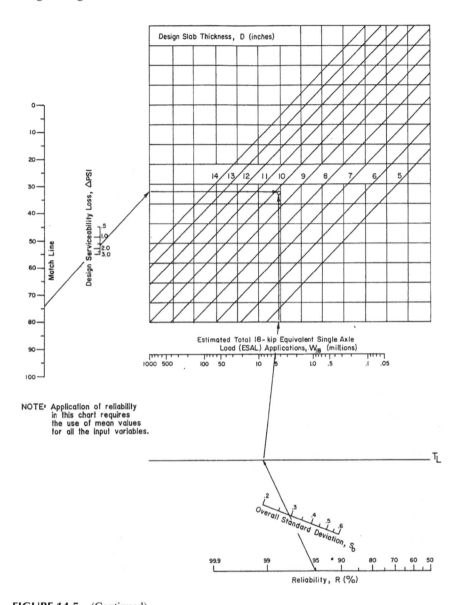

**FIGURE 14.5** (Continued)

- Estimated future traffic, $W_{18}$ (Section 13.2.2), for the performance period.
- Reliability, R (Section 13.2.3).
- Overall standard deviation, $S_o$ (Section 13.2.3).
- Design serviceability loss, $\Delta PSI = p_o - p_t$ (Section 13.3.1).
- Concrete elastic modulus, $E_c$ (Section 13.4.3).
- Concrete modulus of rupture, $S'_c$ (Section 14.4.4).
- Load transfer coefficient, J (Section 14.5.2).
- Drainage coefficient, $C_d$ (Section 14.5.1).

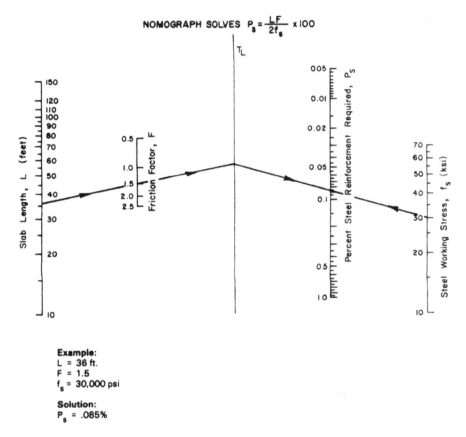

**FIGURE 14.6**   Reinforcement Design Chart for Jointed Reinforced Concrete Pavements.
*Source:* Figure 3.8 of AASHTO Guide, 1993

### 14.7.3   STAGE CONSTRUCTION

A practical maximum performance period (Section 2.2.1) for a given rigid pavement is when it is subjected to some significant level of truck traffic. To consider analysis periods that are longer than this maximum expected performance period or to more rigorously consider the life-cycle costs of rigid pavement designs which are initially thinner, it is necessary to consider the stage construction (planned rehabilitation) approach in the design process. It is also important to recognize the need to compound the reliability for each stage of the strategy. For example, if both stages of a two-stage strategy (an initial PCC pavement with one overlay) have 90% reliability, the overall reliability of the design strategy would be $0.9 \times 0.9$ or 81%. Conversely, if an overall reliability of 95% is desired, the individual reliability for each stage must be $(0.95)^{1/2}$ or 97.5%.

To evaluate secondary stages of such stage construction alternatives, the user should refer to Part III of the AASHTO Guide, 1993 which addresses the design for pavement rehabilitation. That part not only provides a procedure for designing overlays but also provides criteria for the application of other rehabilitation methods that may be used to improve the serviceability and extend the load-carrying capacity of the pavement. The design example in Appendix I illustrates the application of the stage construction approach using a planned future overlay.

### 14.7.4 ROADBED SWELLING AND FROST HEAVE

The approach to considering the effects of swelling and frost heave in rigid pavement design is almost identical to that for flexible pavements (Section 3.2.3). Thus, some of the discussion is repeated here.

Roadbed swelling and frost heave are both important environmental considerations because of their potential effect on the rate of serviceability loss. Swelling refers to the localized volume changes that occur in expansive roadbed soils as they absorb moisture. A drainage system can be effective in minimizing roadbed swelling if it reduces the availability of moisture for absorption.

Frost heave, as it is treated here, refers to the localized volume changes that occur in the roadbed as moisture collects, freezes into ice lenses, and produces distortions on the pavement surface. Like swelling, the effects of frost heave can be decreased by providing some type of drainage system. Perhaps a more effective measure is to provide a layer of nonfrost-susceptible material thick enough to insulate the roadbed soil from frost penetration. This not only protects against frost heave but also significantly reduces or even eliminates the thaw-weakening that may occur in the roadbed soil during early spring.

If either swelling or frost heave is to be considered in terms of their effects on serviceability loss and the need for future overlays, then the following procedure should be applied. It requires the plot of serviceability loss versus time developed in Section 2.2.4.

The procedure for considering environmental serviceability loss is similar to the treatment of stage construction strategies because of the planned future need for rehabilitation. In the stage construction approach, an initial PCC slab thickness is selected and the corresponding performance period (service life) determined. An overlay (or series of overlays) that will extend the combined performance periods past the desired analysis period is then identified.

The difference in the stage construction approach when swelling and/or frost heave are considered is that an iterative process is required to determine the length of the performance period for each stage of the strategy. The objective of this iterative process is to determine when the combined serviceability loss due to traffic and environment reaches the terminal level. This is described with the aid of Table 14.9.

The first step is to select an appropriate slab thickness for the initial pavement. For the relatively small effect slab thickness has on minimizing swelling and frost heave, the maximum initial thickness recommended is derived by assuming no swelling or frost heave. Referring to the example problem presented in Figure 14.5(1) and 14.5(2), the maximum slab thickness is considered as 9.5 inches. Any practical slab thickness less than this value may be justified for swelling or frost heave conditions if it does not violate the minimum performance period (paragraph 13.2.1).

## TABLE 14.9

**Example of Process Used to Predict the Performance Period of an Initial Rigid Pavement Structure Considering Swelling and/or Frost Heave**

Slab Thickness (inches) = 9.5
Maximum Possible Performance Period (years) = 20
Design Serviceability Loss, $\Delta PSI = p_i - p_t = 4.2 - 2.5 = 1.7$

| (1) | (2) | (3) | (4) | (5) | (6) |
|-----|-----|-----|-----|-----|-----|
| Iteration No. | Trial Performance Period (years) | Total Serviceability Loss Due to Swelling and Frost Heave $\Delta PSI_{SW,FH}$ | Corresponding Serviceability Loss Due to Traffic $\Delta PSI_{TR}$ | Allowable Cumulative Traffic (18-kip ESAL) | Corresponding Performance Period (years) |
| 1 | 14.0 | 0.75 | 0.95 | $3.1 \times 10^6$ | 9.6 |
| 2 | 11.8 | 0.69 | 1.01 | $3.3 \times 10^6$ | 10.2 |
| 3 | 11.0 | 0.67 | 1.03 | $3.4 \times 10^6$ | 10.4 |

*Source:* Table 3.4 of the AASHTO Guide, 1993

It is important to note here that for this example, an overall reliability of 90% is desired. Since it is expected that one overlay will be required to reach the 20-year analysis period, the individual reliability that must be used for the design of both the initial pavement and the overlay for two-stage construction is $0.90^{1/2}$ or 95%.

In the second step select a trial performance period that is expected under the swelling/frost heave conditions and enter in Column 2. This number should be less than the maximum possible performance period corresponding to the selected initial slab thickness. The greater environmental loss results smaller the performance period.

In the third step by using the graph of cumulative environmental serviceability loss versus time described in paragraph 13.2.4 Ref. Figure 13.2 is used as an example), obtain the corresponding total environmental serviceability loss due to swelling and frost heave ($\Delta PSI_{sw,FH}$) which is expected in the trial period in the first step and enter in Column 3.

In the fourth step Subtract this environmental serviceability loss (referring to the third step) from the desired total serviceability loss (4.2–2.5 = 1.7 as used in the example) to establish the corresponding traffic serviceability loss and enter in Column 4.

$$\Delta PSI_{TR} = \Delta PSI - \Delta PSI_{swFH}$$

In the fifth use Figure 14.5(1) and 14.5(2) to estimate the allowable cumulative 18-kip ESAL traffic corresponding to the traffic serviceability loss determined in the fourth step and enter in Column 5. It is important to use the same levels of reliability, effective modulus of subgrade reaction, etc., when applying the rigid pavement design chart to estimate the allowable traffic.

| Column No. | Description of Procedures |
|---|---|
| 2 | Estimated by the designer (Step 2). |
| 3 | Using the estimated value from Column 2 with Figure 13.2, the total serviceability loss due to swelling and frost heave is determined (Step 3). |
| 4 | Subtract environmental serviceability loss (Column 3) from design total serviceability loss to determine corresponding serviceability loss due to traffic. |
| 5 | Determined from Figure 14.3 keeping all inputs constant (except for use of traffic serviceability loss from Column 4) and applying the chart in reverse (Step 5). |
| 6 | Using the traffic from Column 5, estimate net performance period from Figure 13.1 (Step 6). |

In the sixth step estimate the corresponding year at which the cumulative 18-kip ESAL traffic (determined in the fifth step) will be reached and enter in Column 6. This should be accomplished with the aid of the cumulative traffic versus time plot developed in Section 13.2.2. (Figure 13.1 as Figure 2.1 of the AASHTO Guide, 1993 is used as an example.)

In the seventh step compare the trial performance period with that calculated in the sixth step. If the difference is greater than 1 year, calculate the average of the two and use this as the trial value for the start of the next iteration (restarting from the second step). If the difference is less than 1 year, convergence is reached, and the average is said to be the predicted performance period of the initial pavement structure corresponding to the selected design slab thickness. In the example, convergence was reached after three iterations and the predicted performance period is about 10.5 years.

The basis of this iterative process is the same for the estimation of the performance period of any subsequent overlays. The major differences in actual application are:

- The overlay design methodology presented in Part III, AASHTO Guide, 1993 is used to estimate the performance period of the overlay.
- Any swelling and/or frost heave losses predicted after overlay should restart from the point when the overlay was placed.

## 14.8 RIGID PAVEMENT JOINT DESIGN

The design considerations for the different types of joints in Portland cement concrete pavements are discussed here. This criterion applies to the design of joints in both jointed and continuous pavements.

### 14.8.1 JOINT TYPES

To allow for the expansion and contraction of the pavement, joints are provided in concrete pavements, these release stresses from environmental changes (i.e., temperature and moisture), friction, and facilitate construction. There are three general types of joints: contraction, expansion, and construction. These joints and their functions are as follows:

- Contraction joints are provided to prevent the tensile stresses due to temperature, moisture, and friction, thus controlling cracking. If contraction joints were not provided, random cracking would occur on the surface of the pavement.
- An expansion joint is to provide space for the expansion of the pavement, thus preventing the development of compressive stresses, which can cause the pavement to buckle.
- Construction joints are required to facilitate multi-session construction. The spacing between longitudinal joints is dictated by the width of the paving machine and by the pavement thickness.

### 14.8.2  JOINT GEOMETRY

The joint geometry is considered in terms of the spacing and general layout.

**Joint Spacing** of both transverse and longitudinal contraction joints depends on local conditions of materials and environment, whereas expansion and construction joints are primarily dependent on layout and construction capabilities.

For contraction joints, the spacing to prevent intermediate cracking decreases as the thermal coefficient, temperature change, or subbase frictional resistance increases; and the spacing increases as the concrete tensile strength increases. The spacing also is related to the slab thickness and the joint sealant capabilities. Local experience must be considered since a change in coarse aggregate type may have a significant effect on the concrete thermal coefficient and consequently, the acceptable joint spacing. As a general guideline, the joint spacing (in feet) for plain concrete pavements should not greatly exceed twice the slab thickness (in inches). Therefore, the maximum joint spacing for an 8-inch slab is 16 feet. The ratio of slab width to length should not exceed 1.25.

Expansion joints are considered less in a project due to cost, complexity, and performance problems. They are used where pavement types change for example, from CRCP to jointed, in prestressed pavements, and at intersections.

The construction joints have spacing which is generally decided for field placement and equipment capabilities. Longitudinal construction joints should be placed at lane edges to maintain pavement smoothness and minimize load transfer problems. Transverse construction joints occur at the end of a day's work or in connection with equipment breakdowns.

## 14.9  RIGID PAVEMENT REINFORCEMENT DESIGN

The purpose of steel reinforcement in reinforced concrete pavement is not to prevent cracking, but to hold the cracks tightly closed, and maintaining the pavement as an integral structural unit. The physical mechanism through which cracks develop is adversely affected by (1) temperature and/or moisture related slab contractions, and (2) frictional resistance from the underlying material.

As temperature or moisture content decreases, the slab tends to contract. This contraction is resisted by the underlying material through its friction and shear with the slab. The restraint to this contraction results in tensile stresses which are maximum at the mid-slab. If these tensile stresses exceed the tensile strength of the concrete, a crack will develop, and all the stresses are transferred to the steel reinforcement. Thus, the reinforcement must be designed for these stresses without any significant elongation which may cause excessive crack width.

The longitudinal steel reinforcement requirements between jointed reinforced (JRCP) and continuously reinforced concrete pavement (CRCP) are significantly different, the reinforcement designs are treated separately. The design for transverse steel in CRCP is the same as the design for longitudinal and transverse steel reinforcement in JRCP. In general, the amount of reinforcement required is specified as a percentage of the concrete cross-sectional area.

### 14.9.1 JOINTED REINFORCED CONCRETE PAVEMENTS

The nomograph for estimating the percent of steel reinforcement required in a jointed reinforced concrete pavement is presented in Figure 14.6. The inputs required include:

- Slab length, L (paragraph 14.6.1).
- Steel working stress, fs (paragraph 14.6.1).
- Friction factor, F (paragraph 14.6.1).

This chart applies to the design of transverse steel reinforcement in both jointed and continuously reinforced concrete pavements, as well as to the design of longitudinal steel reinforcement in JRCP. Normally for joint spacing, less than 15 feet transverse cracking is not anticipated; therefore, the steel reinforcement would not be required.

### 14.9.2 CONTINUOUSLY REINFORCED CONCRETE PAVEMENTS

The design of longitudinal reinforcing steel in continuously reinforced concrete pavements is discussed in here. The design procedure presented here may be systematically performed using the worksheet in Table 14.10. In this table, space is provided for entering the relevant design inputs, intermediate results, and calculations for determining the required percentage for the longitudinal steel. A separate worksheet, presented in Table 14.11, is used for design revisions. The examples used reinforcing bars; the use of deformed wire fabric (DWF) is also acceptable.

The design inputs required by this procedure are as follows:

| | |
|---|---|
| Concrete indirect tensile strength, $f_t$ | (Section 2.6.2) |
| Concrete shrinkage at 28 days, Z | (Section 2.6.2) |
| Concrete thermal coefficient, $\alpha_c$ | (Section 2.6.2) |
| Reinforcing bar or wire diameter, $\phi$ | (Section 2.6.2) |
| Steel thermal coefficient, $a_s$ | (Section 2.6.2) |
| Design temperature drop, $DT_D$ | (Section 2.6.2) |

## TABLE 14.10
### Worksheet for Longitudinal Reinforcement Design

#### Design Inputs

| Input Variable | Value | Input Variable | Value |
|---|---|---|---|
| Reinforcing bar/wire diameter, $\phi$ (inches) | | Thermal coefficient ratio, $\alpha_s/\alpha_c$ (in./in.) | |
| Concrete shrinkage Z (in.in.) | | Design temperature drop, $DT_D$ (°F) | |
| Concrete tensile strength, $f_t$ (psi) | | Wheel load stress, $\sigma_w$ (psi) | |

#### Design Criteria and Required Steel Percentage

| | | Crack Spacing, $\times$ (feet) | AllowableCrack Width, $CW_{max}$ (inches) | Allowable Steel Stress, $(\sigma s)_{max}$ (ksi) | |
|---|---|---|---|---|---|
| Value of liming criterial | | Max. 8.0 Min. 3.5 | | | |
| Minimum required steel percentage | | | | | $(P_{min})$* |
| Maximum allowable steel percentage | | | | | $(P_{max})$** |

*Source:* Table 3.5 of the AASHTO Guide, 1993
\* Enter the largest percentage across line
\*\* If $P_{max} < P_{min}$ then reinforcement criteria are in conflict, design not feasible

## TABLE 14.11
### Worksheet for Revised Longitudinal Reinforcement Design

| | Change in Value from Previous Trial | | | | |
|---|---|---|---|---|---|
| Parameter | Trial 2 | Trial 3 | Trial 4 | Trial 5 | Trial 6 |
| Reinforcing bar/wire diameter, $\phi$ (inches)[a] | | | | | |
| Concrete shrinkage,Z (in./in.) | | | | | |
| Concrete tensile strength,$F_t$ (psi)[a] | | | | | |
| Wheel load stress,$\sigma_w$ (psi) | | | | | |
| Design temperature drop, $DT_D$ (°F)[b] | | | | | |
| Thermal coefficient ratio, $a_s/a_c$ | | | | | |
| Allowable crack width criterion, $CW_{max}$ (inches) | | | | | |
| Allowable steel stress criterion, $(\sigma_s)$ max (ksi) | | | | | |
| Required steel % for crack spacing (min.) | | | | | |
| Required steel % for crack spacing (max.) | | | | | |
| Minimum required steel % for crack width | | | | | |
| Minimum required steel % for steel stress | | | | | |
| Minimum % reinforcement, $P_{min}$ | | | | | |
| Minimum % reinforcement, $P_{max}$ | | | | | |

*Source:* Table 3.6 of the AASHTO Guide, 1993
[a] Change in this parameter will affect steel stress criterion
[b] Change in this parameter will affect crack width criterion

These data should be entered in the space provided in the upper part of Table 14.10.

The input for wheel load tensile stress developed during initial loading of the constructed pavement by either construction equipment or truck traffic is also required. Figure 14.7 may be used to estimate this wheel load stress based on the design slab thickness, the magnitude of the wheel load, and the effective modulus of subgrade reaction. This value should also be entered in the space provided in Table 14.10.

The design of longitudinal reinforcing steel has three limiting criteria which must be considered for the design of longitudinal reinforcing steel: crack spacing, crack

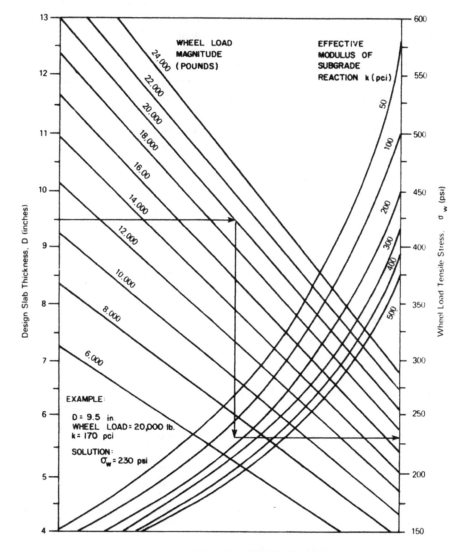

**FIGURE 14.7**   Chart for Estimating Wheel Load Tensile Stress.
*Source:* Figure 3.9 of AASHTO Guide, 1993

width, and steel stress. Acceptable limits of these are mentioned below to ensure that the pavement will respond satisfactorily under the anticipated environmental and vehicular loading conditions.

The limits on crack spacing are determined by considering spalling and punchouts. To minimize the incidence of crack spalling, the maximum spacing between consecutive cracks should not be more than 8 feet. To minimize the potential for the development of punchouts, the minimum desirable crack spacing should be 3.5 feet. These limits are already given in Table 14.10.

The limiting criterion on crack width is based on a consideration of spalling and water percolation. The allowable crack width should not exceed 0.04 inches. In the final determination of the longitudinal steel percentage, the consideration for crack width should be reduced as much as possible through the selection of either a higher percentage of steel or smaller diameter reinforcing bars.

Limiting criteria on steel stress are to prevent steel fracture and excessive permanent deformation. To prevent steel fracture, limiting the stress of 75% of the ultimate tensile strength is considered. The conventional limit on steel stress is 75% of the yield point to ensure that the steel does not undergo any plastic deformation.

AASHTO mentioned that, based on experience, many miles of CRC pavements have performed satisfactorily even though the steel stress was predicted to be above the yield point. This led to reconsideration of this criteria and allowance for a small amount of permanent deformation

Values of allowable mean working stress for steel are listed in Table 14.12, as a function of reinforcing bar size and concrete strength. The indirect tensile strength should be that as determined in paragraph 14.6.2. The limiting working stresses in steel in Table 14.12, are for the Grade 60 steel (meeting **ASTM** A 615 specifications) recommended for longitudinal reinforcement in CRC pavements. Once the allowable steel working stress is determined, it should be entered in the space provided in Table 14.10.

**Design Procedure:** The procedure to determine the required amount of longitudinal reinforcement is described below:

---

### TABLE 14.12
### Allowable Steel Working Stress, ksi (10)

| Indirect Tensile Strength of Concrete at 28 Days, psi | Reinforcing Bar Size* | | |
|---|---|---|---|
| | No. 4 | No. 5 | No.6 |
| 300 (or less) | 65 | 57 | 54 |
| 400 | 67 | 60 | 55 |
| 500 | 67 | 61 | 56 |
| 600 | 67 | 63 | 58 |
| 700 | 67 | 65 | 59 |
| 800 (or greater) | 67 | 67 | 60 |

*Source:* Table 3.7 of the AASHTO Guide, 1993

\* For DWF proportional adjustments may be made using the wire diameter to bar diameter

---

Refer to Table 14.11, Worksheet for Revised Longitudinal Reinforcement
    Design
Refer to Table 14.12, Allowable Steel Working Stress, in ksi

The first step is determining the required amount of steel reinforcement to satisfy
each limiting criterion by using the design charts in Figures 14.8, 14.9, and 14.10.
Enter the resulting steel percentages in the spaces provided in Table 14.10. The second step is to check that if Pmax is greater than or equal to Pmin, then go to the third
step. If Pmax is less than Pmin, then:

- Check the design inputs and identify which input is to be revised.
- Indicate the revised design inputs in the worksheet in Table 14.11. Make any
  corresponding change in the limiting criteria as influenced by the change in
  design parameter and enter that in Table 14.11. Check to see if the revised
  inputs affect the subbase and slab thickness design. It may be necessary to re-
  evaluate the design thickness of the subbase and slab.
- Rework with the design nomographs by entering the resulting steel percentages in Table 14.11.
- If Pm, is greater than or equal to Pmin, go to the third step. If Pmax is less than
  Pmin, then repeat this second step by using the space provided in Table 14.11
  for additional trials.

The third step is to determine the range in the number of reinforcing bars or wires
required:

$$N_{min} = 0.01273 \times P_{min} \times W_s \times D/\phi^2$$

$$N_{max} = 0.01273 \times P_{max} \times W_s \times D/\phi/^2$$

where

$N_{min}$ = minimum required number of reinforcing bars or wires
$N_{max}$ = maximum required number of reinforcing bars or wires
$P_{min}$ = minimum required percent steel
$P_{max}$ = maximum required percent steel
$W_s$ = total width of pavement section (inches)
$D$ = thickness of the concrete layer (inches)
$\phi$ = reinforcing bar or wire diameter (inches), which may be increased if
        the loss of cross section is anticipated due to corrosion.

The fourth step is to determine the final steel design by selecting the number of
reinforcing bars or wires $N_{Design}$, in the final design section, such that $N_{Design}$, is a
whole integer number between $N_{min}$ and $N_{max}$. The correctness of these final design
alternatives may be checked by converting the whole integer number of bars or wires
to percent steel and working backward through the design charts to obtain the corresponding crack spacing, crack width, and steel stress.

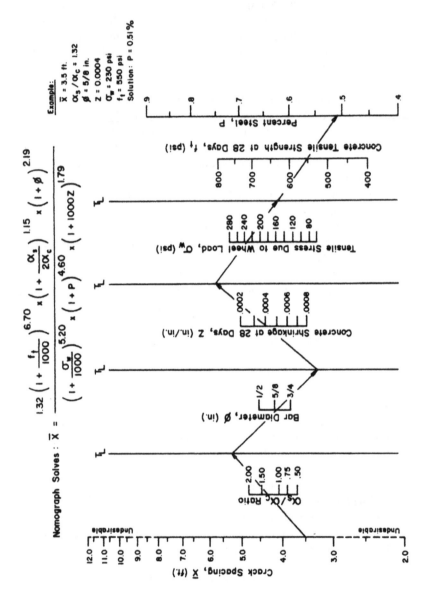

**FIGURE 14.8**  Percent of Long. Reinforcement to Satisfy Crack Spacing Criteria.
*Source:* Figure 3.10 of AASHTO Guide, 1993

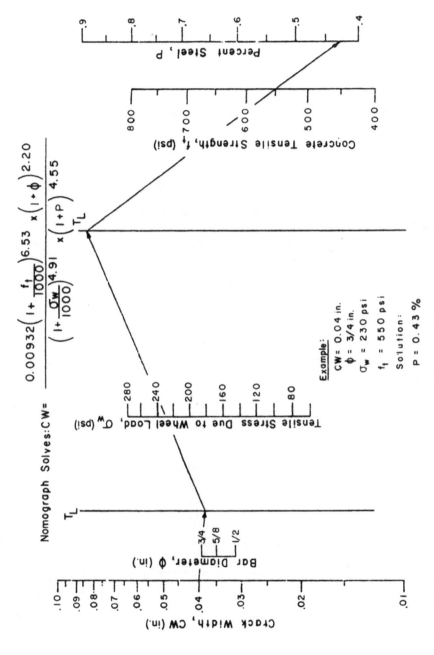

**FIGURE 14.9**   Minimum Percentage Long. Reinforcement to Satisfy Crack Width Criteria.
*Source:* Figure 3.11 of AASHTO Guide, 1993

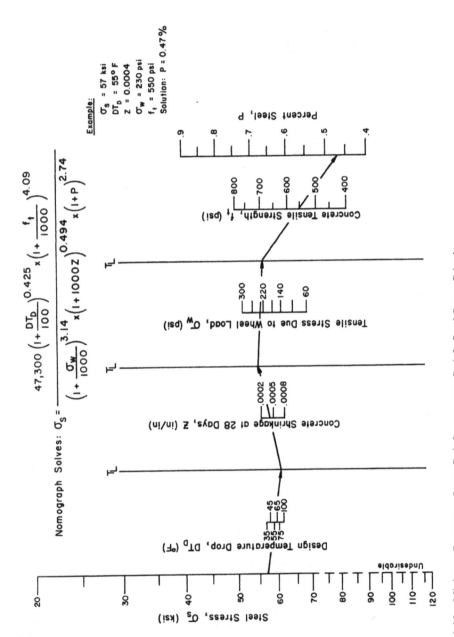

**FIGURE 14.10**  Minimum Percentage Long. Reinforcement to Satisfy Steel Stress Criteria.
*Source:* Figure 3.12 of AASHTO Guide, 1993

**Design Example:** The following example is provided to demonstrate the CRCP longitudinal reinforcement design procedure. Two trial designs are evaluated; the first considers 5/8-inch (No. 5) reinforcing bars and the second trial design examines 3/4-inch (No. 6) bars. Below are the input requirements selected for this example. These values are also recorded for both of the trial designs in the example worksheets presented in Tables 14.13 and 14.14.

- Concrete tensile strength, $f_t$ = 550 psi. This is approximately 86% of the modulus of rupture used in the slab thickness design example, in Figure 14.5(1) and 14.5(2).
- Concrete shrinkage, Z = 0.0004 in./in., This corresponds to the concrete tensile strength, in Table 14.3.
- Wheel load stress, $\sigma_W$ = 230 psi. This is based on the earlier slab thickness design example, a 9.5-inch slab with a modulus of subgrade reaction equal to 170 pci, in Figure 14.7.
- The ratio of steel thermal coefficient to that of Portland cement concrete, $\sigma_s/\sigma_c$ = 1.32, For steel, the thermal coefficient is $5 \times 10^{-6}$ in./ in./15°F. (See paragraph 14.6.2.) Assume limestone coarse aggregate in concrete, therefore, the thermal coefficient is $3.8 \times 10^{-6}$ in./in./°F in Table 14.5.
- Design temperature drop, $DT_D$ = 55°F. (Assume high temperature is 75°F and low is 20°F.)

The limiting criteria corresponding to these design conditions are as follows:

---

**TABLE 14.13**

**Example Application of Worksheet for Longitudinal Reinforcement Design**

| Design Inputs | | | |
|---|---|---|---|
| **Input Variable** | **Value** | **Input Variable** | **Value** |
| Reinforcing bar/wire diameter, $\phi$ (inches) | 5/8 (No. 5) | Thermal coefficient ratio, $\sigma_s/\sigma_c$ (in./in.) | 1.32 |
| Concrete shrinkage Z (in. in.) | 0.0004 | Design temperature drop, $DT_D$ (°F) | 55 |
| Concrete tensile strength, $f_t$ (psi) | 550 | Wheel load stress, $\sigma_w$ (psi) | 230 |
| Design Criteria and Required Steel Percentage | | | |
| | Crack Spacing, x (feet) | Allowable Crack Width, $CW_{max}$ (inches) | Allowable Steel Stress, $(\sigma_s)_{max}$ (ksi) |
| Value of liming criteria | Max. 8.0 Min. 3.5 | 0.04 | 62 |
| Minimum required steel percentage | <0.40% | <0.40% | <0.43% $(P_{min})^*$ |
| Maximum allowable steel percentage | 0.51% | | 0.51% $P_{max}^{**}$ |

*Source:* Table 3.8 of the AASHTO Guide, 1993

* Enter the largest percentage across line

** If $P_{max} < P_{min}$, then reinforcement criteria are in conflict, design not feasible

---

**TABLE 14.14**

**Example Application of Worksheet for Revised Longitudinal Reinforcement Design**

| Parameter | Change in Value from Previous Trial | | | | |
|---|---|---|---|---|---|
| | Trial 2 | Trial 3 | Trial 4 | Trial 5 | Trial 6 |
| Reinforcing bar/wire diameter, $\phi$ (inches)[a] | ¾ (No. 6) | | | | |
| Concrete shrinkage, Z (in./in.) | 0.004 | | | | |
| Concrete tensile strength, $F_t$ (psi)[a] | 550 | | | | |
| Wheel load stress, $\sigma$w (psi) | 230 | | | | |
| Design temperature drop, DTD (°F)[b] | 550 | | | | |
| Thermal coefficient ratio, $a_s/a_c$ | 1.32 | | | | |
| Allowable crack width criterion, $CW_{max}$ (inches) | 0.04 | | | | |
| Allowable steel stress criterion, $(\sigma_s)$max (ksi) | 57 | | | | |
| Required Steel % for Crack Spacing     Min. | <0.04% | | | | |
| Max. | 0.57% | | | | |
| Minimum required steel % for crack width | 0.43% | | | | |
| Minimum required steel % for steel stress | 0.47% | | | | |
| Minimum % reinforcement, $P_{min}$ | 0.47% | | | | |
| Minimum % reinforcement, $P_{max}$ | 0.57% | | | | |

*Source:*  Table 3.9 of the AASHTO Guide, 1993
[a]  Change in this parameter will affect steel stress criterion
[b]  Change in this parameter will affect crack width criterion

- Allowable crack width, CW: 0.04 inch for both trial designs. (See Section 3.4.2, "Continuously reinforced concrete pavements; Limiting criteria.")
- Allowable steel stress, $\sigma_s$: 62 ksi for ⁵/₈-inch bars (Trial 1) and 57 ksi for ³/₄-inch bars. Table 14.12 (Table 3.7 of the AASTHO Guide, 1993) using tensile strength of 550 psi.

Application of the design nomographs in Figures 14.8, 14.9, and 14.10 yields the following limits on steel percentage for the two trial designs:

$$\text{Trial Design} 1 : P_{min}, = 0.43\%, P_{max}, = 0.51\%$$

$$\text{Trial Design} 2 : P_{min} = 0.47\%, P_{max} = 0.57\%$$

The range ($N_{min}$ to $N_{max}$) of the number of reinforcing bars requires (assuming a 12-foot-wide lane) for each trial design is

$$\text{Trial Design} 1 \left(\text{No.5 bars}\right): \quad N_{min} = 19.2, \quad N_{max} = 22.7$$

$$\text{Trial Design} 2 \left(\text{No.6 bars}\right): \quad N_{min} = 14.6, \quad N_{max} = 17.6$$

Using twenty No. 5 bars for Trial 1 (P = 0.45%) and fifteen No. 6 bars for Trial 2 (P = 0.48%), the longitudinal reinforcing bar spacings would be 7.2 and 9.6 inches, respectively. The predicted crack spacing, crack width, and steel stress for these two trial designs are:

| Predicted Response | Trial Design 1 (20 No. 5 Bars) | Trial Design 2 (15 No. 6 Bars) |
|---|---|---|
| Crack Spacing, × (feet) | | |
| Crack Width, | | |
| Steel Stress, | | |

Inspection of these results indicates that there is no significant difference in the predicted response of these two designs such that one should be selected over the other. Thus, in this case, the selection should be based on economics and/or ease of construction.

### 14.9.3 TRANSVERSE REINFORCEMENT

Transverse steel is included in either jointed or continuous pavements for conditions where soil volume changes (due to changes in either temperature or moisture) can result in longitudinal cracking. Steel reinforcement will prevent the longitudinal cracks from opening excessively, thereby maintaining maximum load transfer and minimizing water entry.

If transverse reinforcement and/or tie bars are desired, then the information collected under Section 2.6.1, "Reinforcement Variables for Jointed Reinforced concrete pavements," is applicable. In this case, the "slab length" should be considered as the distance between free longitudinal edges. If tie bars are placed within a longitudinal joint, then that joint is not a free edge. For normal transverse reinforcement, Figure 14.4 may be used to determine the percent transverse steel. The percent transverse steel may be converted to spacing between reinforcing bars as follows:

$$Y = \frac{A_s}{P_t D} \times 100$$

where

$Y$ = transverse steel spacing (inches), (in.2)
$A_s$ = cross-sectional area of transverse reinforcing steel
$P_t$ = percent transverse steel
$D$ = slab thickness (inches)

Figures 14.11 and 14.12 may be used to determine the tie bar spacing for $1/_2$- and $5/_8$-diameter deformed bars, respectively. The designer enters the figure on the horizontal with the distance to the closest free edge axis and proceeds vertically to the pavement thickness obtained from Section 3.3.2, "Determine Required Slab Thickness." From the pavement thickness, move horizontally and read the tie bar spacing from the vertical scale. These nomographs are based on Grade 40 steel and a subgrade friction factor of 1.5.

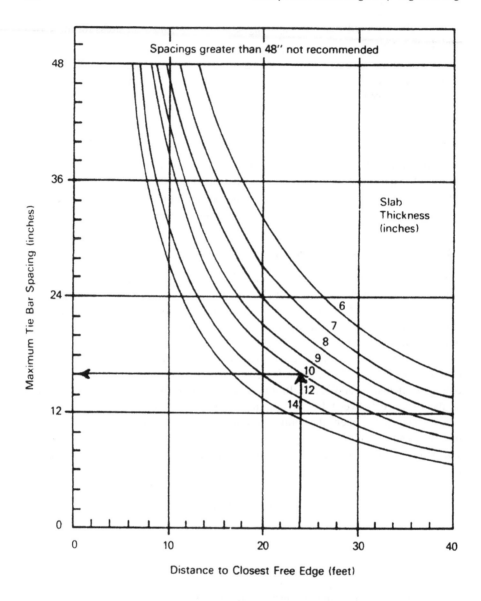

**FIGURE 14.11**   Recommended Maximum Tie Bar Spacings for PCC Pavements Assuming ½-inch Diameter Tie Bars, Grade 40 Steel, and Subgrade Friction Factor of 1.5.
*Source:* Figure 3.13 of AASHTO Guide, 1993

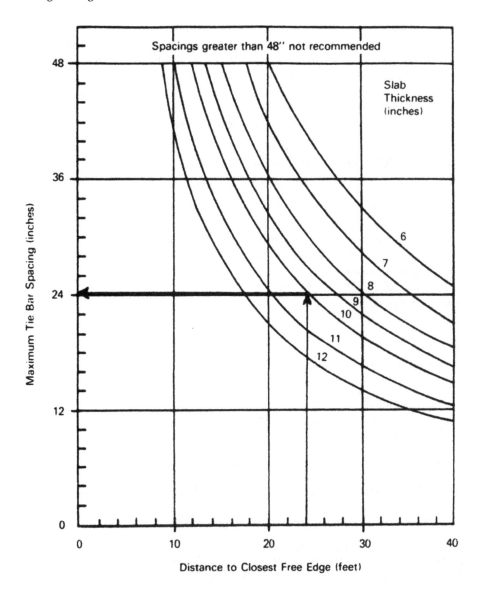

**FIGURE 14.12** Recommended Maximum Tie Bar Spacings for PCC Pavements Assuming 5/8-inch Diameter Tie Bars, Grade 40 Steel, and Subgrade Friction Factor of 1.5.
*Source:* Figure 3.14 of AASHTO Guide, 1993

Note that since steel stress decreases from a maximum near the center of the slab (between the free edges) to zero at the free edges, the required minimum tie bar spacing increases. Thus, to design the tie bars efficiently, the designer should first select the layout of the longitudinal construction joints.

Finally, if the bending of the tie bars is to be permitted during construction, then to prevent steel failures, the use of brittle (high carbon content) steels should be avoided and an appropriate steel working stress level selected.

## 14.10 COMPUTER-AIDED DESIGN OF RIGID CONCRETE PAVEMENT IN AASHTO METHOD

The computer applications for the "Design of Rigid Pavement" are explained in Chapter 14 in the "The Guide for Computer Application Tutorials" part of the book available for download from the website: www.roadbridgedesign.com.

To download the items under book tutorials, visit the website at: www. roadbridgedesign.com.

For downloading the other items, visit the website at: www.techsoftglobal.com/download.php.

### BIBLIOGRAPHY

*AASHTO Guide for Design of Pavement Structures 1986*, AASHTO.
*AASHTO Guide for Design of Pavement Structures 1993*, AASHTO.
*Manual on Pavement Design*, Arahan Teknik (Jalan) 5/85, JKR, Malaysia.
*Guidelines for the Design of Flexible Pavements*, IRC 37-2001, IRC.
*Guidelines for the Design of Plain Jointed Rigid Pavements for Highways*, IRC 58-2002, IRC.
*Tentative Guidelines for Strengthening of Flexible Pavements Using Benkelman Beam Deflection Technique*, IRC 81-1997, IRC.
*Guidelines for the Design of Flexible Pavements for Low Volume Rural Roads*, IRC SP 72-2007, IRC.
*Pavements and Surfacings for Highways and Airports* by Michel Sargious, by Applied Science Publishers Ltd.
*Thickness Design, Asphalt Pavement Structures for Highways and Streets*, The Asphalt Institute.
*Asphalt Pavements for Airports*, The Asphalt Institute.
*Highway Engineering* by Paul H. Wright and Karen K. Dixon, by John Wiley & Sons. Inc.
*Principles of Highway Engineering and Traffic Analysis* by Fred L. Mannering, Walter and Scott, by Wiley India.

# 15 Design of Highway Drainage and Drainage Structures

## 15.1 GENERAL

An effective drainage system is essential on a highway, not only to prevent the accumulation of storm water on the carriageway and causing hazards to the traffic, but also to protect the pavement structure from damage by submerging into the stagnant water.

## 15.2 SURFACE DRAINAGE

The features of a highway structure makes surface drainage effective in non-urban locations, including the roadway crown, shoulder and side slopes, longitudinal side drains, culverts, and bridges. Highways with a divided carriageway in urban areas are provided with inlets (underground pipes) for storm water drains, as the cover drains under the sidewalks handle the surface flow.

### 15.2.1 PAVEMENT AND SHOULDER CROSS-SLOPES

The cross slope of the carriageway is provided in two ways, as a normal camber on straight sections and as super elevation on curved sections. The cross slope on super elevated sections are always either equal to or more than on sections with a normal camber. The cross slope on super elevated sections is determined with respect to the road geometrics and traffic speed, whereas a cross slope with a normal camber is provided by considering the intensity of rainfall and terrain classification in the region. Cumulative lengths of straight sections are much longer than the cumulative length of curved sections. Therefore, the cross slope on straight sections with a normal camber mainly governs the effective drainage of storm water from the surface of the road carriageway.

If the terrain is flat, then in most cases the longitudinal slope of the road section is nearly flat to provide easy access to the users from roads joining the highway. In such cases the cross slope of the carriageway is important to ensure effective surface drainage and the drain must be meticulously designed along the slope.

On curved sections the storm water from the outer carriageway is collected by the drain at the median and from there the storm water is further taken to the side drain on the inner carriageway side. Here the drain on the outer carriageway side is not considered for collecting the storm water from the carriageway surface.

## TABLE 15.1
## Recommended Ranges of Cross-Slopes for Pavements and Shoulders

| Carriageway Element | Range of Cross Slope (%) |
|---|---|
| High-type surface (Concrete/Mastic carriageway) | |
| • Two lanes | 1.5–2.0 |
| • Three or more lanes in each direction | 1.5 minimum; increase 0.5–1.0% per lane, 4.0 max. |
| Intermediate surface | 1.5–3.0 |
| • Low-type surface | 2.0–6.0 |
| • Urban arterials | 1.5–3.0; increase 1.0% per lane |
| Shoulders | |
| • Bituminous or concrete | 2.0–6.0 |
| • With kerbs | ≥4.0 |

*Source:* Drainage of Highway Pavements, Highway Engineering Circular No. 12,Federal Highway Administration, Washington, DC (March 1984).

Recommended ranges of pavement cross-slopes are given in Table 15.1. The amount of cross slope varies with the type of surface, being generally small for relatively impervious surfaces such as concrete pavements and large for pervious surfaces such as bituminous with gravel on earth.

Shoulders are normally sloped to drain away from the pavement surface. Precipitation that occurs on the shoulder area largely flows to the side ditches. The Table 15.1 recommends that shoulder cross-slopes vary from about 2 to 6% depending on the type of surface, and whether curbs are provided.

### 15.2.2 SIDE SLOPES AND SIDE DITCHES

Open side ditches are generally provided in cut sections at highway locations in rural areas or hill stretches to provide for surface drainage. Side ditches may also be constructed along embankment sections when natural drainage channels are inadequate. Both flat-bottomed and V-section ditches are used, with preference being given to the former type, with slope changes in the ditch section provided to improve appearance and prevent erosion. In either case, side slopes are made as flat as possible, consistent with drainage requirements and limiting widths of right-of-ways. Deep and narrow side ditches are avoided whenever possible because of the increased hazards in the adjoining areas and for stability reasons.

In case such ditches are used, adequate measures should be taken for safeguarding traffic through use of traffic barriers. Warrants for traffic barriers, as well as recommended front slopes and back slopes for ditches, are to be provided as per relevant design standard.

Water flowing along side drains is generally in a direction parallel to the roadway centerline. Grades used in open side drains are usually the same as the grades of the highway centerline. Sometimes flat roadway grades and side drains with steeper grades are very frequently used. In very flat countries, ditch grades as low as V1–2 m:H1000 m may be used, while in rolling or mountainous terrain the maximum grade may be determined only by the necessity for preventing erosion.

Side drains are for the removal of surface water from within the limits of the highway right of way. In some situations, storm water from the ground adjacent to the right of way may also come into the drains. The water is suggested to be carried to an outlet, in the form of either a natural or an artificial drainage channel. The side drains must be hydraulically capable of handling the estimated flow of surface water in such a way that the roadway structure is not endangered, or the safety of the motorist threatened. The hydraulic design of side ditches and other open channels is described later in this chapter.

## 15.3 DESIGN OF SURFACE DRAINAGE SYSTEMS

The design of surface drainage systems for a highway may be divided into three major phases:

- Phase one – Estimating the quantity of water that is expected to reach any element of the system.
- Phase two – Hydraulic design of each element of the system.
- Phase three – Comparison of the design system with alternative systems, alternative materials, and other variables to select and provide the most economical system, by considering the lowest annual cost when all variables are taken into consideration.

### 15.3.1 HYDROLOGICAL APPROACHES AND CONCEPTS

There are various methods used to estimate the quantity of runoff for a drainage design. In the case of some culverts and bridges, to handle the flow of an existing stream, the flow used for a hydraulic design is commonly based on the available records for that stream by using the statistical analyses on the recorded stream flow to provide an estimated peak design flow for a given "return period".

The term "return period" means the estimated frequency of rare occurrences such as floods. Selection of the frequency of occurrence in the design for storms is largely a matter of experience and judgment, despite existing guidelines to establish the interval to be used for a given situation. The return period is normally based on statistics. For example, if the system is designed for a return period of 25 years, the statistical assumption is by considering the most severe storm occurred in last 25 years. Considering a return period of 100 years instead of 25 would include a more severe storm in the last 100 years and, hence the design would be for a more costly system. Conversely, if the return period is 10 years, the design for intensity of the less serious storm would result a less costly drainage system. In such cases, although there would be savings in construction costs using the short return period, there would be a risk of more economic losses during severe storms.

For interstate highways and principal arterials, design is done for a return period of 50 years. Collector roads and streets are commonly designed for a return period of 10–25 years, while low traffic roads may be designed for a return period of ten or even 5 years.

Generalized flood-frequency studies are required by the various state highway or irrigation departments and are to be continuous. The studies are to be based on statistical analysis of stream flow records. Usually, the mean annual flood as a function of the size of the drainage area for each hydrologic region within the state. Factors for computing floods of a return period in terms of the mean annual flood should be provided as a graph. The regional flood curves usually cover areas for surface drainage for about 100–2000 square kilometers. Separate flood curves should be given for the larger rivers, which cross regional and state boundaries as these rivers hardly follows the behavior of the smaller rivers.

If the required stream flow information is not available, then those may be obtained by observations of existing culverts, bridges, drains, and the natural stream. Drainage installations in the proposed location may be studied and a design based on those is expected to give satisfactory service for their purposes.

Stream gage data may be used for particular regions to develop statistical regression equations for most areas in the country. These equations normally require parameters such as the drainage area and the average slope of the stream. With such data, the estimate of peak design flows for "ungaged" sites may be determined, within the hydrologic region. This approach is explained in paragraph 15.3.6 of this chapter.

Studies are also required on urban regions where growth and development has resulted in unexpected failures of highway drainage facilities. Regression models for urban runoff should include, in general order of significance, equivalent rural discharge, a "basin development factor", the area contributing toward drainage, slope of terrain, rainfall intensity, storage, and the amount of impervious area. Estimating the runoff from urban watersheds is described in paragraph 15.3.7 of this chapter.

### 15.3.1.1  Flood Hydrograph

A flood wave passing a point along a stream follows a common pattern known as a flood hydrograph or synthetic unit hydrograph (SUH) as presented in Figure 15.1. The flow increases to a maximum point and then recedes.

Traditionally engineers used to design the highway drainage facilities to accommodate the peak flow. If the upstream storage is taken into account, the required size of the hydraulic structure may be significantly reduced.

In urban areas, storage is considered as an integral part of the drainage design. Storage has the effect of broadening and flattening the flood hydrograph and decreasing the potential for downstream flooding resulting in the significant reduction of the sizes of the hydraulic structures.

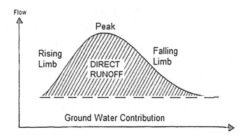

**FIGURE 15.1**  Flood Hydrograph.
*Source:* Federal Highway Administration, Washington

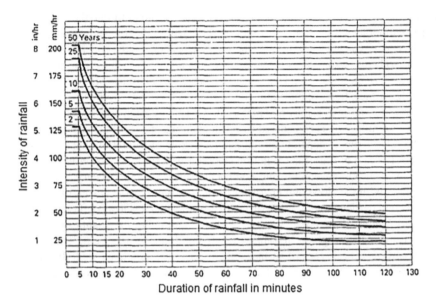

**FIGURE 15.2** Typical Rainfall Intensity – Duration Curves.
*Source:* Federal Aviation Administration, Washington

## 15.3.2 RAINFALL INTENSITY

The estimation of peak runoff for a drainage design is determined by the consideration of severe storms that occur at intervals of the return period and during which the intensity of rainfall and runoff of surface water are much greater than at other periods. Rainfall intensity during the design storm is a function of occurrence, duration, and intensity.

The intensity of rainfall for a particular return period varies greatly with the duration of rainfall. This is illustrated in Figure 15.2. The average rate for a short time – such as five min – is much greater than for a longer period, such as 1 hr. The duration chosen corresponds to the time of concentration in the design of many highway drainage systems, this is explained in paragraph 15.3.4.

A near to accurate estimate of the probable intensity, frequency, and duration of rainfall in a particular location can be made only if sufficient data is possible to be collected over a specified period of time. If such information is available, standard curves may be developed to express rainfall intensity relationships with an accuracy sufficient for drainage problems. The concerned departments should have developed such curves for use in specific areas.

## 15.3.3 SURFACE RUNOFF

When rain falls on a pervious surface, part of it passes into the soil and the remainder disappears over a period of time either by evaporation, runoff, or both. In the design of highway drainage systems, the amount of water lost by evaporation is negligible and drainage must be provided for all rainfall that does not infiltrate the soil, or is not stored temporarily in depressions on the surface within the drainage area.

The rate at which water infiltrates the soil is dependent on the type and gradation of the soil, soil covers, moisture content of the soil, amount of organic material in the soil, temperature of the air, soil, and water, and the presence or absence of impervious layers under the surface. Rates of infiltration on bare soil is less than on surfaces by vegetation turf. Frozen soil is impervious, and rain infiltrates very little until the frozen layer melts. The rate of infiltration is considered as constant during any specific design storm.

The rate of runoff depends on the nature, degree of saturation, and slope of the surface. The rate of runoff is greater on smooth surfaces and is slower where vegetation is present. On pavements and compacted surfaces runoff occurs at a high rate that varies with the slope and character of the surface. For the design of a drainage system for a particular area, these variables are considered, and a coefficient of runoff selected. Values of the coefficient of runoff for estimating the quantity of flow (the rational method) are given in Table 15.2. If the drainage area being considered is composed of several types of surfaces, then separate coefficients are to be taken for each surface and the coefficient for the entire area computed as a weighted average of the individual areas.

### 15.3.4  TIME OF CONCENTRATION

After obtaining the return period for design storm frequency, computations are made to determine the duration of the rainfall that produces the maximum rate of runoff. The duration of rainfall which results in the maximum rate of runoff is the "time of concentration". The time of concentration usually consists of the time of flow on the

---

### TABLE 15.2
### Coefficients of Runoff for the Rational Formula

| Type of Surface of the Drainage Area | Coefficients of Runoff 'C' |
|---|---|
| Rural Areas | |
| • Concrete or mastic asphalt pavement | 0.8–0.9 |
| • Asphalt macadam pavement | 0.6–0.8 |
| • Gravel roadways or shoulders | 0.4–0.6 |
| • Bare earth | 0.2–0.9 |
| • Steep grassed areas (IV:2H) | 0.5–0.7 |
| • Turf with grass | 0.1–0.4 |
| • Forested areas | 0.1–0.3 |
| • Cultivated fields | 0.2–0.4 |
| Urban Areas | |
| • Flat residential, with about 30% of area impervious | 0.40 |
| • Flat residential, with about 60% of area impervious | 0.55 |
| • Moderately steep residential, with about 50% of area impervious | 0.65 |
| • Moderately steep residential, with about 70% of area impervious | 0.80 |
| • Flat commercial, with about 90% of area impervious | 0.80 |

*Source:* Table 12.2, "Highway Engineering" by Paul H. Wright and Karen K. Dixon, by John Wiley & Sons, Inc.

surface and the time of flow in the drainage system. The time of flow on the surface is the time required for water to flow from the farthest point in any section of the drainage area being considered to the point of the drainage system, and the time of flow in the drainage system from the intake to the point being considered must also be added. The time of surface flow varies with the slope, type of surface, length, and other factors. When the particular drainage area consists of several types of surfaces, the time of surface flow is determined by adding together the respective times computed for flow over the lengths of the various surfaces from the most remote point to the inlet. Estimates of time of flow in the drainage system can be made from observed or computed velocities of flow.

## 15.3.5 The Rational Method

For estimating runoff from a drainage area, the rational method is one of the most common methods because it combines engineering judgment with calculations made from analysis, measurement, or estimation. The method is based on the direct relationship between rainfall and runoff.

In conventional U.S. units, the equation is:

$$Q = C \times I \times A \tag{15.1}$$

where

$Q$ = runoff (ft$^3$/sec)
$C$ = a coefficient representing the ratio of runoff to rainfall; typical value of C are given in Table 15.2.
$I$ = intensity of rainfall (in. /hr for the estimated time of concentration)
$A$ = drainage area in acres; the area may be determined from field surveys, topographic maps, or aerial photographs

In metric units, the equation becomes:

$$Q\left(m^3/\sec\right) = \frac{CIA}{360} \tag{15.2}$$

where

$I$ is expressed in millimeters per hour
$A$ is in hectares

The limitation of this method is that it is confined to relatively small drainage areas of up to 200 acres (approximately 80 hectares), according to America's Federal Highway Administration (FHWA).

## 15.3.6 Estimating Runoff from Large Rural Drainage Basins

The process summarizes regression equations for estimating peak flows at "ungaged" sites for a specified return period based on certain hydrological, topographical, and meteorological characteristics of the region. The most frequently used watershed and climatic characteristics are drainage area, main-channel slope, and main annual rain fall. The regression equations are generally reported in the following form:

$$Q_r = a \times A^b \times S^c \times P^d \qquad (15.3)$$

where

    $Q_r$ = rural flood peak discharge for a recurrence interval of "T"
    $A$ = drainage area
    $S$ = channel slope
    $P$ = mean annual precipitation
    $a, b, c, d$ = regression coefficients

Examples of these equations are given in Table 15.3. These equations are intended only for rural locations and should not be used in urban areas unless the effects of urbanization are insignificant. The equations should not be used where dams and other flood-detention structures have a significant effect on peak discharge.

## 15.4   DESIGN OF SIDE DITCHES AND OTHER OPEN CHANNELS

### 15.4.1   The Manning's Formula

The design of side ditches, gutters, stream channels and similar facilities is based on established principles of flow in open channels, which also apply to flow in conduits with a free water surface. Most commonly used for design is Manning's formula, which applies to conditions of steady flow in a uniform channel and has the following form:

$$V = \frac{1.486 R^{2/3} S^{1/2}}{n} \qquad (15.4)$$

where

    $V$ = mean velocity (m/sec)
    $R$ = hydraulic radius (m); this is equal to the area of the cross section of flow
        ($m^2$) divided by the water perimeter (m)
    $S$ = slope of the channel (m per foot)
    $n$ = Manning's roughness coefficient; typical values of n are given in Table 15.3

The conversion factor of 1.186 does not appear, in SI units, the Manning's formula becomes:

$$V = \frac{R^{2/3} S^{1/2}}{n} \qquad (15.5)$$

where

    $V$ = mean velocity (m/sec)
    $R$ = hydraulic radius (m); this is equal to the area of the cross section of flow
        ($m^2$) divided by the wetted perimeter (m)

S = slope of the channel (m/m)
n = Manning's roughness coefficient

Where the equation of continuity is applicable as:

$$Q = V \times A \qquad (15.6)$$

where

Q = discharge (m³/sec)
A = area of the flow cross section (m²)

A set of charts has been published by America's Federal Highway Administration for the solution of Manning's equation for various common channel cross sections.

---

## TABLE 15.3
## Values of Manning's Roughness Coefficient

| Type of Channel or Structure | Values of 'n' |
|---|---|
| Open Channels for Type of Lining Shown | |
| • Smooth Concrete | 0.013 |
| • Rough Concrete | 0.022 |
| • Riprap | 0.03–0.04 |
| • Asphalt, smooth texture | 0.013 |
| • Good stand, any grass – depth of flow more than 6 in. | 0.09–0.30 |
| • Good stand, any grass – depth of flow less than 6 in. | 0.07–0.20 |
| • Earth, uniform section, clean | 0.016 |
| • Earth, fairy uniform section, no vegetation | 0.022 |
| • Channels not maintained, dense weeds | 0.08 |
| Natural Stream Channels (Surface Width at Flood Stage in 100 ft) | |
| Fairly regular section | 0.030–0.035 |
| Some grass and weeds, little or no brush | 0.035–0.05 |
| Dense growth of weeds, depth of flow materially greater than weed height | 0.035–0.05 |
| Some weeds, light brush on banks | 0.035–0.05 |
| Some weeds, heavy brush on banks | 0.05–0.07 |
| Some weeds, dense willows on banks | 0.06–0.08 |
| For trees within channel, with branches submerged at high stage, increase all above values by | 0.01–0.02 |
| Irregular sections, with pools, slight channel meander, increase values given above about | 0.01–0.02 |
| Culverts | |
| Concrete pipe and boxes | 0.012 |
| Corrugated Metal | |
| Unpaved | 0.024–0.027 |
| Paved 25% | 0.021–0.026 |
| Fully paved | 0.012 |

---

*Source:*  Table 12.5 (part), "Highway Engineering" by Paul H. Wright and Karen K. Dixon, by John Wiley & Sons. Inc.

## 15.4.2  ENERGY OR HEAD OF FLOW

The flowing water in open channels and culverts contains energy in two forms, potential and kinetic. The potential energy at a given point is represented by the depth of water plus the elevation of the bottom of the channel above some convenient datum plane, expressed in meters. The kinetic energy is represented by the velocity head, $V_1{}^2/2g$ (m). Bernoulli's energy equation states that the total energy or "head" at one point in an open channel carrying a flow of water is equal to the total energy "head" when allowance is made for any energy added to or taken away from the system. The energy equation is stated as:

$$d_1 + V_1{}^2/2g + Z_1 - \Sigma H_L = d_2 + V_1{}^2/2g + Z_1 \qquad (15.7)$$

where

d          = depth of flow, meters
$V_1^2/2g$ = velocity head or kinetic energy, meters
Z          = elevation of channel bottom, meters
$\Sigma H_L$ = summation of head losses, meters

In open channel problems, the energy content is considered with respect to the channel bottom. This is equal to the depth of water plus the velocity head, $d = V^2/2g$, and it is called the specific energy or "head". Figure 15.3 shows the relationship of the specific head and the depth of a given discharge in a given channel that can be placed with different slopes. The specific head values are plotted along vertical or Y-axis, and the corresponding depths are shown along horizontal or X-axis. The straight diagonal line is drawn at 45 degrees through the points where depth and specific head are equal. The line, therefore, represents the potential energy. The ordinate interval between this line and the specific head curve is the velocity head for the particular depth. The low point in the specific head curve gives the flow with minimum energy content, the depth at this point is the critical depth (dc), and the corresponding velocity is the critical velocity (Vc).

The following equation allows the calculation of critical depth of flow in nonrectangular channels:

$$\frac{Q^2}{g} = \frac{A^3}{b} \qquad (15.8)$$

where

Q  = flow in m³/sec
A  = the cross-sectional area in m²
b  = width of the channel at the surface in m
g  = 9.8 m/sec²

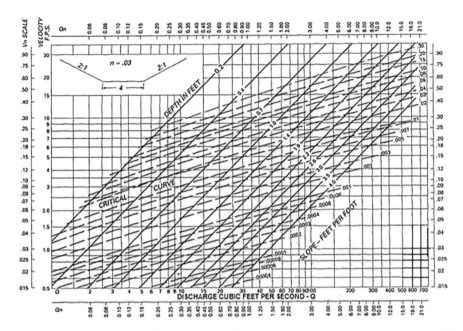

**FIGURE 15.3** Graphical Solution of Manning's Equation for a Channel with Trapezoidal Section Having 2H:1V Side Slopes, Bottom Width of "b = 4 ft", Manning's Roughness Coefficient of 0.03.
*Source:* Design Charts for Open Channel Flow, Federal Highway Administration, Washington, DC, 2001

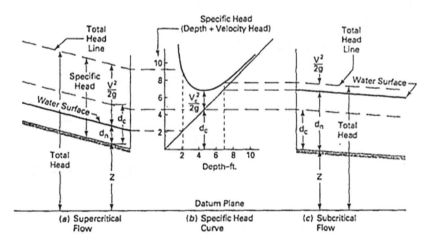

**FIGURE 15.4** Design Chart for Open Channel Flow.
*Source:* Design Charts for Open Channel Flow, Federal Highway Administration, Washington, DC, 2001

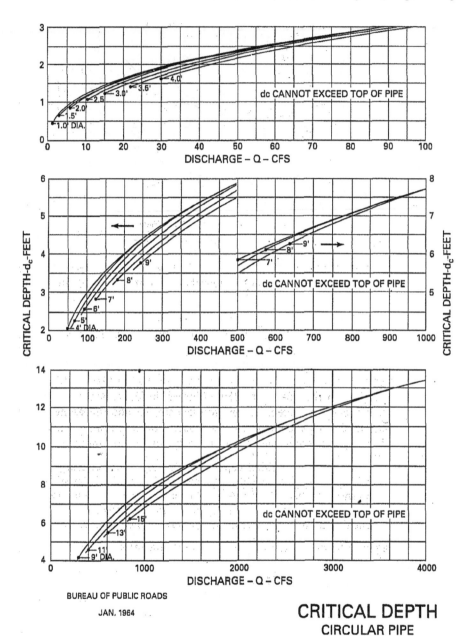

FIGURE 15.5  Critical Depths in Circular Pipes for Various Discharges.
*Source:* Hydraulic Design of Highway Culverts, Federal Highway Administration, Washington, DC, 2001

## 15.5  DESIGN OF CULVERTS

More frequently, culverts are differentiated from bridges on the basis of span length. On an arbitrary basis, structures having a span of 6 meters (approx. 20 ft) or less are called culverts, whereas those having spans of more than 6 meters (approx. 20 ft) are called bridges.

A culvert is differentiated from a bridge by the fact that the top of the culvert does not always form part of the traveled roadway, in most cases these have an earth cushion and the pavement structure is on top, to provide the carriageway for traffic.

Culverts are provided at locations: (i) at the bottom of depressed natural ground falling in the alignment of the road, where no natural watercourse exists, (ii) where natural streams intersect the roadway and (iii) at locations required to discharge the storm water from side drains coming as surface drainage, of the roads to.

Commonly used culvert shapes are illustrated in Figure 15.6, while four standard inlet types are shown in Figure 15.7.

### 15.5.1  Guidelines to Decide Culvert Location

The culverts are commonly installed in natural watercourses that cross the roadway, either at right angles or on a skew. In addition to selecting the proper location for the culvert crossing the centerline of the road, the alignment, grade, and invert level of the culvert are also important.

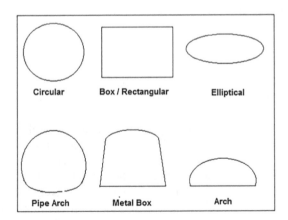

**FIGURE 15.6**  Common Configuration of Culverts.
*Source:* Introduction to Highway Hydraulics, Federal Highway Administration, Washington, DC

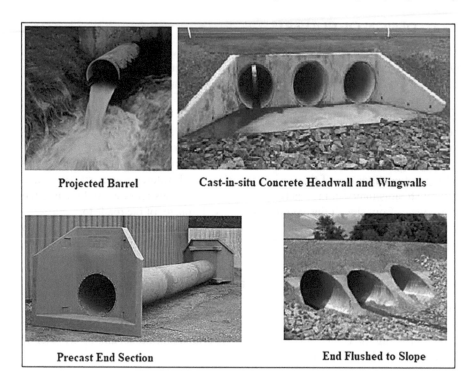

**FIGURE 15.7**   Four Standard Inlet Types.

*Source:* Web Links are as Below; Projected Barrel. (Source unavailable); Cast-in-situ www.theconstruc-tor.org/construction/types-of-culverts-material-construction/12926/; Precast End Section www.thegreen-pipesydney.com.au/culverts-headwells-crossings.htm; End Flushed to Slope www.conteches.com/knowledge-center/pdh-article-series/practical-factors-related-to-the-inspection-evaluation-and-load-rating-of-installed-culverts

The location of the centerline of the culvert on the centerline of the road may be determined by inspection in the field. The locations are generally on the centerline of an existing watercourse or at the bottom of a depression if there is no natural watercourse.

The alignment of the culvert normally follows the alignment of the natural stream, and the culvert should cross the roadway at right angles for economic reasons. However, skewed culverts, located at an angle to the centerline of the road, are needed in many instances. The selection of the natural direction of the stream is somewhat difficult where the stream bed shifts with the passage of time. In such a case, judgment must be exercised in selecting the most desirable location for the culvert. Improvements may be necessary for the streams and water courses to ensure the proper functioning of the culvert in cases where it is already built. Any improvement in the water course that are necessary in the direction of the culvert itself should be done gradually to avoid excessive head losses and consequent reduction in flow.

The grade of the culvert normally follows the grade of the existing water course. If the grade in the culvert is less than the grade of the stream, the velocity of flow may be reduced, in such case sediment carried in the water will be deposited at the mouth or inside the culvert, and the capacity of the culvert will be reduced. If the grade in the culvert is greater than the grade of the stream, velocity of flow through the culvert may increase and at the outlet end scour or erosion of the channel, beyond the culvert may take place, this may require costly protective devices, like apron. Therefore, changes in grade within the length of the culvert should be avoided.

## 15.5.2  HYDRAULIC DESIGN OF CULVERTS

We have already discussed concepts and design procedures related to the estimation of the quantity of runoff from a drainage basin. In this section, the principles and techniques for the hydraulic design of culverts are explained.

The hydraulic design is essential to make a drainage facility or system that will be adequate and economical for the estimated flow throughout the design life without any risk to the roadway structure or nearby property.

The essential steps for hydraulic design of culverts are the following:

- After collecting the required site data, a plot of the roadway cross section at the culvert site, including a profile of the stream channel is done.
- By following the grade of the stream, establish the culvert invert levels at the inlet and outlet. By determining the length of the culvert, the slope is calculated.
- The allowable headwater depth and the probable depth of tail water during the design flood is determined.
- The type and size of culvert and the design features of its appurtenances are selected that will accommodate the design flow under the service conditions. The design of the culvert inlet, discussed in last section, is especially important for the overall hydraulic efficiency of the structure.
- The need for energy dissipaters is checked and, where needed, appropriate protective devices to prevent destructive channel erosion is to be provided.

Whenever a culvert is placed in a natural open channel, there is an increase in the depth of water at the upstream of it. The allowable level of the headwater upstream of the culvert inlet is generally the main controller of the culvert size and inlet geometry. The allowable headwater depth depends on the topography and the nature of land use around the culvert location. Possible harmful effects like flooding causing damages to the pavement, nearby property, and interruptions to traffic should also be considered.

### 15.5.2.1  Types of Culvert Flow

The type of flow in a culvert depends on the total energy available in between the inlet and outlet. It consists primarily of the potential energy or the difference of elevations in the head water and tail water. The flow that occurs naturally will completely utilize all of the available energy, which is expended at entrances, in friction, in velocity head, and in depth.

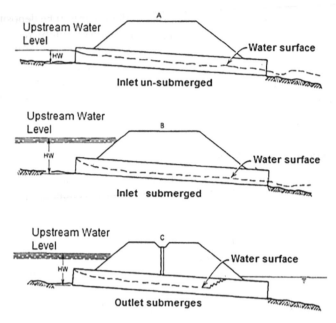

**FIGURE 15.8**   Inlet Controls for Culverts.
*Source:* Federal Highway Administration

The flow characteristics and capacity of a culvert are determined by the location of the control section, which is the section of the culvert that operates at maximum flow. The other parts of the system have a greater capacity than is actually needed.

Laboratory tests and field studies have shown that highway culverts operate with two major types of control – inlet control and outlet control. Examples of flow with inlet control and outlet control are shown in Figures 15.8 and 15.9 respectively.

### 15.5.2.2   Culverts Flowing with Inlet Control

Inlet control means the discharge capacity of a culvert related to the depth of headwater at the entrance and the entrance geometry in terms of barrel shape, cross-sectional area, and type of inlet edge. Inlet control commonly occurs when the slope of the culvert is steep, and the outlet is not submerged.

Analytical relationships have been developed for culverts with inlet control, based on experimental work. For a given entrance shape and submerged or un-submerged flood condition, the relationship is applicable within a specified range of discharge factors. The Federal Highway Administration of the USA has prepared a set of nomographs and design charts, some of them are shown in Figures 15.10 and 15.11.

### 15.5.2.3   Culverts Flowing with Outlet Control

In a culvert operating with outlet control, the maximum flow depends on the depth of headwater and entrance geometry and the additional considerations of the elevation of the tail water at the outlet and the slope, roughness, and length of the culvert. This

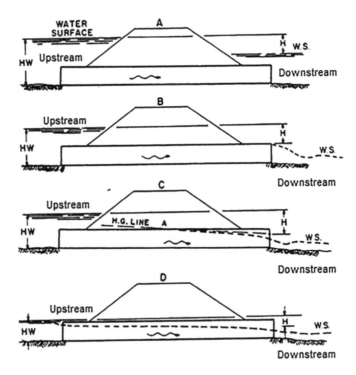

**FIGURE 15.9**   Inlet Controls for Culverts.
*Source:* Federal Highway Administration

type of flow commonly occurs on flat slopes, especially where downstream conditions cause the tail water depth to be more than the critical depth.

For culverts flowing full, the difference between the water surfaces at the upstream
and downstream, is the head "H" which is equal to the velocity head plus the loss of
energy at the entrance and in the culvert.

In conventional U.S. units:

$$H = \left[1 + Ke + (29\,n_2L)/(R_{4/3})\right]\left(V^2/2g\right) \tag{15.9a}$$

In metric (S1) units:

$$H = \left[1 + Ke + (19.6\,n_2L)/(R_{4/3})\right]\left(V^2/2g\right) \tag{15.9b}$$

where

   $H$ = difference in elevation between the head water and tail water surfaces,
   is shown in Figure 15.9a, or $H$ = difference in elevation between the
   headwater surface and the crown of the culvert at the outlet, as shown in
   Figure 15.9b, ft (m)

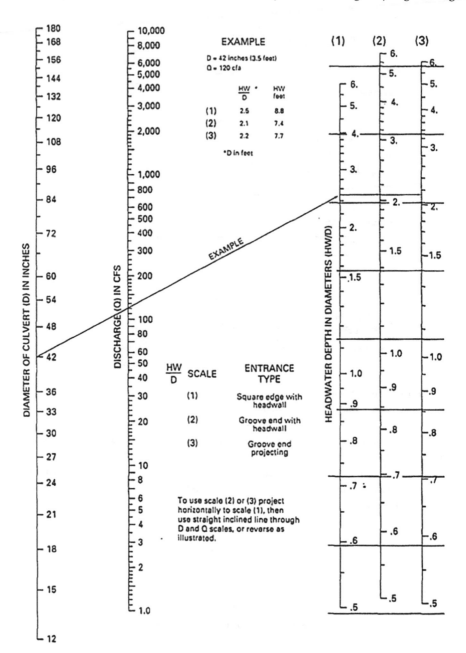

**FIGURE 15.10**  Nomograph to Determine Headwater Depth for Concrete Pipe Culverts with Inlet Control.

*Source:* Hydraulic Design of Highway Culverts, Federal Highway Administration, Washington, DC, 2001

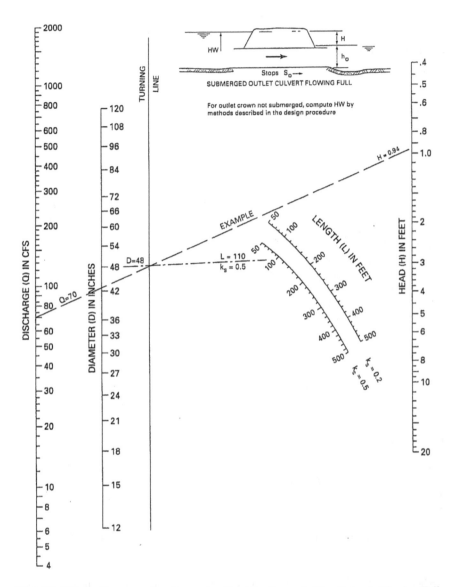

**FIGURE 15.11** Nomograph to Determine Head for Concrete Pipe Culverts Flowing Full.
*Source:* Hydraulic Design of Highway Culverts, Federal Highway Administration, Washington, DC, 2001

$K_e$ = entrance loss coefficient; see Table 15.4
n  = Manning's roughness coefficient; see Table 15.3
L  = length of culvert, ft (m)
R  = hydraulic radius = area/wetted perimeter, ft (m)
V  = velocity, ft/sec (m/sec)
G  = gravity constant, 32.2 ft/sec² (9.8 m/sec²)

## TABLE 15.4
## Entrance Loss Coefficients, $K_e$ for Outlet Control, Full or Partly Full

| Type of Structure and Design of Entrance | Coefficient $K_e$ |
|---|---|
| Pipe, Concrete | 0.2 |
| • Projecting from fill, socket end (groove-end) | 0.5 |
| • Projecting from fill, sq cut end | |
| • Headwall or headwall and wing walls | |
| • Socket end of pipe (groove end) | 0.2 |
| • Square-edged | 0.5 |
| • Rounded (radius = D/12) | 0.2 |
| • Mitered to conform to fill slope | 0.7 |
| • End-section conforming to fill slope | 0.5 |
| • Beveled edges, 33.7" or 45" bevels | 0.2 |
| • Side – or slope-tapered inlet | 0.2 |
| Pipe or Pipe-Arch, *Corrugated Metal* | |
| • Projecting from fill (no headwall) | 0.9 |
| • Headwall or headwall and wing walls square-edged | 0.5 |
| • Mitered to conform to fill slope, paved or unpaved slope | 0.7 |
| • End-section conforming to fill slope | 0.2 |
| • Beveled edges, 33.7" or 45" bevels | 0.2 |
| • Side – or slope-tapered inlet | 0.2 |
| Box, Reinforced Concrete | |
| Headwall parallel to embankment (no wing walls) | |
| • Square – edged on 3 edges | 0.5 |
| • Rounded on 3 edges to radius of D/12 or B/12 | |
| • Or beveled edges on 3 sides | 0.2 |
| Wing walls at 30" to 75" to barrel | |
| • Square-edged at crown | 0.4 |
| • Crown edge rounded to radius of D/12 or beveled top edge | 0.2 |
| Wing wall at 10" to 25" to barrel | |
| • Square-edged at crown | 0.5 |
| Wing walls parallel (extension of sides) | |
| • Square-edged at crown | 0.7 |
| • Side or slope-tapered inlet | 0.2 |

*Source:* Hydraulic Design of Highway Culverts, Federal Highway Administration, Washington, DC, (September, 2001).

When the critical depth falls below the crown at the outlet, then analytically determining the headwater is very tedious via time-consuming backwater computation, as shown in Figures 15.9c or 15.9d. Here the computation is facilitated by using the design charts published by the America's Federal Highway Administration. An example of a nomograph for culverts operating with outlet control is presented in Figure 15.11. The illustrations are given in Figures 15.10 and 15.11, where the headwater depths for both inlet and outlet control can be determined for practically all combinations of culvert size, material, entrance geometry, and discharge (Table 15.1).

To understand the use of Figures 15.10 and 15.11, the reader should study the examples on the nomographs. To further illustrate the use of available US Federal

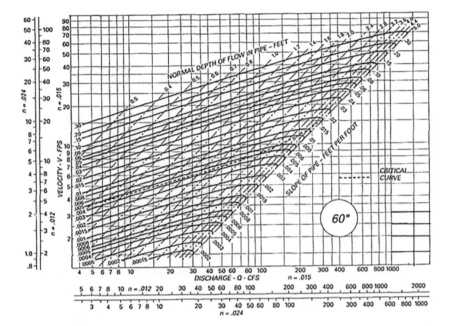

**FIGURE 15.12**   Pipe Flow Chart for a 60-inch Diameter Culvert with Uniform Flow.
*Source:* Design Charts for Open Channel Flow, Federal Highway Administration, Washington, DC, 1961

Highway Administration (FHWA) design nomographs, consider the following example.

## 15.6   DRAINAGE OF CITY STREETS

Accumulation of storm water that occurs on city streets and adjacent areas must be rapidly removed before it becomes a hazard to traffic. The removal of surface water in municipal areas is accomplished by methods like those employed in the drainage of rural highways, except that the surface water is commonly carried to its defined disposal point by means of underground covered drains under sidewalk or pipe drains which are called storm drains. The storm drains are designed to provide for the flow of groundwater as well as surface water, sub-drains may also be provided in certain areas, if required additionally. Construction of sub-drains in urban areas is similar to those done in the rural areas.

The surface water drainage system in a city is composed of the following basic elements: pavement crown, curb and gutter, and the storm drains themselves, including such commonly used appurtenances like inlets, catch basins, and manholes.

### 15.6.1   Pavement Crowns, Curbs and Gutters

The storm water falling on the pavement is removed from the surface and collected in the side drains or gutters via the cross slope on either side of the carriageway, this forms a crown in the center. The height of the crown on any road or city street should be made as small as possible, in the interest of appearance and safety, but also in ensuring proper drainage.

The channel on the road surface formed by the combined curb and gutter, forms the drainage function of a side ditch on a rural highway. It must be designed to adequately transfer the runoff from the pavement and adjacent areas to a suitable collection point. The longitudinal grade of the gutter must be sufficient to ensure rapid removal of the water, the minimum adopted grades are generally recommended as 0.2–0.3%. The grade of the gutter is usually the same as that of the road surface, in exceptional situations it may be different if required for proper drainage. Hydraulically, gutters are provided as an open channel with a special shape.

Flow in the gutters generally continues uninterrupted along the road up to an intersection, unless a low point or sag is reached. At that point an inlet is provided to take the water into the storm sewer system. In the case of long blocks, inlets may also be provided between intersections.

Provision for drainage on the paved area formed by a street intersection is done with adjustment to the grades of the city streets at intersections and is a complex problem. The removal of storm water from the intersection area by adjusting the differences in grades present on the streets that approach and leave the intersection, obtaining the elevations required within the intersection, have been successfully done by various available analytical methods.

## 15.6.2 Inlets

Intersections with improper drainage are a source of hazards to both vehicular and pedestrian traffic. Inlets at intersections are provided to intercept the water flowing in the gutters before it can reach the pedestrian sidewalks. A number of inlets of sufficient size are to be provided to rapidly remove collected storm water.

The selection of the type of inlets to be used, and their positioning in the area adjacent to the intersection, are to be done carefully, in actual practice considerable variation occurs in selection and positioning. The proper positioning of inlets at the intersection of two streets of normal width is described in Figure 15.13.

On streets with less traffic, it may sometimes be desirable to provide for the flow of surface water across the intersection by means of a shallow paved gutter or "dip". Such a provision is not generally recommended but may be necessary in case of budgetary limits.

Inlets of various designs are used by different departments. The most used type is the "combined curb and gutter inlet." This is a concrete or brick box with an opening on one side, near or at the bottom, and accommodates a circular pipe forming one of the laterals of the storm sewer system. Water enters the box through a metal grate placed over a horizontal opening at the top in the gutter and through a relatively narrow opening in the face of the curb. The latter opening at the face of the curb may or may not be protected by a grating of some kind. Inlets of this type are either cast in place or composed of precast units. Inlets are also designed without the opening in the curb, the water flowing in only through a grated opening in the gutter or with only curb openings.

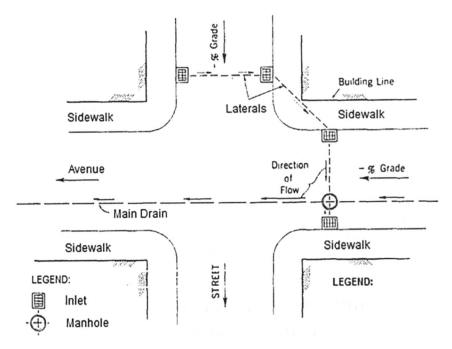

**FIGURE 15.13**   Drainage at an Intersection of Urban Streets.
*Source:* Figure 12.19, "Highway Engineering" by Paul H. Wright and Karen K. Dixon, by John Wiley & Sons. Inc.

While inlets of these types have only an outlet in the form of a circular pipe placed so that the invert of the pipe is flush with the bottom of the box, they are also constructed with a pipe inlet and a pipe outlet. Sometimes, the inlet may be a connecting link in the storm sewer system. Another design, which may be called a "drop inlet," in which the water entering from the gutter must drop some considerable distance to the bottom of the box or may simply be transferred directly into the storm sewer by means of a vertical pipe.

Inlets are susceptible to clogging with debris, ice, etc., and they must be subjected to continual inspection and maintenance to keep them effective. Recent research has increased the understanding of the hydraulic principles involved and the development of various design methods. Curb openings for inlets, on the average, are from 0.6 to 1.2 m in length and from 150 to 200 mm high.

## 15.6.3   CATCH BASINS

Catch basins are like inlets in respect to their function and design. The principal difference between a catch basin and an inlet is that in a catch basin the outlet pipe and inlet pipe if present, is placed some distance above the bottom of the chamber. The catch basins are less important these days because of the universal construction of permanent-type street surfaces and more efficient street-cleaning.

### 15.6.4 MANHOLES

In modern sewerage practice, manholes are generally placed at points where the sewer line changes its grade or direction, where junctions are made, and at intermediate points, usually at intervals of 90–150 m. The opening provided must, be large enough to permit a person to enter the manhole chamber while having room to work. They are usually about 1.2 m in diameter and have a depth sufficient to perform the purpose for which they are intended. The pipes generally enter and leave the manhole at the bottom of the manhole chamber, and essential arrangements are made for carrying the flow through the manhole with a minimum loss in head. Manholes are built of brick masonry, concrete block, or brick, and a department usually adopts a standard design of manhole to be used in all installations. Manhole covers (and frames) are generally of cast iron and the covers are circular in shape with a diameter of about 0.6 m. When the entrance to the manhole is on the traveled way, as it usually does, special care must be taken in the design and placing of the cover relative to the finished street surface so that the opened manhole during work must not endanger the traffic on the road.

## 15.7 COMPUTER-AIDED DESIGN OF HIGHWAY DRAINAGE

### 15.7.1 COMPUTER-AIDED DESIGN OF HIGHWAY SURFACE DRAINAGE

The computer applications for the "Design of Highway Drainage and Drainage Structures" are explained in Chapter 15 in the "The Guide for Computer Application Tutorials" part of the book available for download from the web site www. roadbridgedesign.com.

The following section may be referred for the design applications: "15.6.1 Computer Aided Design of Highway Surface Drainage".

For downloading the items under "Book Tutorials," visit the website at: www. roadbridgedesign.com

For downloading the other items visit the website at: www.techsoftglobal.com/ download.php

### 15.7.2 COMPUTER-AIDED STREAM HYDROLOGY AND SYNTHETIC UNIT HYDROGRAPH

This process determines the adequacy of the bridge opening on a river during a high flood situation.

This study uses the facilities of drawing the alignment of the river centerline, region outline, and downloading the ground elevation data within and outside the river for the length of about 50 km from satellite by using strong internet connections. These are done with the help of software programs such as Google Earth, Global Mapper, and MS-Excel.

The alignment of the river centerline is drawn in Google Earth by using the "My Path" facility, starting from the bridge location (Chainage 0.0) and drawing up to say, 50 km on the upstream side of the river. An outline of the region (specific to relevant zone and sub-zone) is also drawn on the Google Earth by using "My Path". The drawing may be saved as KML/KMZ file and relevant "North" (in the tutorial video "17 Google Earth Video" it is 47 N) is to be noted in Google Earth.

Next, the KMZ drawing is to be opened in Global Mapper, the noted "North" is to be mentioned, drawing is to be saved as DXF file, which may be saved as DWG file later on. The ground elevation data is to be downloaded from the satellite.

The downloaded data is to be opened in MS-Excel, to be formatted with data in five columns as "Serial Number, Easting, Northing, Ground Elevation and Feature Code", which is similar to total station data.

For detail descriptions of various processes user may refer to, "Book Tutorials" of the web site at: "www.roadbridgedesign.com", "www.techsoftglobal.com/download. php", containing the folder "HEADS Pro Tutorials", which further contains the folder "HEADS Pro Tutorial Videos". There are videos with the names "17 Google Earth Video, 18 Global Mapper Video, and 19 Ground Data Formatting." Watching these videos will help the understanding of the procedure of drawing alignment of a river centerline, region outline, and downloading the ground elevation data within and outside the river for the length of about 50 km.

The website of the book contains the folder "Book Tutorials", which further contains the folder "Chapter 15 Drainage Design" and then folder "Stream Hydrology" inside, in which the files SURVEY.TXT and Boundary String.DWG are available and will be used to describe the procedure in this session.

The computer applications for the "Design of Highway Drainage and Drainage Structures" are explained in Chapter 15 in the "The Guide for Computer Application Tutorials" part of the book available for download from the web site www. roadbridgedesign.com.

The following section may be referred for the design applications: "15.6.2 Computer-Aided Stream Hydrology and Synthetic Unit Hydrograph".

For downloading the items under "Book Tutorials" visit the web sites at: www. roadbridgedesign.com.

For downloading the other items visit the web sites at: www.techsoftglobal.com/ download.php.

## BIBLIOGRAPHY

*Guidelines for Road Drainage*, IRC SP 42-1994, IRC.
*Guidelines for Urban Drainage*, IRC SP 50, IRC-1999, IRC.
*Highway Engineering* by Paul H. Wright and Karen K. Dixon, by John Wiley & Sons. Inc.

# Index